# Advances in Intelligent Systems and Computing

## Volume 377

**Series editor**

Janusz Kacprzyk, Polish Academy of Sciences, Warsaw, Poland
e-mail: kacprzyk@ibspan.waw.pl

## About this Series

The series "Advances in Intelligent Systems and Computing" contains publications on theory, applications, and design methods of Intelligent Systems and Intelligent Computing. Virtually all disciplines such as engineering, natural sciences, computer and information science, ICT, economics, business, e-commerce, environment, healthcare, life science are covered. The list of topics spans all the areas of modern intelligent systems and computing.

The publications within "Advances in Intelligent Systems and Computing" are primarily textbooks and proceedings of important conferences, symposia and congresses. They cover significant recent developments in the field, both of a foundational and applicable character. An important characteristic feature of the series is the short publication time and world-wide distribution. This permits a rapid and broad dissemination of research results.

## Advisory Board

More information about this series at http://www.springer.com/series/11156

Jaime Gil-Aluja · Antonio Terceño-Gómez
Joan Carles Ferrer-Comalat
José M. Merigó-Lindahl
Salvador Linares-Mustarós
Editors

# Scientific Methods for the Treatment of Uncertainty in Social Sciences

Springer

*Editors*
Jaime Gil-Aluja
Royal Academy of Economic and Financial
Sciences (RACEF)
Barcelona
Spain

Antonio Terceño-Gómez
Faculty of Business and Economics,
Department of Business Management
Universitat Rovira i Virgili
Reus
Spain

Joan Carles Ferrer-Comalat
Department of Business Administration
University of Girona
Girona
Spain

José M. Merigó-Lindahl
Department of Management Control
and Information Systems
University of Chile
Santiago
Chile

Salvador Linares-Mustarós
Department of Business Administration
University of Girona
Girona
Spain

ISSN 2194-5357          ISSN 2194-5365   (electronic)
Advances in Intelligent Systems and Computing
ISBN 978-3-319-19703-6          ISBN 978-3-319-19704-3   (eBook)
DOI 10.1007/978-3-319-19704-3

Library of Congress Control Number: 2015940743

Springer Cham Heidelberg New York Dordrecht London

Printed on acid-free paper

Springer International Publishing AG Switzerland is part of Springer Science+Business Media
(www.springer.com)

# Preface

The XVIII SIGEF Congress, entitled *Scientific methods for the treatment of uncertainty in social sciences* took place in Girona, Catalonia (Spain) from 6 to 8 July 2015, and was organized by the International Association for Fuzzy-Set Management and Economy (SIGEF) and the University of Girona, through the Department of Business Administration and the research group on Accounting and Mathematics for Economics and Business.

The appearance of the first applications on multivalent logics and the theory of fuzzy sets in economics and business problems, has suggested grouping all those people, businesses, teachers and researchers wishing to collaborate in developing and promoting the new techniques emerging from these achievements that modern science puts in the hands of researchers, professionals and executives. The treatment of uncertainty in the economic and business analysis is fundamental and requires instruments compatible with the uncertain environment of economics and business, because most of the traditional models have been overtaken by this reality when trying to make decisions when the information and the data are uncertain and/or subjective.

To achieve this goal, in 1993 was born the Association for Fuzzy-Set Management and Economy (SIGEF), whose creation has driven the research, study and treatment of all problems associated with the economy in general and business administration in particular.

At this moment, the main objectives of SIGEF are:

(a) To promote research and development of new modelling for the treatment of uncertainty in Economics and Business Administration.
(b) To promote training of postgraduates by organizing courses, workshops and conferences.

(c) To promote the interest of society in the use of new techniques for the treatment of uncertainty in the different areas of Economics and Social Sciences.

(d) To disseminate research, developments and innovations of its members, teachers, researches and businesses.

To achieve these objectives, among other activities, SIGEF publishes the scientific journal *Fuzzy Economic Review* and regularly organizes International Conferences. In this way, this XVIII International SIGEF Congress, entitled *Scientific methods for the treatment of uncertainty in social sciences*, means to introduce these new techniques in social science fields, and it continues and extends the Series of International Conferences organized by SIGEF. Previous Conferences were held in Reus, Spain (1994), Santiago de Compostela, Spain (1995), Buenos Aires, Argentina (1996), Santiago de Cuba, Cuba (1997), Lausanne, Switzerland (1998), Morelia, México (1999), Chania, Greece (2000), Napoli, Italy (2001), Mérida, Venezuela (2002), León, Spain (2003), Reggio Calabria-Messina, Italy (2004), Bahía Blanca, Argentina (2005), Hammamet, Tunisia (2006), Poiana Brasov, Romania (2007), Lugo, Spain (2009), Morelia, México (2010) and Reus, Spain (2012). This international congress aims to stimulate scientific exchanges, to promote international cooperation between the academic community and companies, and disseminate results of theoretical and applied research.

This book includes a selection of papers received by the Scientific Committee, covering the main topics discussed at the Congress. We believe it is a good example of the excellent work of the associates and the significant progress about this line of research in recent times.

The book is organized according to seven thematic groups, although these groups do not offer a disjoint classification: Decision Making, Expert Systems and Forgotten Effects Theory, Forecasting Models, Fuzzy Logic and Fuzzy Sets, Modelling and Simulation Techniques, Neural Networks and Genetic Algorithms and Optimization and Control.

Finally, we would like to address our warmest thanks to all the authors and reviewers, participants at the conference, organizations and institutional sponsors for their help and support and for creating a pleasant connection point for the collaborative efforts and exchange of ideas and information among persons from varied disciplines.

Girona
July 2015

Jaime Gil-Aluja
Antonio Terceño-Gómez
Joan Carles Ferrer-Comalat
José M. Merigó-Lindahl
Salvador Linares-Mustarós

# Honorary Committee

Special thanks to the members of the Honorary Committee for their support in the organization of SIGEF 2015.

Lotfi A. Zadeh, Researcher and professor emeritus of computer science at the University of California, Berkeley

Sergi Bonet Marull, Chancellor at the University of Girona, Spain

Maria Carme Saurina Canals, Dean in the Faculty of Business and Economics at the University of Girona, Spain

Josep Viñas Xifra, Chief in the Department of Business Administration at the University of Girona, Spain

Joan Bonet Amat, Researcher and former professor at the University of Girona, Spain

# Scientific Committee

Thanks to all the members of the Scientific Committee for their kind support in the organization of SIGEF 2015 Girona.

Jaime Gil-Aluja (President)
Antonio Terceño-Gómez (Vice-President)
Joan Carles Ferrer-Comalat (Vice-President)
Xavier Bertran Roure (Vice-President)
M. Glòria Barberá-Mariné
Lisana Belén Martínez
José Manuel Brotons
Christer Carlsson
Sergio-Gerardo de-los-Cobos-Silva
Dolors Corominas Coll
María-Ángeles Fernández-Izquierdo
Emma Fernández Loureiro
Vasile Georgescu
Anna M. Gil-Lafuente
Federico González-Santoyo
M. Belén Guercio
Madan M. Gupta
Emmanuel Haven
Kaoru Hirota
Luca Iandoli
Mariano Jiménez
Janusz Kacprzyk
George Klir
Viktor V. Krasnoproshin
Luisa L. Lazzari
Vicente Liern
Salvador Linares-Mustarós
Tomás Lorenzana
Florica Luban

José M. Merigó
Francesco Carlo Morabito
M. Àngels Farreras Noguer
Witold Pedrycz
Jaime Tinto Arandes
Enric Trillas
I. Burhan Turksen
José Luis Verdegay Galdeano
Hernán P. Vigier
Ronald R. Yager
Hans J. Zimmermann
Giuseppe Zollo
Constantin Zopounidis

# Organizing Committee

Special thanks to all the members of the Organizing Committee for their support during the preparation of the SIGEF 2015 Annual Conference.

**Chairs of the Organizing Committee**

Joan Carles Ferrer-Comalat, Spain
Dolors Corominas Coll, Spain
Salvador Linares-Mustarós, Spain

**Organizing Committee**

Xavier Bertran Roure, Spain
Xavier Càmara Turull, Spain
Elvira Cassú Serra, Spain
Marc Carreras Pijuan, Spain
M. Àngels Farreras Noguer, Spain
Aurelio Fernández-Bariviera, Spain
José M. Merigó-Lindahl, Chile
Elena Rondós Casas, Spain
M. Teresa Sorrosal Forradellas, Spain

# Contents

# Part I
# Decision Making

# New Aggregation Methods for Decision-Making in the Selection of Business Opportunities

Fabio R. Blanco-Mesa, Anna M. Gil-Lafuente and José M. Merigó

**Abstract** We analyse decision-making process in the selection of business opportunities. A new mathematical application based on OWA operator and selection index is developed. We consider the use of the OWA distance (OWAD), the OWA adequacy coefficient (OWAAC) and the OWA index of maximum and minimum level OWAIMAM operators. The study proposes a fuzzy significance vector (FS), which can aggregate information according to the importance of its characteristics. The introduction of the selection OWA operator using fuzzy significance vector can reflect decision with different degrees of optimism through normalization process where the maximum value of the aggregated information can be higher than 1. These methods are called FS-OWAAC, FS-OWAD and FS-OWAIMAM operator. By using FS-OWA operator, we can parameterize attitudinal character of decisor and the importance of characteristics of the information. A numerical example is developed in decision-making process for the selection of opportunities to start a new business within different sectors according to preference of decisor and environmental factors.

**Keywords** Group decision making · OWA operator · Probability · Weighted average · Induced aggregation operators · Strategic management

F.R. Blanco-Mesa (✉) · A.M. Gil-Lafuente
Department of Business Administration, University of Barcelona,
Av. Diagonal 690, 08034 Barcelona, Spain
e-mail: frblamco@gmail.com

A.M. Gil-Lafuente
e-mail: amgil@ub.edu

J.M. Merigó
Department of Management Control and Information Systems,
University of Chile, Av. Diagonal Paraguay, 257, 8330015 Santiago, Chile
e-mail: jmerigo@fen.uchile.cl

© Springer International Publishing Switzerland 2015
J. Gil-Aluja et al. (eds.), *Scientific Methods for the Treatment of Uncertainty in Social Sciences*, Advances in Intelligent Systems and Computing 377,
DOI 10.1007/978-3-319-19704-3_1

# 1 Introduction

Entrepreneurship is considered a key factor for the economic performance because it is able to introduce innovation, enhancing rivalry, contribute employment generation and creating competition (Wong et al. 2005; Praag and Versloot 2007). Furthermore, the entrepreneurship is considered as process of the change (Audretsch and Keilbach 2004) where the best agent for this change is the entrepreneur (Acs and Amorós 2008). The creation of business opportunities are conditioned by the market and the industrial structure which in turn, are influenced by technological development, globalization and economic development (Verheul et al. 2002). A potential entrepreneur considers risk and reward of these opportunities and makes a decision with any degree of uncertainty. In fact, the change market conditions and a possible increase of the opportunity costs can cause more uncertainty and hesitation in the individual when making decision regarding your preferences (Wennekers and Thurik 1999; Verheul et al. 2002). In the literature, there are a wide range of decision making methods capable of processing personal preferences and attitudinal character (Herrera 1995; Martinez and Herrera 2000; Merigó 2009; Merigó and Gil-Lafuente 2007, 2008a, c, 2010, 2013; Merigó et al. 2013; Peng and Ye 2014; Wei 2009), which aid to decisor maker to make a decision to invest and to create a new firm in the most appropriate market (Cuervo 2005; Merigó and Peris-Ortiz 2014; Merigó et al. 2014). The Hamming distance, the adequacy coefficient and index of maximum and minimum level are some of these effective techniques to allow us to lead to a selection process and to solve decision problem (Merigó and Gil-Lafuente 2011). These techniques are developed in Gil-Aluja (1999), Gil-Lafuente (2001), Kaufmann and Gil-Aluja (1986), Merigó (2009) and Merigó and Gil-Lafuente (2010). These techniques let entrepreneur or investor team comparing different characteristics of each new business alternatives from ideal condition to optimal solution. However, according to Merigó and Gil-Lafuente (2011) these techniques cannot take into account attitudinal character of the decisor maker. In order to estimating more or less the attitudinal character (optimism or pessimisms), they propose the ordered weighted averaging (OWA) operator (Yager 1988), as method for aggregation information according to decision attitude. The OWA operator has been studied by several authors (Gil-Lafuente and Merigó 2007, 2012b; Herrera 1995; Martinez and Herrera 2000; Merigó 2009; Merigó and Gil-Lafuente 2008a–c, 2013; Peng and Ye 2014; Wei 2009; Xu 2004; Yager 1993; Yager et al. 2011; Zhao et al. 2010; Zhou and Chen 2010), which have developed new aggregation operators providing a parameterized family of them (Merigó and Gil-Lafuente 2012a). The combination between selection index and OWA operator has developed one of these families, which allow changing neutrality of selection methods (Merigó and Gil-Lafuente 2011). The methods introduced are: the ordered weighted averaging distance (OWAD) operator (Merigó and Gil-Lafuente 2007, 2010), the ordered weighted averaging adequacy coefficient (OWAAC) (Merigó and Gil-Lafuente 2008c, 2010) and the ordered weighted averaging index of maximum and minimum level (OWAIMAM) (Merigó and Gil-Lafuente 2011, 2012).

The main aim of the paper is to developing a new mathematical application based on OWA operator and their selection indices extension (see Merigó 2009). This application consists in that selection OWA operators in combination with fuzzy vector can also be aggregated information when the maximum value is higher than 1 using normalization process. This proposition allows taking into account at a certain level all characteristics. The main advantage of this proposition is that it takes into account the importance of the information of each characteristic in the decision-making process and parameterization of interest of the decisor maker, which are combined and considered to make a decision.

Numerical example is developed in order to see the usefulness of new proposition. Selection of business opportunities within specific sector is considered, where entrepreneur is seeking more optimal option to starting up new business. Depending of aggregation operator and particular attitude, the results might be different, which implies that decision-making might be different too. Furthermore, application of these methods allows obtaining abroad view of the decision problem due to considering different scenarios. The structure of this paper is as follows: In Sect. 2, basic concepts of OWA operator and their extensions with selection index are briefly reviewed and new proposition is presented. In Sect. 3, decision-making approach is explained. In Sect. 4, numerical application process for starting new business using FS-OWAAC, FS-OWAD and FS-OWAIMAM operators are exposed. In Sect. 4, the numerical example of new proposition is developed focusing on business opportunities example in several industries. In Sect. 5, summary and main conclusion are presented.

# 2 Preliminares

In this Section, we briefly review of OWA operator and their extensions with selection index OWAAC operator, OWAD operator and OWAIMAM operator and a new method for dealing with the weight of the OWA operator is presented.

## 2.1 The OWA Operator

The OWA operator (Yager 1988) provides a parameterized class of mean type of aggregation operators. It can be defined as follows.

**Definition 1.** An OWA operator of dimension n is a mapping $OWA: R^n \rightarrow R$ that has an associated weighting vector $W$ of dimension $n$ with $w_j \in [0, 1]$ and $\sum_{j=1}^{n} w_j = 1$, such that:

$$OWA(a_1, a_2, \ldots, a_n) = \sum_{j=1}^{n} w_j b_j,$$ (1)

where $x_i$ and $y_i$ are the *ith* arguments of the sets $X$ and $Y$.

## 2.2 The OWAAC Operator

The OWAAC operator (Merigó and Gil-Lafuente 2008b, 2010; Gil-Lafuente and Merigó 2009) is an aggregation operator that uses the adequacy coefficient and the OWA operator in the same formulation. It can be defined as follows for two sets X and Y.

**Definition 2.** An OWAAC operator of dimension n is a mapping $OWAAC$: $[0,1]^n x[0,1]^n \rightarrow [0,1]$ that has an associated weighting vector W, with $w_j \in [0,1]$ and $\sum_{j=1}^{n} w_j = 1$, such that:

$$OWAAC(\langle x_1, y_1 \rangle, \ldots, \langle x_n, y_n \rangle) = \sum_{j=1}^{n} w_j K_j, \qquad (2)$$

where $K_j$ represents the *jth* largest of $[1 \wedge (1 - x_i + y_i)], x_i, y_i \in [0,1]$.

## 2.3 The OWAD Operator

The OWAD operator (Gil-Lafuente and Merigó 2007; Merigó 2009; Merigó and Gil-Lafuente 2010) is an aggregation operator that uses OWA operators and distance measures in the same formulation. It can be defined as follows for two sets X and Y.

**Definition 3.** An OWAD operator of dimension n is a mapping $OWAD: R^n x R^n \rightarrow R$ that has an associated weighting vector $W$, $\sum_{j=1}^{n} w_j = 1$ and $w_j \in [0,1]$ such that:

$$OWAD(\langle x_1, y_1 \rangle, \ldots, \langle x_n, y_n \rangle) = \sum_{j=1}^{n} w_j D_j, \qquad (3)$$

where $D_j$ represents the *jth* largest of the $|x_i - y_i|$.

## 2.4 The OWAIMAM Operator

The OWAIMAM operator (Merigó 2009; Merigó and Gil-Lafuente 2011, 2012b) is an aggregation operator that uses the Hamming distance, the adequacy coefficient and the OWA operator in the same formulation. It can be defined as follows.

**Definition 4.** An OWAIMAM operator of dimension n, is a mapping $OWAIMAM: [0,1]^n x[0,1]^n \rightarrow [0,1]$ that has an associated weighting vector $W$, with $w_j \in [0,1]$ and the sum of the weights is equal to 1, such that:

$$OWAIMAM(\langle x_1, y_1 \rangle, \langle x_2, y_2 \rangle, \ldots, \langle x_n, y_n \rangle) = \sum_{j=1}^{n} w_j K_j, \qquad (4)$$

where $K_j$ represents the *jth* largest of all the $|x_i - y_i|$ and the $[0 \vee (x_i, y_i)]$.

## 2.5 A New Method for Dealing with the Weight of the OWA Operator

Following (Gil-Aluja 1999) OWA operator can also be aggregated according to the significance of its characteristics. This proposition allows taking into account at a certain level all characteristics, i.e., the importance of the relationship of each characteristic is considered to make a decision. In order to deal with a decision-making problem, it is made the following proposition.

**Proposition 1.** An OWAAC operator with fuzzy significance can have a weighting vector $W$, with $w_j \in [0, 1]$ and $\sum_{j=1}^{n} \frac{w_j}{max(w_j)}$, such that:

$$FS - OWAAC(\langle x_1, y_1 \rangle, \ldots, \langle x_n, y_n \rangle) = \sum_{j=1}^{n} \frac{w_j}{max(w_j)} K_j, \tag{5}$$

where $K_j$ represents the *jth* largest of $[1 \wedge (1 - x_i + y_i)], x_i, y_i \in [0, 1]$.

**Proposition 2.** An OWAD operator with fuzzy significance can have a weighting vector $W$, $\sum_{j=1}^{n} \frac{w_j}{max(w_j)}$, and $w_j \in [0, 1]$ such that:

$$FS - OWAD(\langle x_1, y_1 \rangle, \ldots, \langle x_n, y_n \rangle) = \sum_{j=1}^{n} \frac{w_j}{max(w_j)} D_j, \tag{6}$$

where $D_j$ represents the *jth* largest of the $|x_i - y_i|$.

**Proposition 3.** An OWAIMAM with fuzzy significance can have a weighting vector $W$, with $w_j \in [0, 1]$ and $\sum_{j=1}^{n} \frac{w_j}{max(w_j)}$, such that:

$$FS - OWAIMAM(\langle x_1, y_1 \rangle, \ldots, \langle x_n, y_n \rangle) = \sum_{j=1}^{n} \frac{w_j}{max(w_j)} K_j, \tag{7}$$

where $K_j$ represents the *jth* largest of all the $|x_i - y_i|$ and the $[0 \vee (x_i, y_i)]$.

Note that the boundary and idempotency condition is not accomplished in the operator because it is easy to obtain a result higher than the maximum. This happens because the sum of the weights may be higher than one. In order to accomplish the boundary condition, a normalization process could be used where the sum of the weights become one (Merigó, Gil-Lafuente 2009). A practical method for doing so is by dividing the weights by the sum of all of them. Therefore, this operator is very useful as a method for decision-making but it does not represent the information from a classical point of view.

These OWA operators are commutative, monotonic, non-negative and reflexive. They are commutative from the OWA perspective because $f(\langle x_1, y_1 \rangle, \ldots, \langle x_n, y_n \rangle) = f(\langle c_1, d_1 \rangle, \ldots, \langle c_n, d_n \rangle)$ where $(\langle x_1, y_1 \rangle, \ldots, \langle x_n, y_n \rangle)$ is any permutation of the arguments $(\langle c_1, d_1 \rangle, \ldots, \langle c_n, d_n \rangle)$. They are also commutative from the distance measure perspective because $f(\langle x_1, y_1 \rangle, \ldots, \langle x_n, y_n \rangle) = f(\langle y_1, x_1 \rangle, \ldots, \langle y_n, x_n \rangle)$. They are monotonic because if $|x_i - y_i| \geq |c_i - d_i|$, for all $i$, then $f(\langle x_1, y_1 \rangle, \ldots, \langle x_n, y_n \rangle) \geq f(\langle c_1, d_1 \rangle, \ldots, \langle c_n, d_n \rangle)$. Non-negativity is also accomplished always, that is, $f(\langle x_1, y_1 \rangle, \ldots, \langle x_n, y_n \rangle) \geq 0$. Finally, they are also reflexive because $f(\langle x_1, x_1 \rangle, \ldots, \langle x_n, x_n \rangle) = 0$.

Another important issue is to consider the formulation for the characterization of the weighting vector under this framework: The degree orness/andness (degree of optimist or pessimist) (Yager 1988) is defined in the following way with this new approach:

$$\alpha(w) = \sum \frac{n-j}{n-1} * \frac{w_j}{max(w_j)}, \tag{10}$$

where min $\rightarrow 0$ although the maximum can be higher than 1. By the introduction of significance degree in more than one variable is where the result is not contained in [0,1].

The entropy of dispersion (Shannon 1948; Yager 1988) can be defined as follows:

$$H(w) = - \sum_{j=1}^{n} \left( \frac{w_j}{max(w_j)} \right) ln \left( \frac{w_j}{max(w_j)} \right). \tag{11}$$

Finally, let us briefly study some key particular cases of these operators. If one of the sets is empty, we get the FS-OWA operator in the FS-OWAD. In the FS-OWAAC, if X is empty all the individual comparisons are 1, so the result is 1 or higher. The FS-OWAIMAM is a combination of the previous ones.

If $w_1 = 1$ and $w_j = 0$ for all $j \neq 1$, we get the maximum distance, FS-OWAD = Max$\{D_j\}$ and if $w_n = 1$ and $w_j = 0$ for all $j \neq n$, the minimum distance, FS-OWAD = Min$\{D_j\}$. Note that similar results are also found with the FS-OWAAC and the FS-OWAIMAM operator. If $w_j = 1/n$, for all $j$, the result is the total operator, which is also the sum of all the individual distances (Merigó et al. 2014). This occurs because in this situation all the weights are $(1/n)/(1/n) = 1$ so the sum is $n$. The result obtained is the absolute distance.

The step-FS-OWA aggregation (Yager 1993) provides the same results than the classical OWA approach. The median-FS-OWA aggregation (Merigó and Gil-Lafuente 2009) gives the same results only when it is odd. If it is even, the results of the FS-OWAD are twice as high as they should in the classical framework. The Olympic-FS-OWAD (Merigó and Gil-Lafuente 2009; Yager 1993) is used when $w_1 = w_n = 0$, and for all others $w_{j*} = 1/(n-2)$. In this case, the result is $n - 2$ which is close to the total and has close connections with the arithmetic distance.

## 3 Decision Making Approach for Starting New Business

The using OWA operator allow aggregating information obtaining a single value representative of the information according to the parameters of optimist and pessimist of entrepreneur in the selection of business. Entrepreneurs use personal preferences and knowledge (Wenneker and Thurick 2002) in order to perceive opportunities and lead to a decision-making process. The introduction of the selection OWA operator using fuzzy significance vector can reflect decision with different degrees of optimism and pessimism according to the importance of the characteristics.

Similar models that use the OWA operator have been developed for other selection problems (Merigó 2009; Merigó and Gil-Lafuente 2007, 2008a–c, 2010, 2011). Likewise, similar process has been developed in Gil-Aluja (1999); Gil-Lafuente (2001, 2002); Kaufmann and Gil-Aluja (1986) with instruments that can be applied in the selection process with OWA operator. The five steps of decision-making process (Merigó 2011) are described as follows:

*Step 1.* Analyse and determine the characteristics of potential business. Theoretically, it will be represented as: $C = \{C_1, C_2, ..., C_i, ..., C_n\}$, where $C_i$ is the *ith* characteristic of potential business to be considered.

*Step 2.* Fixation of the ideal levels of each characteristic in order to set up the ideal conditions to starting up new business (see Table 1) where P is the ideal conditions to entrepreneur expressed by a fuzzy subset, $C_i$ is the ith characteristic to be considered and $l_i \in [0, 1]$; $i = 1, 2, ..., n$, is the valuation between 0 and 1 for the ith characteristic.

*Step 3.* Fixation of the real level of each characteristic for all different potential business (see Table 2) with $k = 1, 2, ..., m$; where $P_k$ is the kth potential business expressed by a fuzzy subset, $C_i$ is the ith characteristic to be considered and $l_i^{(k)} \in [0, 1]$; $i = 1, ..., n$ is the valuation between 0 and 1 for the ith characteristic of the kth potential business.

*Step 4.* Comparison between the ideal conditions to starting up new business, different potential business and determination of the level of significance is considered through the use of the FS-OWA operator. In this step, we use the new method in combination with selection OWA operators such as FS-OWAD, FS-OWAAC and FS-OWAIMAM. Thus, it is to express numerically the significance between ideal conditions to starting up new business and different potential businesses.

*Step 5.* It is adopted of decisions according to the results found in the previous steps. Noticeably, the decision is based on choosing the business that best fits with entrepreneur interests.

## 4 Numerical Example

In this section we present an application on new approach suggested above. The main advantage on using OWA operators is that they can overestimate information according to attitudinal character of the decision-making. The application is focused on entrepreneurial activity example in several industries. The new method describes the procedure for the development of the application allowing a comparison of selection between its alternatives. The design of approaching is formed by four steps, which are presented as follow:

Step 1.  We have assumed that potential entrepreneurs want to start a new business between six possible options $P_1$, $P_2$, $P_3$, $P_4$, $P_5$ and $P_6$, with different characteristics (see Table 3). Each characteristic of the sector is considered a property: $P_1$: Shoe Manufacture; $P_2$: Clothing Manufacture; $P_3$: Telecommunications; $P_4$: Business Services; $P_5$: Trade in goods; $P_6$: Health Services.

Step 2.  We have analyzed and determined significant characteristics of business opportunities according to key entrepreneurial framework conditions (EFC's) (GEM 2014). Based on EFC's, the ideal level for each characteristic is fixed: $C_1$ = finances; $C_2$ = national policy-general policy; $C_3$ = national policy-regulation; $C_4$ = government programs; $C_5$ = education-primary and second; $C_6$ = education-post-school; $C_7$ = R&D transfer; $C_8$ = commercial infrastructure; $C_9$ = internal market-dynamics; $C_{10}$ = internal market-openness; $C_{11}$ = physical infrastructure; $C_{12}$ = culture and social norms. In Table 1 is shown the ideal conditions to start new business.

Table 1  Ideal conditions to start new business (IC)

|      | $C_1$ | $C_2$ | $C_3$ | $C_4$ | $C_5$ | $C_6$ | $C_7$ | $C_8$ | $C_9$ | $C_{10}$ | $C_{11}$ | $C_{12}$ |
|------|-------|-------|-------|-------|-------|-------|-------|-------|-------|----------|----------|----------|
| ICSB | 0.46  | 0.56  | 0.52  | 0.6   | 0.46  | 0.64  | 0.48  | 0.56  | 0.58  | 0.56     | 0.66     | 0.62     |

Step 3.  The real level is fixed for each characteristic for each different sectors considered. It is also remarkable that real level is identified as necessary conditions. Here, each level could be composed by objective or subjective information according to expert specifications and its advices. In Table 2 is shown necessary conditions to start a business and its degree of importance.

Table 2  Necessary conditions

| NC    | $C_1$ | $C_2$ | $C_3$ | $C_4$ | $C_5$ | $C_6$ | $C_7$ | $C_8$ | $C_9$ | $C_{10}$ | $C_{11}$ | $C_{12}$ |
|-------|-------|-------|-------|-------|-------|-------|-------|-------|-------|----------|----------|----------|
| $P_1$ | 0.1   | 0.6   | 0.4   | 0.8   | 0.4   | 0.7   | 0.7   | 0.8   | 0.7   | 0.7      | 0.9      | 0.9      |
| $P_2$ | 0.4   | 0.6   | 0.4   | 0.2   | 0.7   | 0.6   | 0.3   | 0.6   | 0.6   | 0.6      | 0.7      | 0.6      |
| $P_3$ | 0.6   | 0.6   | 0.5   | 0.5   | 0.7   | 0.6   | 0.3   | 0.6   | 0.1   | 0.3      | 0.9      | 0.9      |
| $P_4$ | 0.1   | 0.6   | 0.5   | 0.3   | 0.7   | 0.9   | 0.2   | 0.2   | 0.5   | 0.7      | 0.5      | 0.3      |
| $P_5$ | 0.1   | 0.6   | 0.7   | 0.2   | 0.3   | 0.6   | 0.3   | 0.4   | 0.1   | 0.1      | 0.7      | 0.6      |
| $P_6$ | 0.1   | 0.6   | 0.2   | 0.2   | 0.6   | 0.6   | 0.5   | 0.8   | 0.6   | 0.7      | 0.7      | 0.9      |

*Step 4.* For making a technical comparison between ICSB and NC are considered the use of the FS-OWA operator, expressly the FS-OWAAC, the FS-OWAIMAM and the FS-OWAD operators. In this application, we assumes that the entrepreneur decide to take into account *the most problematic factor for doing business* (WEF 2014) as weight vector ($W_{PF}$) (see Table 3). The main idea is to analyze and compare how these conditions can influence at the time of deciding to start a new business.

If we use FS-OWAAC, we get the following results shown in Table 4. In this case the highest result obtained by its average, it is selected as optimal option.

If we use FS-OWAD, we get the following results shown in Table 5. In this case the lowest result obtained by its average, it is selected as optimal option.

**Table 3** Problematic Factors for doing business rates

| $w_{PF}$ | Rate | N | IN | $w_{PF}$ | Rate | N | IN | $w_{PF}$ | Rate | N | IN |
|---|---|---|---|---|---|---|---|---|---|---|---|
| 1 | 20.2 | 1 | 0 | 7 | 7.4 | 0.36 | 0.63 | 12 | 2.3 | 0.11 | 0.88 |
| 2 | 14.6 | 0.72 | 0.27 | 8 | 6.3 | 0.31 | 0.68 | 13 | 1.8 | 0.08 | 0.91 |
| 3 | 12.2 | 0.60 | 0.39 | 9 | 3.3 | 0.16 | 0.83 | 14 | 1.7 | 0.08 | 0.91 |
| 4 | 8.1 | 0.40 | 0.59 | 10 | 2.8 | 0.13 | 0.86 | 15 | 0.9 | 0.04 | 0.95 |
| 5 | 8.1 | 0.40 | 0.59 | 11 | 2.4 | 0.11 | 0.88 | 16 | 0.2 | 0.01 | 0.99 |
| 6 | 7.7 | 0.38 | 0.61 | | | | | | | | |

Source: Own elaboration based on Global Competitiveness Report, 2012–2013. Note that resulting data has been Normalized (N) to establish the weight of each factor and inverse normalized (IN) for showing positive factor effects

We use FS-OWAIMAM I and FS-OWAIMAM II, which are combinatorial of Hamming distance and adequacy coefficient. Thus, it is assumed that characteristics $C_1$, $C_4$, $C_7$, $C_8$, $C_{11}$ and $C_{12}$ are dealt with adequacy coefficient and $C_2$, $C_3$, $C_5$, $C_6$, $C_9$ and $C_{10}$ are dealt with Hamming distance for FS-OWAIMAM I, and characteristics $C_1$, $C_4$, $C_7$, $C_8$, $C_{11}$ and $C_{12}$ are dealt with Hamming distance and $C_2$, $C_3$, $C_5$, $C_6$, $C_9$ and $C_{10}$ are dealt with adequacy coefficient for FS-OWAIMAM II. Their results are shown in Tables 6 and 7. Finally, it is shown in the Table 8 the order matrix of each business alternative obtained. Furthermore, It is important to take into account that results lead to different decision depending on the type of operator of aggregation used.

Then, the order established for all vectors allows us to analyze the individual decision-making to start a new business (see Table 8). Firstly, we look at business opportunity $P_1$ is the most favourable alternative followed by $P_2$. Business opportunities $P_3$ and $P_6$ have some option to be considered, although there are some necessary conditions that are not favourable. Finally, business opportunities $P_4$ and $P_5$ are far less likely to be considered.

**Table 4** Results of vector $W_{PF}$ with FS-OWAAC operator

| Vectors $w_{PF}$ | 1 | 2 | 3 | 4 | 5 | 6 | 7 | 8 | 9 | 10 | 11 | 12 | 13 | 14 | 15 | 16 | Ave |
|---|---|---|---|---|---|---|---|---|---|---|---|---|---|---|---|---|---|
| $P_1$ | 0.0 | 3.82 | 4.08 | 7.17 | 7.17 | 7.95 | 8.21 | 8.98 | 10.01 | 10.01 | 10.53 | 10.53 | 10.79 | 11.05 | 11.10 | 11.46 | 8.30 |
| $P_2$ | 0.0 | 3.72 | 3.97 | 6.98 | 6.98 | 7.73 | 7.98 | 8.74 | 9.74 | 9.74 | 10.24 | 10.24 | 10.49 | 10.75 | 10.80 | 11.15 | 8.08 |
| $P_3$ | 0.0 | 3.65 | 3.90 | 6.86 | 6.86 | 7.60 | 7.85 | 8.59 | 9.58 | 9.58 | 10.07 | 10.07 | 10.32 | 10.56 | 10.61 | 10.96 | 7.94 |
| $P_4$ | 0.0 | 3.39 | 3.62 | 6.36 | 6.36 | 7.05 | 7.28 | 7.97 | 8.88 | 8.88 | 9.34 | 9.34 | 9.57 | 9.80 | 9.84 | 10.16 | 7.37 |
| $P_5$ | 0.0 | 3.28 | 3.50 | 6.17 | 6.17 | 6.83 | 7.05 | 7.72 | 8.61 | 8.61 | 9.05 | 9.05 | 9.27 | 9.49 | 9.54 | 9.85 | 7.14 |
| $P_6$ | 0.0 | 3.60 | 3.84 | 6.76 | 6.76 | 7.49 | 7.73 | 8.46 | 9.43 | 9.43 | 9.92 | 9.92 | 10.16 | 10.41 | 10.45 | 10.79 | 7.82 |

**Table 5** Results of vector $W_{PF}$ with FS-OWAD operator

| Vectors $W_{PF}$ | 1 | 2 | 3 | 4 | 5 | 6 | 7 | 8 | 9 | 10 | 11 | 12 | 13 | 14 | 15 | 16 | Ave |
|---|---|---|---|---|---|---|---|---|---|---|---|---|---|---|---|---|---|
| $P_1$ | 0.0 | 1.22 | 1.30 | 2.28 | 2.28 | 2.53 | 2.61 | 2.86 | 3.19 | 3.19 | 3.35 | 3.35 | 3.43 | 3.51 | 3.53 | 3.65 | 2.64 |
| $P_2$ | 0.0 | 1.43 | 1.52 | 2.68 | 2.68 | 2.97 | 3.07 | 3.36 | 3.74 | 3.74 | 3.94 | 3.94 | 4.03 | 4.13 | 4.15 | 4.28 | 3.10 |
| $P_3$ | 0.0 | 1.35 | 1.44 | 2.53 | 2.53 | 2.81 | 2.90 | 3.17 | 3.54 | 3.54 | 3.72 | 3.72 | 3.81 | 3.90 | 3.92 | 4.05 | 2.93 |
| $P_4$ | 0.0 | 1.64 | 1.75 | 3.08 | 3.08 | 3.41 | 3.52 | 3.85 | 4.29 | 4.29 | 4.52 | 4.52 | 4.63 | 4.74 | 4.76 | 4.91 | 3.56 |
| $P_5$ | 0.0 | 1.88 | 2.01 | 3.54 | 3.54 | 3.92 | 4.05 | 4.43 | 4.94 | 4.94 | 5.19 | 5.19 | 5.32 | 5.45 | 5.47 | 5.65 | 4.10 |
| $P_6$ | 0.0 | 1.48 | 1.58 | 2.77 | 2.77 | 3.07 | 3.17 | 3.47 | 3.87 | 3.87 | 4.07 | 4.07 | 4.17 | 4.27 | 4.29 | 4.43 | 3.21 |

**Table 6** Results of vector $W_{PF}$ with FS-OWAIMAM I operator

| Vectors $W_{PF}$ | 1 | 2 | 3 | 4 | 5 | 6 | 7 | 8 | 9 | 10 | 11 | 12 | 13 | 14 | 15 | 16 | Ave |
|---|---|---|---|---|---|---|---|---|---|---|---|---|---|---|---|---|---|
| $P_1$ | 0.0 | 0.24 | 0.25 | 0.45 | 0.45 | 0.50 | 0.51 | 0.56 | 0.63 | 0.63 | 0.66 | 0.66 | 0.67 | 0.69 | 0.69 | 0.72 | 0.52 |
| $P_2$ | 0.0 | 0.31 | 0.33 | 0.58 | 0.58 | 0.65 | 0.67 | 0.73 | 0.82 | 0.82 | 0.86 | 0.86 | 0.88 | 0.90 | 0.90 | 0.93 | 0.68 |
| $P_3$ | 0.0 | 0.37 | 0.40 | 0.70 | 0.70 | 0.78 | 0.80 | 0.88 | 0.98 | 0.98 | 1.03 | 1.03 | 1.06 | 1.08 | 1.09 | 1.12 | 0.81 |
| $P_4$ | 0.0 | 0.76 | 0.81 | 1.42 | 1.42 | 1.57 | 1.63 | 1.78 | 1.98 | 1.98 | 2.09 | 2.09 | 2.14 | 2.19 | 2.20 | 2.27 | 1.64 |
| $P_5$ | 0.0 | 0.72 | 0.77 | 1.36 | 1.36 | 1.50 | 1.55 | 1.70 | 1.89 | 1.89 | 1.99 | 1.99 | 2.04 | 2.09 | 2.10 | 2.16 | 1.57 |
| $P_6$ | 0.0 | 0.45 | 0.48 | 0.84 | 0.84 | 0.93 | 0.96 | 1.05 | 1.18 | 1.18 | 1.24 | 1.24 | 1.27 | 1.30 | 1.30 | 1.35 | 0.98 |

**Table 7** Results of vector $W_{PF}$ with FS-OWAIMAM II operator

| Vectors $W_{PF}$ | 1 | 2 | 3 | 4 | 5 | 6 | 7 | 8 | 9 | 10 | 11 | 12 | 13 | 14 | 15 | 16 | Ave |
|---|---|---|---|---|---|---|---|---|---|---|---|---|---|---|---|---|---|
| $P_1$ | 0.0 | 0.53 | 0.56 | 0.99 | 0.99 | 1.10 | 1.14 | 1.24 | 1.39 | 1.39 | 1.46 | 1.46 | 1.49 | 1.53 | 1.54 | 1.59 | 1.15 |
| $P_2$ | 0.0 | 0.25 | 0.27 | 0.47 | 0.47 | 0.52 | 0.53 | 0.58 | 0.65 | 0.65 | 0.68 | 0.68 | 0.70 | 0.72 | 0.72 | 0.75 | 0.54 |
| $P_3$ | 0.0 | 0.55 | 0.59 | 1.04 | 1.04 | 1.15 | 1.19 | 1.30 | 1.45 | 1.45 | 1.53 | 1.53 | 1.56 | 1.60 | 1.61 | 1.66 | 1.20 |
| $P_4$ | 0.0 | 0.58 | 0.62 | 1.08 | 1.08 | 1.20 | 1.24 | 1.35 | 1.51 | 1.51 | 1.59 | 1.59 | 1.63 | 1.67 | 1.67 | 1.73 | 1.25 |
| $P_5$ | 0.0 | 0.68 | 0.73 | 1.28 | 1.28 | 1.42 | 1.47 | 1.60 | 1.79 | 1.79 | 1.88 | 1.88 | 1.93 | 1.97 | 1.98 | 2.05 | 1.48 |
| $P_6$ | 0.0 | 0.52 | 0.55 | 0.97 | 0.97 | 1.08 | 1.11 | 1.21 | 1.35 | 1.35 | 1.42 | 1.42 | 1.46 | 1.49 | 1.50 | 1.55 | 1.12 |

**Table 8** Order matrix

| Order | 1 | 2 | 3 | 4 | 5 | 6 |
|---|---|---|---|---|---|---|
| FS-OWAAC | $P_1$ | $P_2$ | $P_3$ | $P_6$ | $P_4$ | $P_5$ |
| FS-OWAD | $P_1$ | $P_2$ | $P_3$ | $P_6$ | $P_5$ | $P_4$ |
| FS-OWAIMAM I | $P_2$ | $P_6$ | $P_1$ | $P_3$ | $P_4$ | $P_5$ |
| FS-OWAIMAM II | $P_1$ | $P_3$ | $P_2$ | $P_6$ | $P_4$ | $P_5$ |

# 5   Conclusion

We have analysed the selection of opportunities of entrepreneurship for starting new business. We have proposed new method that uses selection OWA operators and fuzzy significance vector, which allow combining attitudinal character of entrepreneur and importance of characteristics of the environment. Thus, the introduction of the Selection OWA operator using fuzzy significance vector can reflect decision with different degrees of optimism and pessimism according to the significance of the characteristic. These methods are called FS-OWAAC, FS-OWAD and FS-OWAIMAM operator.

We have developed a numerical example where depending on the type index used and weight of vector, the results might be different, which implies that decision-making might be different too. Furthermore, we have analysed the results obtained. Values of each alternative change according to weight of 16 vectors assessed and depending on selection operator used the results are abroad or narrow. Likewise, FS-OWAIMAM is shown a dual version of it. Finally, order matrix has shown opportunities for starting business according to the most favourable alternative, potential alternative and more difficult to develop. Therefore, these methods lead us to select the alternative that fits better the interest of entrepreneur or investor team.

This new extension of the OWA operator can be applied in others families of OWA operators due to the boundary condition is not accomplished in the operator because it is really easy to obtain a result through the maximum. Notwithstanding, in order to accomplish the boundary condition, a normalization process could be used where the sum of the weights become one (Merigó and Gil-Lafuente 2009). Finally, it will be considered to apply in decision-making problems in business as strategy management, investment, marketing and costumer services.

# References

Acs, Z.J., Amorós, J.E.: Entrepreneurship and competitiveness dynamics in Latin America. Small Bus. Econ. **31**(3), 305–322 (2008)

Audretsch, D., Keilbach, M.: Entrepreneurship capital and economic performance. Reg. Stud. **38** (8), 949–959 (2004)

Cuervo, A.: Individual and environmental determinants of entrepreneurship. Int. Entrep. Manag. J. **1**(3), 293–311 (2005)

GEM.: Global Entrepreneurship Monitor 2013 Global Report, Global Entrepreneurship Monitor. http://gemconsortium.org/docs/3106/gem-2013-global-report (2014)

Gil-Aluja, J.: Elements for a Theory of Decision in Uncertainty. Kluwer Academic Publishers, Dordrecht (1999)

Gil-Lafuente, A.M., Merigó, J.M.: The ordered weighted averaging distance operator. Lect. Modell. Simul. **8**(1), 84–95 (2007)

Gil-Lafuente, A.M., Merigó, J.M.: On the use of the OWA operator in the adequacy coefficient. Modell. Meas. Control **30**(1), 1–17 (2009)

Gil-Lafuente, J.: Index of maximum and minimum level in the optimization of athlete signing. In: Proceeding of X Internacional Congress A.E.D.E.M, Reggio Calabria (Italy), 4–6 Sept 2001, pp. 439–443 (2001) (In Spanish)

Herrera, F.: A sequential selection process in group decision making with a linguistic assessment approach. Inf. Sci. **85**(4), 223–239 (1995)

Kaufmann, A., Gil-Aluja, J.: Introduction of the fuzzy sub-sets theory to the business administration. Milladoiro Santiago de Compostela (1986) (In Spanish]

Martinez, L., Herrera, F.: A 2-tuple fuzzy linguistic representation model for computing with words. IEEE Trans. Fuzzy Syst. **8**(6), 746–752 (2000)

Merigó, J.M.: New extentions to the OWA operators and their applications in decision making. Doctoral thesis. University of Barcelona, Spain (2009) (In Spanish)

Merigó, J.M., Gil-Lafuente, A.M.: Unification point in methods for the selection of financial products. Fuzzy Econ. Rev. **12**(1), 35–50 (2007)

Merigó, J.M., Gil-Lafuente, A.M.: On the use of the OWA operator in the Euclidean distance. Int. J. Comput. Sci. Eng. **2**(4), 170–176 (2008a)

Merigó, J.M., Gil-Lafuente, A.M.: The generalized adequacy coefficient and its application in strategic decision making. Fuzzy Econ. Rev. **13**(2), 17–36 (2008b)

Merigó, J.M., Gil-Lafuente, A.M.: Using the OWA operator in the Minkowski distance. Int. J. Comput. Sci. **3**, 147–157 (2008c)

Merigó, J.M., Gil-Lafuente, A.M.: New decision-making techniques and their application in the selection of financial products. Inf. Sci. **180**(11), 2085–2094 (2010)

Merigó, J.M., Gil-Lafuente, A.M.: Decision-making in sport management based on the OWA operator. Expert Syst. Appl. **38**(8), 10408–10413 (2011)

Merigó, J.M., Gil-Lafuente, A.M.: Decision making techniques with similarity measures and OWA operators. SORT–Stat. Oper. Res. Trans. **36**, 81–102 (2012a)

Merigó, J.M., Gil-Lafuente, A.M.: A method for decision making with the OWA operator. Comput. Sci. Inf. Syst. **9**(1), 357–380 (2012b)

Merigó, J.M., Gil-Lafuente, A.M.: A method for decision making based on generalized aggregation operators. Int. J. Intell. Syst. **28**(5), 453–473 (2013)

Merigó, J.M., Gil-Lafuente, A.M., Xu, Y.J.: Decision making with induced aggregation operators and the adequacy coefficient. Econ. Comput. Econ. Cybern. Stud. Res. **1**, 185–202 (2013)

Merigó, J.M., Peris-Ortiz, M.: Entrepreneurship and decision- making in Latin America. INNOVAR—J. Admin. Soc. Sci. **24**(1), 101–111 (2014)

Merigó, J.M., Peris-Ortiz, M., Palacios-Marqué, D.: Entrepreneurial fuzzy group decision-making under complex environments. J. Intell. Fuzzy Syst. **27**(2), 901–912 (2014)

Peng, B., Ye, C.: An approach based on the induced uncertain pure linguistic hybrid harmonic averaging operator to group decision making. Econ. Comput. Econ. Cybern. Stud. Res. **47**(4), 275–296 (2014)

Praag, C.M., Versloot, P.H.: What is the value of entrepreneurship? A review of recent research. Small Bus. Econ. **29**(4), 351–382 (2007)

Shannon, C.E.: A mathematical theory of communication. Bell Syst. Tech. J. **27**(3), 379–423 (1948)

Verheul, I., Wennekers, S., Audretsch, D., Thurik, R.: An eclectic theory of entrepreneurship: policies, instittutions and culture. In: Audretsch, B., Thurik, R., Verheul, I., Wennekers, S. (eds.) Entrepreneurship : Determinants and Policy in a European-US Comparison, pp. 11–81. Kluwer Academic, Boston (2002)

WEF.: The global competitiveness report 2013–2014. World Economic Forum. http://www. weforum.org/reports/global-competitiveness-report-2013-2014 (2014)

Wei, G.: Uncertain linguistic hybrid geometric mean operator and its applications to group decision making under uncertain linguistic environment. Int. J. Uncertain. Fuzziness Knowl.-Based Syst. 17(02):251–267 (2009)

Wennekers, S., Thurik, R.: Linking entrepreneurship and economic growth. Small Bus. Econ. 13 (1), 27–56 (1999)

Wong, P.K., Ho, Y.P., Autio, E.: Entrepreneurship, innovation and economic growth: evidence from GEM data. Small Bus. Econ. 24(3), 335–350 (2005)

Xu, Z.S.: EOWA and EOWG operators for aggregating linguistic labels based on linguistic preference relations. Int. J. Uncertain. Fuzziness Knowl.-Based Syst. 12(06), 791–810 (2004)

Yager, R.R.: On ordered weighted averaging aggregation operators in multicriteria decision making. IEEE Trans. Syst. Man Cybern. 18(1), 183–190 (1988)

Yager, R.R.: Families of OWA operators. Fuzzy Sets Syst. 59(2), 125–148 (1993)

Yager, R.R., Kacprzyk, J., Beliakov, G.: Recent Developments on the Ordered Weighted Averaging Operators: Theory and Practice. Springer, Berlin (2011)

Zhao, H., Xu, Z., Ni, M., Liu, S.: Generalized aggregation operators for intuitionistic fuzzy sets. Int. J. Intell. Syst. 25(1), 1–30 (2010)

Zhou, L.G., Chen, H.: Generalized ordered weighted logarithm aggregation operators and their applications to group decision making. Int. J. Intell. Syst. 25(7), 683–707 (2010)

# Credit Analysis Using a Combination of Fuzzy Robust PCA and a Classification Algorithm

Onesfole Kurama, Pasi Luukka and Mikael Collan

**Abstract** Classification is a key part of credit analysis and bankruptcy prediction and new powerful classification methods coming from artificial intelligence are often applied. Most often classification methods require pre-processing of data. This paper presents a two-part classification process that combines a pre-processing step that uses fuzzy robust principal component analysis (FRPCA) and a classification step. Combinations of three FRPCA algorithms and two different classifiers, similarity classifier and fuzzy k-nearest neighbor classifier, are tested to find the combination that gives the most accurate mean classification result. Tests are run with a small Australian credit data set that can be considered "rough" and to require "robust" methods, due to the small number of observations. The created principal components are used as inputs in the classification methods. Results obtained indicate a mean classification accuracies of over 80 % for all combinations. It becomes clear that parameters of the used methods clearly affect the results and emphasis is put on finding suitable parameters.

**Keywords** Credit decision-making · Australian credit screening dataset · Fuzzy robust PCA · Similarity classifier · Fuzzy k-nearest neighbor

## 1 Introduction

Business and academic communities have paid a lot of attention to bankruptcy prediction and credit scoring within the fields of accounting and finance. Many methods that are commonly used in bankruptcy prediction stem from the broad

O. Kurama (✉) · P. Luukka
Laboratory of Applied Mathematics, Lappeenranta University of Technology, P.O Box 20, 53851 Lappeenranta, Finland
e-mail: pasi.luukka@lut.fi

P. Luukka · M. Collan
School of Business and Management, Lappeenranta University of Technology, P.O Box 20, 53851 Lappeenranta, Finland
e-mail: mikael.collan@lut.fi

© Springer International Publishing Switzerland 2015
J. Gil-Aluja et al. (eds.), *Scientific Methods for the Treatment of Uncertainty in Social Sciences*, Advances in Intelligent Systems and Computing 377, DOI 10.1007/978-3-319-19704-3_2

discipline of artificial intelligence (AI). The problem is real, failed credit decision-making in financial institutions may run them into financial difficulty, while simultaneously it is not good business to not lend to good borrowers – the accuracy of the ability of financial institutions to *select good borrowers* from the group of potential borrowers is matter of competitive advantage.

The grand assumption in modern bankruptcy prediction and credit analysis, since the late 1960s, is that financial information, extracted from financial statements, such as financial ratios, contains large amounts of relevant information (Altman 1968, Beaver 1968). A variety of information about the status of client companies is available from financial and accounting data and in addition to this "hard" data, information may be based on normative judgments of experts. When it comes to individual borrowers the data may be in a different format, but the same principles for credit analysis hold. Results from these analyses can also be used as "early warning signs" that can be acted on to prevent business failure by stakeholder intervention (such as, forcing mergers of distressed firms, liquidation, or reorganization). Another, perhaps a more commonly used, way of using the results is to use them in credit decision-making that is, in the credit approval process (Dimitras et al. 1996).

Bankruptcy prediction and credit analysis in general can be thought of as a classification problem, where by using multiple inputs, we try to classify potential lenders into two categories: ones with a high bankruptcy risk and the ones with no or low bankruptcy risk that is, to the group to grant a loan to and the group not to grant a loan to. As the information systems era has brought with it large data sets of relevant data for credit analysis, both with regards to companies and individuals as clients, the main focus of research has shifted to using modern data mining techniques and a variety of these methods has been applied to bankruptcy prediction, for overviews on different methods see, e.g., (Back et al. 1996; Dimitras et al. 1996; Huang et al. 2004; Lensberg et al. 2006; Kumar and Ravi 2007). It has been shown that data mining and AI techniques can outperform classical statistical approaches in bankruptcy prediction (Zhang et al. 1999; Min and Lee 2005; Shin et al. 2005; Kumar and Ravi 2007). Recently, models that incorporate more than just client specific that is, "environmental" information in the bankruptcy prediction process have started to emerge see, e.g., (Karas and Reznakova 2014; López Iturriaga and Pastor Sanz 2015). Bankruptcy prediction methods from fuzzy and artificial intelligence have recently grown in popularity see, e.g., (Zopounidis et al. 1999; Ahn et al. 2000; Mckee 2000; Quek et al. 2009; Terceno and Vigier 2011).

In this paper we use a two-step data classification procedure for credit analysis, where in the first step, the data undergoes a fuzzy robust principal component analysis (FRPCA) in line with (Yang and Wang 1999) and by using (testing) three different FRPCA algorithms. The resulting principal components are then used in the second step as inputs in classification. In the classification step we use two different classification methods, the fuzzy K-nearest neighbor classifier (FKNN) by (Keller et al. 1985) and the similarity classifier presented in (Luukka et al. 2001, Luukka and Leppälampi 2006). A blueprint of the process is visible in Fig. 1.

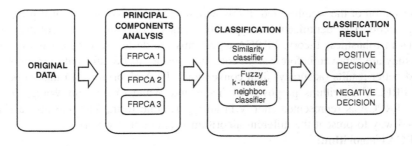

**Fig. 1** Blueprint of the two-step classification process with the tested methods indicated. The known correct values are compared to the resulting classification

Combinations of the three first-step FRPCA methods and the two second-step classification methods are all tested for parameter values to find combinations that result in good classification accuracy. A two-step system with these characteristics is, to the best of our knowledge, a new approach.

The next section presents the details of the two-step classification procedure by first going through the three different fuzzy robust principal component analysis algorithms and how they are used, the two classification algorithms are then presented in detail. This is followed by a walk-through of classification results with the procedure, obtained with test data. The paper closes with discussion and conclusions.

## 2 Classification Procedure

In this section we first briefly review the applied methods, then summarize how they are used, and then the "combinations" are compared. Methods include three fuzzy robust principal component analysis methods (Yang and Wang 1999) in forming robust linear combinations from the data and the fuzzy k-nearest neighbor (Keller et al. 1985) and the similarity classifier (Luukka and Leppälampi 2006) as classification methods. The task is to find the best combination through selecting the most suitable FRPCA method with the proper principal components, and parameter values and to run with the two classifiers.

### 2.1 *Fuzzy Robust Principal Component Analysis (FRPCA)*

In standard PCA, principal components are often influenced by outliers that can in a bad case distort the weights applied in forming these linear combinations. This distortion can affect the accuracy of the results, or even lead to misleading results. The underlying idea in fuzzy robust PCA methods is that these outliers are removed in formation of the weights, and in this way we can produce more robust principal components. The identification and possible removal of outliers is one of the main topics in robustness theory. FRPCA algorithms are designed to detect outliers so that they are removed before weight computations are made, this makes them often yield better results (accuracy) than conventional PCA algorithms. In this work we use FRPCA algorithms proposed by Yang and Wang (Yang and Wang 1999). Next FRPCA is represented shortly in an algorithmic form, allowing also a simplified way to present the different algorithm versions that we apply.

**FRPCA1 algorithm:**

Step 1: Initially set the iteration count to $t = 1$, iteration bound T, learning coefficient $\alpha_0 \in (0, 1]$, soft threshold $\lambda$ to a small positive value, and randomly initialize the weight $\phi$.

Step 2: While $t$ is less than T, perform the next steps 3–9.

Step 3: Compute $\alpha_t = \alpha_0 (1 - t/T)$, set $i = 1$ and $\sigma = 0$.

Step 4: While $i$ is less, than $n$ perform steps 5–8.

Step 5: Compute $y = \phi^T x_i$, $v = y\phi$, and $u = \phi^T v$

Step 6: Update the weight with: $\varnothing^{new} = \varnothing^{old} + \alpha_t \beta(x_i)[y(x_i - v) + (y - u)x_i]$

Step 7: Update the temporary count, $\sigma = \sigma + \theta_1(x_i)$

Step 8: Add 1 to $i$.

Step 9: Compute $\lambda = \sigma/n$ and add 1 to $t$.

**FRPCA2 algorithm:**

The procedures above are followed, except for steps 6 and 7, where we change the way we update the weight and the count:

Step 6: Update the weight with: $\varnothing^{new} = \varnothing^{old} + \alpha_t \beta(x_i) \left( x_i y - \frac{\varnothing}{\varnothing^T \varnothing} y^2 \right)$

Step 7: Update the temporary count: $\sigma = \sigma + \theta_2(x_i)$

**FRPCA3 algorithm:**

Follow the same procedure as in FRPCA1, except for steps 6 and 7:

Step 6: Update the weight with: $\varnothing^{new} = \varnothing^{old} + \alpha_t \beta(x_i)(x_i y - \varnothing y^2)$

Step 7: Update the temporary count: $\sigma = \sigma + \theta(x_i)$

## 2.2 Fuzzy k-Nearest Neighbor Classifier

The k-nearest neighbor algorithm is used to assign patterns of unknown classification to the class of the majority of its k-nearest neighbors of known classification (Keller et al. 1985). The classical model assumes that each sample belongs to a specific (single) class, the fuzzy extension however allows a sample to belong to more than one class simultaneously. Membership values are assigned to the samples in relation to their contribution to the classification process. Starting from the closest neighbor, the algorithm runs until all the clusters in the neighborhood are

covered. Initialization determines to which local minima the algorithm will termi-
nate. Thus the algorithm will always search for the labeled patterns in the nearest
class (Keller et al. 1985). To compute membership values for different samples, the
expression below is useful:

$$\mu_i(x) = \frac{\sum_{j=1}^{K} \mu_{i,j} (1 / \|x - x_j\|^{2/(n-1)})}{\sum_{j=1}^{K} (1 / \|x - x_j\|^{2/(n-1)})} \tag{1}$$

where $\mu_{i,j}$ is the membership in the $i^{th}$ class of the $j^{th}$ sample of the labeled
pattern and $n_1$ is the scaling factor for fuzzy weights.

## 2.3 Similarity Classifier

Similarity classifier was introduced in (Luukka and Leppälampi 2006). Main
components in the classifier are the computation of ideal vectors for each class and
computing similarity between samples and the ideal vectors. Ideal vectors, $v_i$ for
class $i$ are given as:

$$v_i = (v_{i1}, v_{i2}, \dots v_{in}) \tag{2}$$

In current computations we compute ideal vectors by using the generalized
mean, by fixing a parameter $m$, and the ideal vector is then computed as:

$$v_i(d) = (\frac{1}{\#X_i} \sum_{x \in X_i} x_d^m)^{1/m}, \quad \forall d = 1, 2, \dots, n \tag{3}$$

Generalized Łukasiewicz similarity (Luukka and Leppälampi 2006) have been
found to be useful in comparisons between samples and ideal vectors. A weight
parameter $w_d$, which varies feature by feature and depends on feature importance, is
also applied and makes the resulting similarity:

$$S\langle x, y \rangle = \frac{1}{n} \sum_{d=1}^{n} w_d (1 - |x_d^p - v_d^p|)^{m/p} \tag{4}$$

In experiments with real data it is important to calibrate the parameters $m$ and $p$,
see the Eq. (4), since they affect the classification accuracy.

# 3 Classification Results with an Australian Credit Screening Dataset

This section presents the results obtained by using the proposed classification process, by using the "Australian credit screening dataset" as a basis. The data set is available for download at the UCL machine learning data repository (URL: http://archive.ics.uci.edu/ml/). Key properties of the data are summarized in Table 1.

| Table 1 Summary of properties of Australian credit screening data set | | |
|---|---|
| Number of variables: | 14(continuous: 6, categorical:8) |
| Number of classes: | 2 |
| Number of samples: | 690 |
| Class distribution: | Bad credit: 307, Good credit 383 |

Emphasis is put on the performance of the classifiers, as well as, on the preprocessing procedures utilized. Parameters used have also been "optimized" in the sense that they have been pre-tested to obtain acceptable ranges for each parameter.

## 3.1 Results from the Australian Credit Scoring Data with FRPCA and the Similarity Classifier

In this experiment, ideal vectors are computed for each class across all features and compared with the rest of the samples by calculating the similarities between them. Generalized Łukasiewicz based similarity, with generalized mean is used for computation of the similarities. Matlab software has been utilized in designing and implementing the algorithms. Since not all the principal components necessarily hold important information related to classification process, we study how many principal components need to be considered, in order to obtain the best accuracy. This is straight-forward, because the first principal components should hold the most information, and in an ideal scenario the last component should mostly consist of noise. This way we also get a ranking order for principal components. Different FRPCA-algorithms (FRPCA1, FRPCA2, and FRPCA3) are used in forming these principal components. Random division of the data to training and testing sets is done several times (30 times in this case) – that is we run 30 runs with different sets selected from the data. Mean classification accuracies are computed. In these experiments we set the fuzziness variable set to n = 2.

The best mean accuracy of 82.92 % is reached with FRPCA3, when all principal components (14) are used. Since the classification accuracy improves until when all principal components are used, it seems that important information is "still left" even in the last principal component (PC). The other two methods also perform well: FRPCA2 gives a mean classification accuracy of 81.84 % (with 13 PC), and

FRPCA1 results in 81.73 % (also with 13 PC). Figure 2 shows how the number of principal components affects the mean classification accuracies with the similarity classifier.

**Fig. 2** Mean classification accuracies with respect to the number of principal components used for FRPCA1, FRPCA2, and FRPCA3

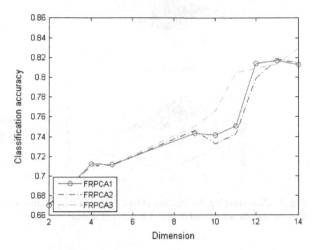

In a follow-up experiment we examine the effect of using different fuzziness variable $n$ values for FRPCA algorithms. The best choice of the fuzziness parameter value for FRPCA1 was found to be around $n = 1.8$ and corresponding to a classification accuracy of 83.22 %. With the fuzziness parameter value around $n = 1.2$ FRPCA2 has a classification accuracy of 82.09 %, and FRPCA3 has an accuracy of 82.97 %. The experiment shows that the highest mean classification accuracy can be reached with FRPCA1. In addition to the parameter values of the FRPCA algorithms also the parameters in the similarity classifier have an effect on the mean classification accuracy of the process, (parameters m and p) were introduced shortly in Eq. (4).

A way to locate the best similarity classifier parameter values is to compute mean classification accuracies for a grid of suitable parameter values. By using figures, where we plot the mean classification accuracy with regards to different parameter values, we can by visual inspection find the best "area" for these parameters.

Example of this is given in Fig. 3, where we have plotted the variance, mean, minimum, and maximum classification accuracies with regards to parameters m and p, for FRPCA1 with 13 PC used, and with the fuzziness parameter set to n = 1.8. By visual inspection one can see that the best value of p should be chosen from within the interval [0.5, 3], with p = 2 seeming to be the best choice. The value of m should be taken from the interval [0.3, 6] with m = 5 looking to be the best. The approximately best parameter pair $(p, m) = (2, 5)$ is chosen.

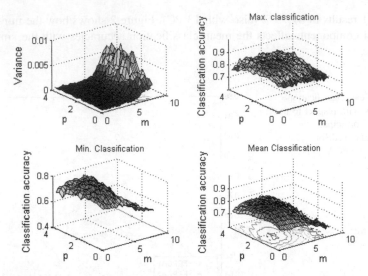

**Fig. 3** Plots showing the location of best values of the parameters $p$ and $m$ with FRPCA1, with the fuzziness parameter fixed at $n = 1.8$ and 13PC used

## 3.2 Results from the Australian Credit Scoring Data with FRPCA and Fuzzy k-Nearest Neighbor Classifier

The experiment is set up as the first one, the number of principal components used for the best mean classification result and a suitable number of nearest neighbors are investigated. We also look into the fuzziness parameter $n$ value for the FRPCA algorithms and the scaling factor value for fuzzy weights $n_1$ used in the FKNN classifier, see Eq. (3). Table 2 summarizes the results for the mean classification accuracy with different numbers of principal components used and the number of nearest neighbors, when using FRPCA2. The same analysis was made for all three FRPCA algorithms, but using FPRCA 2 with 14 PC and k = 13 nearest neighbors resulted in the best mean classification result of 84.55 %. These results were obtained with the fuzziness variable $n$ fixed at 2.

**Table 2** Mean classification results using FRPCA2 with FKNN classifier for principal components PC and the number of nearest neighbors K. Fuzziness variables are fixed to $n = n_1 = 2$. The best mean classification accuracy is given in **boldface**

| K<br>PC | 8 | 9 | 10 | 11 | 12 | 13 | 14 |
|---|---|---|---|---|---|---|---|
| 14 | 0.8366 | 0.8380 | 0.8402 | 0.8414 | 0.8421 | 0.8417 | 0.8427 |
| 13 | 0.8392 | 0.8390 | 0.8419 | 0.8437 | 0.8437 | 0.8446 | **0.8455** |
| 12 | 0.8390 | 0.8410 | 0.8413 | 0.8425 | 0.8420 | 0.8418 | 0.8433 |
| 11 | 0.8344 | 0.8351 | 0.8363 | 0.8381 | 0.8395 | 0.8410 | 0.8422 |
| 10 | 0.8104 | 0.8101 | 0.8120 | 0.8138 | 0.8153 | 0.8167 | 0.8173 |

It can generally be seen that the mean classification accuracy improves with increase in both the number of PC used and the number of nearest neighbors. Next, the proper values for the fuzziness parameter $n$, for FRPCA algorithms is studied. The best classification result, 85.00 %, is obtained by using FRPCA2 with a fuzziness parameter value $n = 2.5$. The fuzziness parameter value for FKNN was fixed to $n_1 = 2$. The best classification accuracy for FRPCA1 is 84.87 % and was obtained, with $n = 2.2$, FRPCA3 reaches a classification accuracy of 84.82 %, when $n = 3.5$. For finding suitable FKNN classifier fuzziness parameter $n_1$ values we fixed FRPCA variable $n$ values to those that return the best classification results. The mean classification accuracy obtained for different $n_1$ parameter values is shown in Table 3.

**Table 3** Mean classification accuracies for several FKNN parameter $n_1$ values. Values of $n$ fixed for FRPCA1 $n = 2.2$, for FRPCA2 $n = 2.5$, and for FRPCA3 $n = 3.5$. The best result in **boldface**

| $n_1$ | FRPCA1 | FRPCA2 | FRPCA3 |
|-------|--------|--------|--------|
| 2.0 | 0.8481 | 0.8485 | 0.8491 |
| 2.8 | 0.8515 | 0.8552 | 0.8565 |
| 3.0 | 0.8527 | 0.8535 | 0.8566 |
| 4.0 | 0.8516 | 0.8531 | 0.8579 |
| 4.5 | 0.8521 | **0.8593** | 0.8574 |
| 5.0 | 0.8546 | 0.8544 | 0.8562 |

The best mean classification accuracy of 85.93 % is achieved with FRPCA2, when $n_1 = 4.5$ and $n = 2.5$. FRPCA3 reaches the accuracy of 85.79 %, when $n_1 = 4.0$ and $n = 3.5$ and FRPCA1 85.46 %, when $n_1 = 5.0$ and $n = 2.2$. The mean classification accuracy did not improve with $n_1 > 5.0$.

## 3.3 Short Discussion About the Results

We ran tests on the proposed credit analysis process by using three FRPCA algorithms with two classification algorithms and tuned the parameters of these methods for better results. Experiments with the similarity classifier show that combining it with FRPCA1 gives the best mean classification accuracy of 83.22 %, when $n = 1.8$, and p & m are chosen around 2 and 5 respectively. Experiments with the FRPCA2 paired with FKNN classifier produced the best results, with a mean classification accuracy of 85.93 % reached. Table 4 summarizes the results.

**Table 4** Summary of the results obtained with suitable choices for fuzziness parameters $n$ (FRPCA) and $n_l$ (FKNN), parameters $p$ (Łukasiewicz structure) and m (generalized mean), number of principal components D, and the number of nearest neighbors K, for the two classifiers with all FRPCA methods used

| Method | Similarity classifier | | FKNN classifier | |
|---|---|---|---|---|
| FRPCA1 | p = 2, | m = 5 | $n_1$ = 2.2 | K = 14 |
| n | 1.8 | | 2.2 | |
| D | 13 | | 13 | |
| Accuracy | 83.22 % | | 85.46 % | |
| FRPCA2 | p = 1.9, | m = 3.5 | $n_1$ = 4.5 | K = 14 |
| n | 1.2 | | 2.5 | |
| D | 13 | | 13 | |
| Accuracy | 82.09 % | | **85.93 %** | |
| FRPCA3 | p = 2.1, | m = 2.5 | $n_1$ = 4.0 | K = 14 |
| n | 1.2 | | 3.5 | |
| D | 14 | | 14 | |
| Accuracy | 82.97 % | | 85.79 % | |

Classification accuracies of above 80 % are achieved with all FRPCA and classifier combinations, when parameters are tuned. The choice of FRPCA-parameters is significant for the results, as is the number of principal components used in the classifiers. Furthermore, tuning of the classifier parameter values also affects the mean accuracy of the classification.

# 4 Summary and Conclusions

In this article we have tackled the problem of classifying potential borrowers with a two-step process that first uses a fuzzy robust principal components analysis and then a classifier that classifies the borrowers based on the principal components. Three FRPCA algorithms were tested for the first step and two classifier algorithms for the second step. All combinations of steps one and two were tested for mean classification accuracy, by using an Australian credit screening data set. The results show that with all combinations, when parameter values are set correctly, an above 80 % classification accuracy can be reached, in fact, the best accuracy obtained was 85.93 %. It is to be noted that the used data set is not very large and contains only 690 observations, making the problem a "robust" one. This is why we feel that the fit of the robust FRPCA method is a valid choice for the first-step of the process. What we can take home from the results of this paper is that in a two-step setup like the one presented, tuning of parameters influences the results and should be emphasized when designing and using systems.

# References

Ahn, B., Cho, S., Kim, C.: The integrated methodology of rough set theory and artificial neural network for business failure prediction. Expert Syst. Appl. **18**, 65–74 (2000)

Altman, E.I.: Financial ratios, discriminant analysis and the prediction of corporate bankruptcy. J. Finance **23**, 589–609 (1968)

Back, B., Laitinen, T., Sere, K.: Neural networks and genetic algorithms for bankruptcy predictions. Expert Syst. Appl. **11**, 407–413 (1996)

Beaver, W.H.: Market prices, financial ratios, and the prediction of failure. J. Account. Res. **6**, 179–192 (1968)

Dimitras, A.I., Zanakis, S.H., Zopounidis, C.: A survey of business failures with an emphasis on prediction methods and industrial applications. Eur. J. Oper. Res. **90**, 487–513 (1996)

Huang, Z., Chen, H., Hsu, C.J., Chen, W.H., Wu, S.: Credit rating analysis with support vector machines and neural networks: a market comparative study. Decis. Support Syst. **37**, 543–558 (2004)

Karas, M., Reznakova, M.: To what degree is the accuracy of a bankruptcy prediction model affected by the environment? the case of the Baltic States and the Czech Republic. In: Gimžauskienė, E. (ed.) 19th International Scientific Conference Economics and Management 2014 ICEM-2014, vol. 156, pp. 564–568. Elsevier, Riga, Latvia (2014)

Keller, J., Gray, M., Givens, J.R.: A fuzzy K nearest neighbor algorithm. IEEE Trans. Syst. Man Cybern. **15**, 580–585 (1985)

Kumar, P., Ravi, V.: Bankruptcy prediction in banks and firms via statistical and intelligent techniques—a review. Eur. J. Oper. Res. **180**, 1–28 (2007)

Lensberg, T., Ellifsen, A., McKee, T.E.: Bankruptcy theory development and classification via genetic programming. Eur. J. Oper. Res. **169**, 677–697 (2006)

López Iturriaga, F.J., Pastor Sanz, I.: Bankruptcy visualization and prediction using neural networks: a study of U.S. commercial banks. Expert Syst. Appl. **42**, 2857–2869 (2015)

Luukka, P., Leppälampi, T.: Similarity classifier with generalized mean applied to medical data. Comput. Biol. Med. **36**, 1026–1040 (2006)

Luukka, P., Saastamoinen, K., Könönen, V.: A classifier based on the maximal fuzzy similarity in the generalized Lukasiewicz-structure. In: FUZZ-IEEE (ed.), pp. 195–198. IEEE, Melbourne, Australia (2001)

Mckee, T.: Developing a bankruptcy prediction model via rough sets theory. Int. J. Intell. Syst. Account. Finance Manage. **16**, 159–173 (2000)

Min, J.H., Lee, Y.C.: Bankruptcy prediction using support vector machine with optimal choice of kernel function parameters. Expert Syst. Appl. **28**, 603–614 (2005)

Quek, C., Zhou, D., Lee, C.: A novel fuzzy neural approach to data reconstruction and failure prediction. Intell. Syst. Account. Finance. Manage. **16**, 165–187 (2009)

Shin, K.S., Lee, T.S., Kim, H.J.: An application of support vector machines in bankruptcy prediction model. Expert Syst. Appl. **28**, 127–135 (2005)

Terceno, A., Vigier, H.: Economic-financial forecasting model of businesses using fuzzy relations. Econ. Comput. Econ. Cybern. Stud. Res. **2011**, 185–203 (2011)

Yang, T.N., Wang, S.D.: Robust algorithms for principal component analysis. Pattern Recogn. Lett. **20**, 927–933 (1999)

Zhang, G., Hu, M.Y., Patuwo, B.E., Indro, D.C.: Artificial neural networks in bankruptcy prediction: general framework and cross-validation analysis. Eur. J. Oper. Res. **116**, 16–32 (1999)

Zopounidis, C., Slowinski, R., Doumpus, M., Dimitras, A., Susmaga, R.: Business failure prediction using rough sets: a comparison with multivariate analysis techniques. Fuzzy Econ. Rev. **4**, 3–33 (1999)

# Fuzzy TOPSIS for an Integrative Sustainability Performance Assessment: A Proposal for Wearing Apparel Industry

Elena Escrig-Olmedo, María-Ángeles Fernández-Izquierdo, María-Jesús Muñoz-Torres and Juana-María Rivera-Lirio

**Abstract** It is particularly important assessing sustainability in an integrative way considering different stakeholder perspectives to overcome the weaknesses of reductionist approaches to measure sustainability. This integration could help organizations to understand and engage with their stakeholders; and could contribute to sustainable development. The objective of this paper is developing stakeholder methodological approach, based on an application of fuzzy multi-criteria decision-making method (MCDM), to improve the contributions to organizations to sustainable development considering the particular sustainability interests of stakeholders into corporate sustainability performance measurement. With the aim of illustrating the proposed methodology in a specific industry, this study is focused on the wearing apparel industry

**Keywords** Fuzzy TOPSIS · Fuzzy inference system (FIS) · Fuzzy multi-criteria decision-making method (MCDM) · Stakeholder · Sustainability · Wearing apparel

The authors wish acknowledge the support received from P1·1B2013-31 and P1·1B2013-48 projects through the Universitat Jaume I and Sustainability and Corporate Social Responsibility Master Degree (UJI–UNED). This research is partially financed by Nils Science and Sustainability Programme (ES07), ABEL – Coordinated Mobility of Researchers (ABEL-CM-2014A), and supported by a grant from Iceland, Liechtenstein and Norway through the EEA Financial Mechanism. Operated by Universidad Complutense de Madrid.

E. Escrig-Olmedo (✉) · M.-Á. Fernández-Izquierdo
M.-J. Muñoz-Torres · J.-M. Rivera-Lirio
Department of Finance and Accounting, Universiat Jaume I,
Campus Del Riu Sec - Avda. Vicent Sos Baynat s/n, 12071 Castellón, Spain
e-mail: eescrig@uji.es

M.-Á. Fernández-Izquierdo
e-mail: afernand@uji.es

M.-J. Muñoz-Torres
e-mail: munoz@uji.es

J.-M. Rivera-Lirio
e-mail: jrivera@uji.es

© Springer International Publishing Switzerland 2015
J. Gil-Aluja et al. (eds.), *Scientific Methods for the Treatment of Uncertainty in Social Sciences*, Advances in Intelligent Systems and Computing 377,
DOI 10.1007/978-3-319-19704-3_3

31

# 1 Introduction

Companies play an important role to contribute to sustainable development of markets, economy and society (Schaltegger and Wagner 2011). In this context, assessing corporate sustainability performance is a challenge. Due to the multidimensional character of the sustainability concept, several questions should be considered in the evaluation process: (i) the qualitative nature of indicators and the complexity of developing a synthetic index; (ii) the difficulty of choosing properly statistical techniques for aggregation (Windolph 2011) and; (iii) the difficulty to introduce stakeholder's preferences in the assessment models.

The integration of different stakeholder perspectives could help organizations to understand and engage with their stakeholders; and could contribute to sustainable development. The goal of contributing to sustainable development will be a necessary logical conclusion from applying stakeholder theory (Hörisch et al. 2014).

This paper is an attempt to address this challenge by developing stakeholder methodological approach, based on an application of fuzzy multi-criteria decision-making method (MCDM), to improve the contributions to organizations to sustainable development, considering the particular sustainability interests of stakeholders into corporate sustainability performance measurement.

With the aim of illustrating the proposed methodology in a specific industry, this study is focused on the wearing apparel industry. This sector is a valuable example of important environmental and social impacts along the supply chain and moreover, it is one of the most global industries in the world.

# 2 Fuzzy Multi-criteria Decision-Making Method (MCDM)

Fuzzy MCDMs represent an alternative for the assessment of sustainability overcoming the limitations of the traditional sustainability assessment tools for many reasons: (i) they are capable of representing uncertain data of complex concepts and emulating skilled humans (Andriantiatsaholiniaina et al. 2004; Muñoz et al. 2008); (ii) they allow for the combination of quantitative and qualitative data (Munda et al. 1995; Zadeh 1975); (iii) they integrate expert knowledge in the evaluation process and; (iv) they can handle stochasticity or uncertainty of the variables without the prior knowledge of their probability distributions, which makes it superior to the existing stochastic methods for assessing sustainability (Prato 2005).

## 2.1 Fuzzy Inference System

Fuzzy logic starts with the concept of a fuzzy set. Fuzzy set theory has been developed for modeling complex systems in uncertain and imprecise environment.

According to Zadeh (1965) a fuzzy set is defined as a 'class' with a continuum of grades of membership. A fuzzy subset A of a universal set is defined by a membership function $fA(x)$ which associates each element into a real number in the interval (0, 1).

One of the most useful tools presented within the context of fuzzy set theory to deal with nonlinear, but ill-defined, mapping of input variables to some output ones is what is known as Fuzzy Inference System (FIS). FIS is a framework, which simulates the behaviour of a given system as IF–THEN rules through knowledge of experts or past available data of the system.

Our FIS is Mamdani-type (Jang 1993), and it is composed of five functional blocks.

**Block 1: A fuzzification inference which transforms the crisp inputs into degree of match with linguistic values.** In our study, the defined membership functions for the input and output variables are triangular membership functions. The defined membership functions are triangular because of their simplicity (Lin et al. 2007), because they can approximate most non-triangular functions (Pedrycz 1994) and because it is most common in the performance measurement (Chan and Qi 2003).

$$\mu_{\tilde{A}_1}(x) = \begin{cases} (x-a)/(b-a), & a \leq x \leq b, \\ (d-x)/(d-b) & b \leq x \leq d, \\ 0 & otherwise, \end{cases} \tag{1}$$

**Block 2: A rule base containing a number of fuzzy IF–THEN rules.** These rules assume the form:

$$\text{IF } \chi \text{ is A THEN y is B}$$

where A and B are the linguistic values defined by fuzzy sets of universes of discourse X and Y, respectively. As De Siqueira Campos Boclin and DeMello (2006) assert, the 'IF' part of the rule defines the situation for which it is designed (antecedent), and the 'THEN' part defines the action of the fuzzy system in this situation (consequent).

**Block 3: A database which defines the membership functions of the fuzzy sets used in the fuzzy rules.** Once the inputs were fuzzified, according to their own membership function and linguistic variables, the "if– then" rules for each group were generated. Subsequently, the individual fuzzy sets were related according to the fuzzy rules to generate a final single membership function

**Block 4: A decision making unit which performs the inference operations on the rules.**

**Block 5: A defuzzification interface which transforms the fuzzy results of the inference into a crisp output.** There are several defuzzification methods in the literature (Lee 1990), and in this research it is used the centroid method due to it is the most frequently used (Wu and Lee 2007). The fuzzy logic toolbox of MATLAB was applied to entering the membership functions and fuzzy rules.

## 2.2 Fuzzy TOPSIS

Fuzzy Set Theory (FST) presents a framework for modelling the stakeholders' selection problem in an uncertain environment. Moreover, FST can be combined with other techniques, such as TOPSIS, to improve the quality of the final result. The Fuzzy TOPSIS method was proposed by Chen (2000) to solve multi-criteria decision making problems under uncertainty.

Fuzzy TOPSIS can be expressed in a series of steps.

**Step 1: Determine the weighting of evaluation criteria.** The Linguistic variables are used by the decision makers, DMs (s = 1, ..., k), in our case stakeholders, to assess the weights of the criteria and the ratings of the alternatives

**Step 2: Establish of decision matrix $\tilde{D}$**, where the number of criteria is $n$ and the count of alternatives is $m$. Fuzzy decision matrix will be obtained with $m$ rows and $n$ columns

**Step 3: Calculate the normalized decision matrix** of the alternatives ($\tilde{D}$) using linear scale transformation

**Step 4: Calculate the weighted normalized fuzzy decision matrix.** The weighted normalized value $\tilde{v}_{ij}$ is calculated by multiplying the weights $\tilde{w}_j$ of criteria with the normalized fuzzy decision matrix $\tilde{r}_{ij}$. The weighted normalized decision matrix $\tilde{R}$ for each criterion is calculated through the following relations:

$$\tilde{R} = \left[\tilde{v}_{ij}\right]_{m \times n} \quad i = 1, \ldots, m, j = 1, \ldots, n$$
$$Where \quad \tilde{v}_{ij} = \tilde{r}_{ij}(\cdot)\tilde{w}_i \tag{2}$$

**Step 5: Determine the fuzzy positive-ideal solution (FPIS) and fuzzy negative-ideal solution (FNIS).** The basic concept of TOPSIS is that the chosen alternative should have the shortest distance from the positive-ideal solution (PIS), i.e., the solution that maximises the benefit criteria and minimises the cost criteria. The chosen alternative should also have the farthest distance from the negative-ideal solution (NIS), i.e., the solution that maximises the cost criteria and minimises the benefit criteria (Sghafian and Hejazi 2005)

$$A^+ = \left(\tilde{v}_1^+, \tilde{v}_2^+, \ldots, \tilde{v}_n^+\right) = \left\{\left(\max \tilde{v}_{ij} | i = 1, \ldots, m\right), j = 1, \ldots, n\right\}$$
$$A^- = \left(\tilde{v}_1^-, \tilde{v}_2^-, \ldots, \tilde{v}_n^-\right) = \left\{\left(\min \tilde{v}_{ij} | i = 1, \ldots, m\right), j = 1, \ldots, n\right\} \tag{3}$$

Here $\tilde{v}_j^+$ (1,1,1) and $\tilde{v}_j^-$ (0,0,0)

**Step 6: Calculate distance from the FPIS and FNIS for each alternative.** According to Bojadziev and Bojadziev (1995), the distance between two triangular fuzzy numbers ($\tilde{A}$) and ($\tilde{B}$) is calculated as:

$$d(\tilde{A}, \tilde{B}) = \sqrt{\frac{1}{3}\left[(a_1 - b_1)^2 + (a_2 - b_2)^2 + (a_3 - b_3)^2\right]} \qquad (4)$$

The separation of each alternative from the ideal solution is given as

$$d_i^+ = \sum_{j=1}^{k} d\left(\tilde{v}_{ij}, \tilde{v}_j^+,\right), i = 1, \ldots, m \qquad (5)$$

Similarly, the separation from the negative ideal solution is given as

$$d_i^- = \sum_{j=1}^{k} d\left(\tilde{v}_{ij}, \tilde{v}_j^-,\right), i = 1, \ldots, m \qquad (6)$$

**Step 7: Calculate the relative closeness coefficient to the ideal solution.** The alternative with the highest closeness coefficient (CC) value will be the best choice

$$CC \quad or \quad \bar{R}_i = \frac{\bar{d}_i^-}{\bar{d}_i^+ + \bar{d}_i^-}, i = 1, \ldots, m \qquad (7)$$

**Step 8: Rank preference order.** The ranking of the alternatives can be determined according to the CC in descending order. The best alternative is closed to the FPIS and farthest to the FNIS.

## 3  Empirical Design

This research contributes to the literature by proposing an application of Fuzzy MCDM to advance in the corporate sustainability evaluation. This is a schematic representation of the same (Table 1).

**Table 1** Schematic representation of the proposed method

| Step | Description |
| --- | --- |
| Step 1. Selection of the sample | 54 world companies belonging wearing apparel industry |
| Step 2. Selection of variables for sustainability definition | 18 indicators grouped in 4 different levels: (i) financial, (ii) social, (iii) environmental and, (iv) governance |
| Step 3. Aggregation method for the 18 indicators into 4 levels | Mamdani-type Fuzzy Inference Systems (FIS) design (Jang 1993). 4 FIS, one for each level (financial, environmental, social and governance) |
| Step 4. The consideration of stakeholders' preferences | Fuzzy Topsis model design with a specific set of rules and variables in order to incorporate the main three external market actors' preferences of wearing apparel sector: legislators, investor's long term and society[1] |

[1]The developed methodology has been empirically applied in a simulated scenario, in order to test its applicability

Note that, Fuzzy Topsis model, the importance weights of various criteria and the ratings of qualitative criteria are considered as linguistic variables. These linguistic variables can be expressed in positive triangular fuzzy numbers. The three stakeholders have expressed their opinions on the importance weights of the four criteria and the ratings of each company regarding the four criteria independently. Tables 2 shows the relative importance weights of the four criteria by the three stakeholders (decision makers- DMs).

**Table 2** The relative importance weights of the four criteria by three stakeholders

| Criterion | $DM_1$ (legislators) | $DM_2$ (investors' long term) | $DM_3$ society | Aggregated fuzzy number |
|-----------|----------------------|-------------------------------|----------------|-------------------------|
| C1 = F    | M                    | H                             | M              | (0.33, 0.58, 0.83)      |
| C2 = E    | VH                   | VH                            | VH             | (0.75, 1.00, 1.00)      |
| C3 = S    | H                    | M                             | VH             | (0.50, 0.75, 0.92)      |
| C4 = G    | L                    | VH                            | H              | (0.42, 0.67, 0.83)      |

# 4 Results

The abovementioned problems regarding the evaluation of corporate sustainability assessment are solved. To carefully analyse these results and the accuracy of the methodology, we have compared the results according to Asset4 vs. Fuzzy MCDM rating (Table 3). Social, Environmental, Financial and Governance indicators have been normalized in order to contextualize the companies' results and to highlight the differences among them −1 is the best in class and 0 is the worst behaved.

Although movements between rankings are not dramatic, this methodological approach overcomes the weaknesses of reductionist approaches to measure corporate sustainability. For example, on the one hand, "company A" lost 7 positions applying our methodological approach because its scores in the environmental and social criteria are LOW, and these are the two most important criteria for legislators, investors' long term and society. On the other hand, "company B" gained 4 positions because its scores in the environmental and social criteria are HIGH.

The TOPSIS methodology relativized scores on each aspect according to the importance of this aspect has to stakeholders. For example, the governance performance has little influence on TOPSIS ranking, because at least there is one decision maker that gives LOW significance. As a result, there are organizations that fall into their positions on ASSET4 when present high scores in governance but medium-low values in all other aspect. Similarly, the environmental aspect is considered by all simulated stakeholders as VERY HIGH, so their relative importance in terms of ranking elaboration will be higher than the rest. Therefore, companies with high environmental performance but modest results in the other scores, they could be benefited in their initial position, and vice versa, companies with low environmental performance could be penalized, although in the rest of scores present acceptable results.

**Table 3** An example of Fuzzy MCDM ranking results vs. Asset4 ranking

| Comp. | Score FUZZY Topsis method | Ranking FUZZY Topsis method | Score ASSET4 | Ranking ASSET4 | Asset4 –FUZZY Topsis method | FIN score ASSET4 | ENV score ASSET4 | SOC score ASSET4 | GOV score ASSET4 |
|---|---|---|---|---|---|---|---|---|---|
| A | 0.176 | 44 | 0.276 | 37 | −7 | 0.110 | 0.237 | 0.426 | 0.332 |
| B | 0.485 | 14 | 0.600 | 18 | 4 | 0.364 | 0.835 | 0.866 | 0.336 |

# 5 Conclusions

The integration of different stakeholder perspectives could help organizations to understand and engage with their stakeholders; and could contribute to sustainable development.

This paper develops stakeholder methodological approach, based on an application of fuzzy multi-criteria decision-making method (MCDM), to improve the contributions to organizations to sustainable development. This methodological approach overcomes the weaknesses of reductionist approaches to measure corporate sustainability, and to integrate in the evaluation process sustainability interests of stakeholders.

This methodological approach advance in corporate sustainability assessment in two main aspects: (i) it aggregates multidimensional information to avoid the information loss problem and; (ii) it allows to integrate the different sensibilities and objectives of different market actors and to reach a consensus about what sustainability is and how to translate it to company behaviour.

The results derived can be used: (i) externally, in the financial markets, in particular by sustainability rating agencies, to advance in the design of their corporate sustainability assessment models as well as to measure corporate contribution to sustainable development of markets and society; (ii) internally, to engage with their stakeholders.

# References

Andriantiatsaholiniaina, L.A., Kouikoglou, V.S., Phillis, Y.A.: Evaluating strategies for sustainable development: fuzzy logic reasoning and sensitivity analysis. Ecol. Econ. **48**(2), 149–172 (2004)

Bojadziev, G., Bojadziev, M.: Fuzzy Sets, Fuzzy Logic, Applications. Advances in Fuzzy Systems-Applications and Theory. World Scientific, Singapore (1995)

Chan, F.T.S., Qi, H.J.: An innovative performance measurement method for supply chain management. Supply Chain Manage. Int. J. **8**(3), 209–223 (2003)

Chen, C.T.: Extensions of the TOPSIS for group decision-making under fuzzy environment. Fuzzy Sets Syst. **114**, 1–9 (2000)

De Siqueira Campos Boclin, A., De Mello, R.A.: Decision support method for environmental impact assessment using a fuzzy logic approach, Ecol. Econ. **58**(1), 170–181 (2006)

Hörisch, J., Freeman, R.E., Schaltegger, S.: Applying stakeholder theory in sustainability management: links, similarities, dissimilarities, and a conceptual framework. Organ. Environ. **27**(4), 328–346 (2014)

Jang, J.-S.R.: ANFIS: adaptative-network-based fuzzy inference system. IEEE Trans. Syst. Man Cybern. **23**(3), 665–685 (1993)

Lee, C.C.: Fuzzy logic in control systems: fuzzy logic controller, part II. IEEE Trans. Syst. Man Cybern. **20**(2), 419–435 (1990)

Lin, H.-Y., Hsu, P.-Y., Sheen, G.-J.: A fuzzy-based decision-making procedure for data warehouse system selection. Expert Syst. Appl. **32**(3), 939–953 (2007)

Munda, G., Nijkamp, P., Rietveld, P.: Qualitative multi-criteria methods for fuzzy evaluation problems: an illustration of economic ecological evaluation. Eur. J. Oper. Res. **82**, 79–97 (1995)

Muñoz, M.J., Rivera, J.M., Moneva, J.M.: Evaluating sustainability in organizations with a fuzzy logic approach. Ind. Manage. Data Syst. **108**(6), 829–841 (2008)

Pedrycz, W.: Why triangular membership functions? Fuzzy Sets Syst. **64**(1), 21–30 (1994)

Phillis, Y.A., Andriantiatsaholiniaina, L.A.: Sustainability: an ill-defined concept and its assessment using fuzzy logic. Ecol. Econ. **37**(3), 435–456 (2001)

Prato, T.A.: fuzzy logic approach for evaluating ecosystem sustainability. Ecol. Model. **187**, 361–368 (2005)

Saghafian, S., Hejazi, S.R.: Multi-criteria group decision making using a modified fuzzy topsis procedure. In: Proceedings of the 2005 International Conference on Computational Intelligence for modeling, Control and Automation. IEEE Computer Society (2005)

Schaltegger, S., Wagner, M.: Sustainable entrepreneurship and sustainability innovation: categories and interactions. Bus. Strat. Environ. **20**, 222–237 (2011)

Windolph, S.E.: Assessing corporate sustainability through ratings: challenges and their causes. J. Environ. Sustain. **1**(1), 61–81 (2011)

Wu, W.-W., Lee, Y.-T.: Developing global managers' competencies using the fuzzy DEMATEL method. Expert Syst. Appl. **32**(2), 499–507 (2007)

Zadeh, L.A.: Fuzzy sets. Inf. Control **8**(3), 338–353 (1965)

Zadeh, L.A.: The concept of a linguistic variable and its application to approximate reasoning. Inf. Sci. **8**, 199–249 (1975)

# On the Orness of SUOWA Operators

Bonifacio Llamazares

**Abstract** There is in the literature a great variety of functions utilized in the aggregation processes. For this reason, numerous indicators have been suggested to understand the behavior of such functions. One of the measures proposed for this purpose is the orness, which allows to know the degree of closeness to the maximum. The aim of this paper is to provide the orness of some specific cases of SUOWA operators, a family of aggregation functions that simultaneously generalize weighted means and OWA operators.

**Keywords** SUOWA operators · Orness · Weighted means · OWA operators · Choquet integral

## 1 Introduction

Aggregation information is a usual activity in many research fields. Weighted means and ordered weighted averaging (OWA) operators (Yager 1988) are well-known functions widely used for this task. Both classes of functions are defined by means of weighting vectors, but their behavior is quite different: Weighted means allow to weight each information source in relation to their reliability while OWA operators allow to weight the values according to their ordering. Nevertheless, there are situations where both weightings are necessary (see, for instance, Torra 1997, 2001; Llamazares 2015a, b, c; and the references therein).

Different aggregation functions have appeared in the literature to deal with this kind of problems. A usual approach is to consider families of functions parametrized by two weighting vectors, one for the weighted mean and the other one for the OWA type aggregation, that generalize weighted means and OWA operators in

B. Llamazares (✉)
Dpto. de Economía Aplicada, Instituto de Matemáticas (IMUVA), Universidad de Valladolid, Avda. Valle de Esgueva 6, 47011 Valladolid, Spain
e-mail: boni@eco.uva.es

© Springer International Publishing Switzerland 2015
J. Gil-Aluja et al. (eds.), *Scientific Methods for the Treatment of Uncertainty in Social Sciences*, Advances in Intelligent Systems and Computing 377,
DOI 10.1007/978-3-319-19704-3_4

the following sense: A weighted mean (or a OWA operator) is obtained when the other weighting vector has a "neutral" behavior; that is, it is $(1/n, \ldots, 1/n)$ (see Llamazares 2013, for an analysis of some functions that generalize the weighted means and the OWA operators in this sense). Two of the solutions having better properties are the weighted OWA (WOWA) operator (Torra 1997) and the semi-uninorm based ordered weighted averaging (SUOWA) operator (Llamazares 2015a).

An important issue in the field of aggregation functions is the behavior of compensative operators regarding the minimum and the maximum. For dealing with this task, the notion of orness was introduced for the case of the root-mean-powers (Dujmović 1974) and, in an independent way, for OWA operators (Yager 1988). Later, it has been generalized by using the notion of average value (Marichal 1998). In this paper we analyze some specific cases of SUOWA operators with respect to this indicator and show how to get a SUOWA operator with a previously fixed degree of orness.

The paper is organized as follows. In Sect. 2 we recall some basic properties of aggregation functions and the concepts of semi-uninorms and uninorms. Section 3 is devoted to Choquet integral, including some of the most interesting particular cases: weighted means, OWA operators and SUOWA operators. In Sect. 4 we give some properties of SUOWA operators related to their orness, and illustrate how to obtain a SUOWA operator having a particular degree of orness. Finally, some concluding remarks are provided in Sect. 5.

## 2 Preliminaries

Throughout the paper we will use the following notation: $N = \{1, \ldots, n\}$; given $A \subseteq N, |A|$, denotes the cardinality of $A$; vectors are denoted in bold and $\eta$ denotes the tuple $(1/n, \ldots, 1/n) \in \mathbb{R}^n$. We write $x \geq y$ if $x_i \geq y_i$ for all $i \in N$. For a vector $x \in \mathbb{R}^n$, $[\cdot]$ and $(\cdot)$ denote permutations such that $x_{[1]} \geq \cdots \geq x_{[n]}$ and $x_{(1)} \leq \cdots \leq x_{(n)}$.

Some well known properties of aggregation functions are the following.

**Definition 1.** Let $F: \mathbb{R}^n \to \mathbb{R}$ be a function.

1. $F$ is symmetric if $F(x_{\sigma(1)}, \ldots, x_{\sigma(n)}) = F(x_1, \ldots, x_n)$ for all $x \in \mathbb{R}^n$ and for all permutation $\sigma$ of $N$.
2. $F$ is monotonic if $x \geq y$ implies $F(x) \geq F(y)$ for all $x, y \in \mathbb{R}^n$.
3. $F$ is idempotent if $F(x, \ldots, x) = x$ for all $x \in \mathbb{R}$.
4. $F$ is compensative (or internal) if $\min(x) \leq F(x) \leq \max(x)$ for all $x \in \mathbb{R}^n$.
5. $F$ is homogeneous of degree 1 (or ratio scale invariant) if $F(rx) = rF(x)$ for all $x \in \mathbb{R}^n$ and for all $r > 0$.

A class of necessary functions in the definition of SUOWA operators are semiuninorms (Liu 2012). These functions are monotonic and have a neutral element in the interval [0,1]. They were introduced as a generalization of uninorms, which, in turn, were proposed as a generalization of t-norms and t-conorms (Yager and Rybalov 1996).

**Definition 2.** Let $U: [0, 1]^2 \rightarrow [0, 1]$.

1. $U$ is a semi-uninorm if it is monotonic and possesses a neutral element $e \in [0, 1]$ (for all $x \in [0, 1]$, $U(e, x) = U(x, e) = x$).
2. $U$ is a uninorm if it is a symmetric and associative (for all $x, y, z \in [0, 1]$, $U(x, U(y, z)) = U(U(x, y), z)$) semi-uninorm.

We denote by $\mathcal{U}^e$ (respectively, $\mathcal{U}_i^e$) the set of semi-uninorms (respectively, idempotent semi-uninorms) with neutral element $e \in [0, 1]$.

SUOWA operators are defined by using semi-uninorms with neutral element $1/n$. Moreover, they have to belong to the following subset (Llamazares 2015a):

$$\tilde{\mathcal{U}}^{1/n} = \{U \in \mathcal{U}^{1/n} \mid U(1/k, 1/k) \leq 1/k \text{ for all } k \in N\}.$$

Obviously $\mathcal{U}_i^{1/n} \subseteq \tilde{\mathcal{U}}^{1/n}$. Notice that the smallest and the largest elements of $\mathcal{U}_i^{1/n}$ are, respectively, the following uninorms (Yager and Rybalov 1996):

$$U_{\min}(x, y) = \begin{cases} \max(x, y), & \text{if } (x, y) \in [1/n, 1]^2, \\ \min(x, y), & \text{otherwise,} \end{cases}$$

$$U_{\max}(x, y) = \begin{cases} \min(x, y), & \text{if } (x, y) \in [0, 1/n]^2, \\ \max(x, y), & \text{otherwise.} \end{cases}$$

## 3 Choquet Integral

The notion of Choquet integral is based on that of capacity (Choquet 1953; Murofushi and Sugeno 1991). The concept of capacity resembles that of probability measure but in the definition of the former additivity is replaced by monotonicity (see also fuzzy measures in Sugeno 1974). A game is then a generalization of a capacity where the monotonicity is no longer required.

**Definition 3.**

1. A game $v$ on $N$ is a set function $v: 2^N \rightarrow \mathbb{R}$ satisfying $v(\emptyset) = 0$.
2. A capacity (or fuzzy measure) $\mu$ on $N$ is a game on $N$ satisfying $\mu(A) \leq \mu(B)$ whenever $A \subseteq B$. In particular, it follows that $\mu: 2^N \rightarrow [0, \infty)$. A capacity $\mu$ is said to be normalized if $\mu(N) = 1$.

A straightforward way to get a capacity from a game is to consider the monotonic cover of the game (Maschler and Peleg 1967; Maschler et al. 1971).

**Definition 4.** Let $v$ be a game on $N$. The monotonic cover of $v$ is the set function $\hat{v}$ given by

$$\hat{v}(A) = \max_{B \subseteq A} v(B).$$

Some basic properties of $\hat{v}$ are given in the sequel.

**Remark 1.** Let $v$ be a game on $N$. Then:

1. $\hat{v}$ is a capacity on $N$.
2. If $v$ is a capacity on $N$, then $\hat{v} = v$.
3. If $v(A) \leq 1$ for all $A \subseteq N$ and $v(N) = 1$, then $\hat{v}$ is a normalized capacity on $N$.

Although the Choquet integral is usually defined as a functional (see, for instance, Choquet 1953; Murofushi and Sugeno 1991), in this paper we consider the Choquet integral as an aggregation function over $\mathbb{R}^n$ (see, for instance, Grabisch et al. 2009, p. 181).

**Definition 5.** Let $\mu$ be a capacity on $N$. The Choquet integral with respect to $\mu$ is the function $C_\mu : \mathbb{R}^n \to \mathbb{R}$ given by

$$C_\mu(x) = \sum_{i=1}^{n} \mu(B_{(i)})(x_{(i)} - x_{(i-1)}),$$

where $B_{(i)} = \{(i), \ldots, (n)\}$, and we use the convention $x_{(0)} = 0$.

It is worth noting that Choquet integral-based operators possess several properties which are useful in certain information aggregation contexts (Grabisch et al. 2009, p. 193 and p. 196).

**Remark 2.** Let $\mu$ be a capacity on $N$. Then $C_\mu$ is continuous, monotonic and homogeneous of degree 1. Moreover, it is idempotent and compensative when $\mu$ is a normalized capacity.

Choquet integral can also be represented by using a decreasing sequences of values (see, for instance, Torra 1998; Llamazares 2015a):

$$C_\mu(x) = \sum_{i=1}^{n} \mu(A_{[i]})(x_{[i]} - x_{[i+1]}), \tag{1}$$

where $A_{[i]} = \{[1], \ldots, [i]\}$, and we use the convention $x_{[n+1]} = 0$.

From the previous expression, it is straightforward to show explicitly the weights of the values $x_{[i]}$ by representing the Choquet integral as follows:

$$C_\mu(x) = \sum_{i=1}^{n} (\mu(A_{[i]}) - \mu(A_{[i-1]}))x_{[i]},$$

where we use the convention $A_{[0]} = \emptyset$.

## 3.1 Weighted Means and OWA Operators

Weighted means and OWA operators (Yager 1988) are well-known functions in the theory of aggregation operators. Both classes of functions are defined in terms of weight distributions that add up to 1.

**Definition 6.** A vector $q \in \mathbb{R}^n$ is a weighting vector if $q \in [0, 1]^n$ and $\sum_{i=1}^{n} q_i = 1$.

**Definition 7.** Let $p$ be a weighting vector. The weighted mean associated with $p$ is the function $M_p \colon \mathbb{R}^n \to \mathbb{R}$ given by

$$M_p(x) = \sum_{i=1}^{n} p_i x_i.$$

**Definition 8.** Let $w$ be a weighting vector. The OWA operator associated with $w$ is the function $O_w \colon \mathbb{R}^n \to \mathbb{R}$ given by

$$O_w(x) = \sum_{i=1}^{n} w_i x_{[i]}.$$

It is well known that weighted means and OWA operators are specific cases of Choquet integral (Fodor et al. 1995; Grabisch 1995a, b; Llamazares 2015a).

**Remark 3.**

1. If $p$ is a weighting vector, then $M_p$ is the Choquet integral with respect to the normalized capacity $\mu_p(A) = \sum_{i \in A} p_i$.
2. If $w$ is a weighting vector, then $O_w$ is the Choquet integral with respect to the normalized capacity $\mu_{|w|}(A) = \sum_{i=1}^{|A|} w_i$.

So, according to Remark 2, weighted means and OWA operators are continuous, monotonic, idempotent, compensative and homogeneous of degree 1. Moreover, in the case of OWA operators, given that the values of the variables are previously ordered in a decreasing way, they are also symmetric.

## 3.2 SUOWA Operators

SUOWA operators (Llamazares 2015a) were introduced in order to consider situations where both the importance of information sources and the importance of values had to be taken into account. These functions are Choquet integral-based operators where their capacities are the monotonic cover of specific games. These games are defined by using semi-uninorms with neutral element $1/n$ and the values of the capacities associated with the weighted means and the OWA operators. To be specific, the games from which SUOWA operators are built are defined as follows.

**Definition 9.** Let $p$ and $w$ be two weighting vectors and let $U \in \tilde{\mathcal{U}}^{1/n}$.

1. The game associated with $p$, $w$ and $U$ is the set function $v_{p,w}^U : 2^N \to \mathbb{R}$ defined by

$$v_{p,w}^U(A) = |A| U \left( \frac{\mu_p(A)}{|A|}, \frac{\mu_{|w|}(A)}{|A|} \right)$$

if $A \neq \varnothing$, and $v_{p,w}^U(\varnothing) = 0$.

2. $\hat{v}_{p,w}^U$, the monotonic cover of the game $v_{p,w}^U$, will be called the capacity associated with $p$, $w$ and $U$.

**Definition 10.** Let $p$ and $w$ be two weighting vectors and let $U \in \tilde{\mathcal{U}}^{1/n}$. The SU-OWA operator associated with $p$, $w$ and $U$ is the function $S_{p,w}^U : \mathbb{R}^n \to \mathbb{R}$ given by

$$S_{p,w}^U(x) = \sum_{i=1}^n s_i x_{[i]},$$

where $s_i = \hat{v}_{p,w}^U(A_{[i]}) - \hat{v}_{p,w}^U(A_{[i-1]})$ for all $i \in N$, $\hat{v}_{p,w}^U$ is the capacity associated with $p$, $w$ and $U$, and $A_{[i]} = \{[1], \ldots, [i]\}$ (with the convention that $A_{[0]} = \varnothing$).

According to expression (1), the SUOWA operator associated with $p$, $w$ and $U$ can also be written as

$$S_{p,w}^U(x) = \sum_{i=1}^n \hat{v}_{p,w}^U(A_{[i]}) (x_{[i]} - x_{[i+1]}).$$

By the choice of $\hat{v}_{p,w}^U$ we have $S_{p,\eta}^U = M_p$ and $S_{\eta,w}^U = O_w$ for any $U \in \tilde{\mathcal{U}}^{1/n}$. Moreover, by Remark 2 and given that $\hat{v}_{p,w}^U$ is a normalized capacity, SUOWA operators are continuous, monotonic, idempotent, compensative and homogeneous of degree 1.

## 4 Orness Measures

An important concept to measure the degree to which the aggregation is disjunctive (i.e. it is like an *or* operation) is the notion of orness. This concept was initially introduced for the case of the root-mean-powers (Dujmović 1974) and, in an independent way, for OWA operators (Yager 1988). Afterwards, Marichal (1998) derived an orness measure for the Choquet integral by using the notion of average value.

**Definition 11.** Let $\mu$ be a capacity on $N$.

1. The average value of $\mathcal{C}_\mu$ is defined by

$$E(\mathcal{C}_\mu) = \int_{[0,1]^n} \mathcal{C}_\mu(x)dx.$$

2. The orness value of $\mathcal{C}_\mu$ is defined by

$$\text{orness}(\mathcal{C}_\mu) = \frac{E(\mathcal{C}_\mu) - E(\min)}{E(\max) - E(\min)}.$$

The orness of $\mathcal{C}_\mu$ can be written in terms of the capacity $\mu$ (Marichal 2004):

$$\text{orness}(\mathcal{C}_\mu) = \frac{1}{n-1} \sum_{t=1}^{n-1} \frac{1}{\binom{n}{t}} \sum_{\substack{T \subseteq N \\ |T|=t}} \mu(T). \tag{2}$$

The degree of orness is known for some Choquet integral-based operators (Marichal 2004). In fact, in the particular cases of weighted means and OWA operators we get the following values:

$$\text{orness}(M_p) = \frac{1}{2}, \qquad \text{orness}(O_w) = \frac{1}{n-1} \sum_{i=1}^{n} (n-i)w_i. \tag{3}$$

However, in the vast majority of cases, the calculation of this value through a closed expression does not seem an easy task (Torra 2001). In the case of SUOWA operators, we can get a very interesting result when we consider a semi-uninorm obtained as a convex combination of two semi-uninorms satisfying that the games associated with them are capacities. In this case, the orness of the SUOWA operator associated with this new semi-uninorm can be gotten through the same convex combination of the orness of the SUOWA operators associated with the former semi-uninorms. This result, together with others concerning the game and the SUOWA operator associated with this new semi-uninorm, are collected in the following theorem.

**Theorem 1.** Let $p$ and $w$ be two weighting vectors, let $U_1, U_2 \in \tilde{\mathcal{U}}^{1/n}$ such that $v_{p,w}^{U_1}$ and $v_{p,w}^{U_2}$ be capacities, let $\lambda \in [0,1]$, and let $U = \lambda U_1 + (1-\lambda)U_2$. Then:

1. $U \in \tilde{\mathcal{U}}^{1/n}$ and if $U_1, U_2 \in \mathcal{U}_i^{1/n}$ then $U \in \mathcal{U}_i^{1/n}$.
2. $v_{p,w}^{U}(A) = \lambda v_{p,w}^{U_1}(A) + (1-\lambda)v_{p,w}^{U_2}(A)$ for any subset $A$ of $N$.
3. $v_{p,w}^{U}$ is a normalized capacity on $N$.
4. $S_{p,w}^{U}(x) = \lambda S_{p,w}^{U_1}(x) + (1-\lambda)S_{p,w}^{U_2}(x)$ for all $x \in \mathbb{R}^n$.
5. $\text{orness}\left(S_{p,w}^{U}\right) = \lambda \, \text{orness}\left(S_{p,w}^{U_1}\right) + (1-\lambda)\,\text{orness}\left(S_{p,w}^{U_2}\right)$.

Among the great variety of semi-uninorms belonging to $\tilde{\mathcal{U}}^{1/n}$ that could be chosen to generate a SUOWA operator, idempotent semi-uninorms are of specific interest owing to their notable properties (Llamazares 2015a). Notice that, since any idempotent semi-uninorm is located between $U_{\min}$ and $U_{\max}$, we straightforwardly obtain the following results.

**Remark 4.** Let $p$ and $w$ be two weighting vectors, and $U \in \mathcal{U}_i^{1/n}$. Then:

1. $S_{p,w}^{U_{\min}}(x) \leq S_{p,w}^{U}(x) \leq S_{p,w}^{U_{\max}}(x)$ for all $x \in \mathbb{R}^n$.
2. $\text{orness}\left(S_{p,w}^{U_{\min}}\right) \leq \text{orness}\left(S_{p,w}^{U}\right) \leq \text{orness}\left(S_{p,w}^{U_{\max}}\right)$.

Given that SUOWA operators generated from idempotent semi-uninorms are located between those obtained from the uninorms $U_{\min}$ and $U_{\max}$, next we establish some results on the games and the SUOWA operators associated with these uninorms when the weighting vector $w$ is an increasing (in the case of considering $U_{\min}$) or decreasing (in the case of considering $U_{\max}$) sequence of weights.

**Theorem 2.** Let $w$ be a weighting vector such that $w_1 \leq w_2 \leq \cdots \leq w_n$ and $w \neq \eta$. Then, for all weighting vector $p$, we have

1. $v_{p,w}^{U_{\min}}$ is a normalized capacity on $N$.
2. $v_{p,w}^{U_{\min}}(A) = \min\left(\mu_p(A), \mu_{|w|}(A)\right)$ for all $A \subseteq N$.
3. $S_{p,w}^{U_{\min}}(x) \leq \min\left(M_p(x), O_w(x)\right)$ for all $x \in \mathbb{R}^n$.
4. $\text{orness}\left(S_{p,w}^{U_{\min}}\right) \leq \text{orness}(O_w) \leq 0.5$.

Similar results can be established for the uninorm $U_{\max}$.

**Theorem 3.** Let $w$ be a weighting vector such that $w_1 \geq w_2 \geq \cdots \geq w_n$ and $w \neq \eta$. Then, for all weighting vector $p$, we get

1. $v_{p,w}^{U_{\max}}$ is a normalized capacity on $N$.
2. $v_{p,w}^{U_{\max}}(A) = \max\left(\mu_p(A), \mu_{|w|}(A)\right)$ for all $A \subseteq N$.
3. $S_{p,w}^{U_{\max}}(x) \geq \max\left(M_p(x), O_w(x)\right)$ for all $x \in \mathbb{R}^n$.
4. $\text{orness}\left(S_{p,w}^{U_{\max}}\right) \geq \text{orness}(O_w) \geq 0.5$.

In the sequel, and using some of the previous results, we illustrate how we can get a SUOWA operator having a particular degree of orness.

**Example 1.** Let us to consider the weighting vectors $p = (0.1, 0.4, 0.4, 0.1)$ and $w = (0.3, 0.3, 0.2, 0.2)$, and the idempotent uninorms $U_{\min}$ and $U_{\max}$. In the case of $U_{\max}$, the first item of Theorem 3 guarantees that $v_{p,w}^{U_{\max}}$ is a capacity. As we can see in Table 1, this is also the case when we consider the uninorm $U_{\min}$.

According to expressions (2) and (3), we have

$$\text{orness}(O_w) = 1.7/3, \quad \text{orness}\left(S_{p,w}^{U_{\min}}\right) = 4.7/9, \quad \text{orness}\left(S_{p,w}^{U_{\max}}\right) = 5.5/9.$$

And by the second item of Remark 4, for any idempotent semi-uninorm $U$ we get

$$4.7/9 \leq \text{orness}\left(S_{p,w}^{U}\right) \leq 5.5/9.$$

**Table 1** Capacities associated with $U_{\min}$ and $U_{\max}$

| Set | $v_{p,w}^{U_{\min}}$ | $v_{p,w}^{U_{\max}}$ |
|---|---|---|
| {1} | 0.1 | 0.3 |
| {2} | 0.4 | 0.4 |
| {3} | 0.4 | 0.4 |
| {4} | 0.1 | 0.3 |
| {1,2} | 0.6 | 0.6 |
| {1,3} | 0.6 | 0.6 |
| {1,4} | 0.2 | 0.6 |
| {2,3} | 0.8 | 0.8 |
| {2,4} | 0.6 | 0.6 |
| {3,4} | 0.6 | 0.6 |
| {1,2,3} | 0.9 | 0.9 |
| {1,2,4} | 0.6 | 0.8 |
| {1,3,4} | 0.6 | 0.8 |
| {2,3,4} | 0.9 | 0.9 |
| N | 1 | 1 |

Now, by the fifth item of Theorem 1, we can easily obtain an idempotent semi-uninorm that allows us to get a SUOWA operator having a particular degree of orness in the range 4.7/9 to 5.5/9. For instance, if we look for an idempotent semi-uninorm $U$ such that $\text{orness}\left(S_{p,w}^{U}\right) = \text{orness}(O_w) = 1.7/3$, then, since $1.7/3 = 0.5 \cdot 4.7/9 + 0.5 \cdot 5.5/9$, it is sufficient to consider $U_{am} = 0.5U_{\min} + 0.5U_{\max}$, that is, the idempotent semi-uninorm obtained through the arithmetic mean:

$$U_{am}(x,y) = \begin{cases} \min(x,y), & \text{if}(x,y) \in [0, 0.25]^2, \\ \max(x,y), & \text{if}(x,y) \in [0.25, 1]^2 \setminus \{(0.25, 0.25)\}, \\ (x+y)/2, & \text{otherwise.} \end{cases}$$

# 5 Conclusion

SUOWA operators have been introduced recently for dealing with situations where combining values by using both a weighted mean and a OWA type aggregation is necessary. Given that they are Choquet integral-based operators with respect to normalized capacities, they have some natural properties such as continuity, monotonicity, idempotency, compensativeness and homogeneity of degree 1. For this reason, it seems interesting to analyze their behavior through indicators proposed in the literature. In this paper we have investigated the orness of some specific cases of SUOWA operators and illustrated how to obtain a SUOWA operator with a given degree of orness.

**Acknowledgments** The partial financial support from the Ministerio de Economía y Competitividad (Project ECO2012-32178) and the Junta de Castilla y León (Consejería de Educación, Project VA066U13) is gratefully acknowledged.

# References

Choquet, G.: Theory of capacities. Annales de l'Institut Fourier **5**, 131–295 (1953)

Dujmović, J.J.: Weighted conjunctive and disjunctive means and their application in system evaluation. J. Univ. Belgrade, EE Dept., Ser. Math. Phys. **483**, 147–158 (1974)

Fodor, J., Marichal, J.L., Roubens, M.: Characterization of the ordered weighted averaging operators. IEEE Trans. Fuzzy Syst. **3**, 236–240 (1995)

Grabisch, M.: Fuzzy integral in multicriteria decision making. Fuzzy Sets Syst. **69**, 279–298 (1995a)

Grabisch, M.: On equivalence classes of fuzzy connectives—the case of fuzzy integrals. IEEE Trans. Fuzzy Syst. **3**, 96–109 (1995b)

Grabisch, M., Marichal, J., Mesiar, R., Pap, E.: Aggregation Functions. Cambridge University Press, Cambridge (2009)

Liu, H.W.: Semi-uninorms and implications on a complete lattice. Fuzzy Sets Syst. **191**, 72–82 (2012)

Llamazares, B.: An analysis of some functions that generalizes weighted means and OWA operators. Int. J. Intell. Syst. **28**, 380–393 (2013)

Llamazares, B.: Constructing Choquet integral-based operators that generalize weighted means and OWA operators. Inf. Fusion **23**, 131–138 (2015a)

Llamazares, B.: SUOWA operators: constructing semi-uninorms and analyzing specific cases. Fuzzy Sets Syst. forthcoming (2015b)

Llamazares, B.: A study of SUOWA operators in two dimensions. Math. Probl. Eng. forthcoming (2015c)

Marichal, J.L.: Aggregation operators for multicriteria decision aid. Ph.D. thesis, University of Liège (1998)

Marichal, J.L.: Tolerant or intolerant character of interacting criteria in aggregation by the Choquet integral. Eur. J. Oper. Res. **155**, 771–791 (2004)

Maschler, M., Peleg, B.: The structure of the kernel of a cooperative game. SIAM J. Appl. Math. **15**, 569–604 (1967)

Maschler, M., Peleg, B., Shapley, L.S.: The kernel and bargaining set for convex games. Int. J. Game Theory **1**, 73–93 (1971)

Murofushi, T., Sugeno, M.: A theory of fuzzy measures. Representation, the Choquet integral and null sets. J. Math. Anal. Appl. **159**, 532–549 (1991)

Sugeno, M.: Theory of fuzzy integrals and its applications. Ph.D thesis, Tokyo Institute of Technology (1974)

Torra, V.: The weighted OWA operator. Int. J. Intell. Syst. **12**, 153–166 (1997)

Torra, V.: On some relationships between the WOWA operator and the Choquet integral. In: Proc. 7th Int. Conf. on Information Processing and Management of Uncertainty in Knowledge-Based Systems (IPMU'98), pp. 818–824, Paris (1998)

Torra, V.: Empirical analysis to determine weighted OWA orness. In: Proc. 4th Int. Conf. on Information Fusion (FUSION 2001), Montreal (2001)

Yager, R.R.: On ordered weighted averaging operators in multicriteria decision making. IEEE Trans. Syst. Man Cybern. **18**, 183–190 (1988)

Yager, R.R., Rybalov, A.: Uninorm aggregation operators. Fuzzy Sets Syst. **80**, 111–120 (1996)

# OWA Operators in Portfolio Selection

Sigifredo Laengle, Gino Loyola and José M. Merigó

**Abstract** Portfolio choice is the process of selecting the optimal proportion of various assets. One of the most well-known methods is the mean-variance approach developed by Harry Markowitz. This paper introduces the ordered weighted average (OWA) in the mean-variance model. The key idea is that the mean and the variance can be extended with the OWA operator being able to consider different degrees of optimism or pessimism in the analysis. Thus, this method can adapt to a wide range of scenarios providing a deeper representation of the available information from the most pessimistic situation to the most optimistic one.

**Keywords** Portfolio selection · Ordered weighted average · Mean · Variance

## 1 Introduction

Portfolio selection is a theory of finance that aims to maximize the expected return of a portfolio considering a specific level of portfolio risk or minimize the risk for a given amount of expected return. The objective is to select a set of investment assets that have collectively a lower risk than any individual asset. Initially, it was introduced by Markowitz with the development of a mean-variance portfolio selection approach (Markowitz 1952). This model was able to find the optimal portfolio but it needed a lot of calculations. In the 50s and 60s this was a significant problem because the computers were not very strong and not able to make huge

S. Laengle (✉) · G. Loyola · J.M. Merigó
Department of Management Control and Information Systems, University of Chile, Av. Diagonal Paraguay 257, 8030015 Santiago, Chile
e-mail: slaengle@fen.uchile.cl

G. Loyola
e-mail: gloyola@fen.uchile.cl

J.M. Merigó
e-mail: jmerigo@fen.uchile.cl

© Springer International Publishing Switzerland 2015
J. Gil-Aluja et al. (eds.), *Scientific Methods for the Treatment of Uncertainty in Social Sciences*, Advances in Intelligent Systems and Computing 377,
DOI 10.1007/978-3-319-19704-3_5

calculations. Therefore, it was not easy to deal with Markowitz's theory. In order to solve this weakness, Sharpe (1963) suggested a new approach that simplified Markowitz's model a lot. He found that most of the investment assets were subject to similar conditions. Thus, he introduced a diagonal model based on regression techniques that could find a solution with a substantial lower number of calculations than Markowitz's model.

During the last decades, Markowitz's approach has received increasing attention due to the development of computers that can make a lot of calculations in a short period of time. A key example is the special issue published in the European Journal of Operational Research celebrating the 60th anniversary of Markowitz's model (Zopounidis et al. 2014) that published several overviews regarding the newest developments in the field (Kolm et al. 2014; Markowitz 2014). Moreover, many other contributions have appeared during the last years in a wide range of journals (Garlappi et al. 2007; Maccheroni et al. 2013; Wu et al. 2014).

A fundamental issue in Markowitz's mean-variance framework is the aggregation of the mean and the variance with the arithmetic mean or the weighted average. However, it is possible to use other aggregation operators for this purpose (Beliakov et al. 2007; Grabisch et al. 2011). A well-known aggregation operator that aggregates the information according to the attitudinal character of the decision maker is the ordered weighted average (OWA) (Yager 1988; Yager et al. 2011). The OWA operator has been extended and generalized by a lot of authors (Emrouznejad and Marra 2014). Yager and Filev (1999) developed a more general framework that used order inducing variables in the reordering process. Fodor et al. (1995) introduced the quasi-arithmetic OWA operator by using quasi-arithmetic means in the analysis. Merigó and Gil-Lafuente (2009) presented the induced generalized OWA operator as an integration of the induced and quasi-arithmetic approaches. Some other studies have focused on the unification between the probability and the OWA operator (Engemann et al. 1996; Merigó 2012) and some other ones on the weighted average and the OWA operator (Merigó et al. 2013; Torra 1997; Xu and Da 2003). Recently, some further generalizations have been developed in this direction with the integration of the probability, the weighted average and the OWA operator in the same formulation (Merigó et al. 2012).

The aim of this paper is to introduce a mean-variance portfolio approach by using the OWA operator in the aggregation of the expected returns and risks. The key idea is to replace the mean and the variance used in Markowitz's model by the OWA operator and the variance-OWA (Var-OWA) (Yager 1996). Note that the OWA and the Var-OWA are extensions of the classical mean and the variance by considering the attitudinal character of the decision maker and any scenario that may occur from the most pessimistic to the most optimistic one. The key motivation for using the OWA operator instead of the arithmetic mean or the weighted average is that for uncertain environments where the available information is very complex, many times it is not possible to assign weights or probabilities to the expected results. Therefore, the expected value and the variance cannot be calculated using the common formulations and a different framework is needed. The main assumption of the OWA operator is that the information can be weighted according to the attitudinal character of the decision maker instead of the possibilities or probabilities that each state of nature or criteria will occur. The motivation for doing so is that often the probabilities or weights of the future events are not available and a different methodology is needed for aggregating the information. The OWA aggregates the information under or overestimating it according to the specific attitude of the decision maker. Additionally, it also considers any result that can occur from the minimum to the maximum providing a complete representation of the information that does not loose information in the analysis.

Markowitz' approach is reformulated by using OWA operators in the mean and the variance. Some key properties of the conceptual implications of this approach are studied including the measures for the characterization of the OWA operator (Yager 1988) and some additional extensions suggested in this paper. A wide range of particular types of OWA operators are also considered in order to see some specific attitudes that the decision maker may adopt.

The rest of the paper is organized as follows. Section 2 presents some basic preliminaries concerning the OWA operator and Markowitz's portfolio model. Section 3 introduces the use of the OWA in the mean-variance portfolio methodology. Section 4 summarizes the main results and conclusions of the paper.

## 2 Preliminares

### 2.1 The OWA Operator

The ordered weighted average (OWA) is an aggregation operator that provides a parameterized family of aggregation operators between the minimum and the maximum (Yager 1988). In decision making under uncertainty, it is very useful for taking decisions with a certain degree of optimism or pessimism. It generalizes the classical methods into a single formulation including the optimistic, pessimistic, Laplace and Hurwicz criterion. It can be defined as follows.

**Definition 1.** An OWA operator of dimension $n$ is a mapping OWA: $R^n \rightarrow R$ that has an associated weighting vector $W$ of dimension $n$ with $\sum_{j=1}^{n} w_j = 1$ and $w_j \in [0, 1]$, such that:

$$OWA (a_1, a_2, \ldots, a_n) = \sum_{j=1}^{n} w_j b_j, \tag{1}$$

where $b_j$ is the $j$th largest of the $a_i$.

An important issue when dealing with the OWA operator is the reordering process. In definition 1 the reordering has been presented in a descending way although it is also possible to consider an ascending order by using $w_j = w^*_{n-j+1}$, where $w_j$ is the $j$th weight of the descending OWA and $w^*_{n-j+1}$ the $j$th weight of the ascending OWA operator. Moreover, it is also possible to adapt the ordering of the arguments to the weights and vice versa (Yager 1988). Note that the OWA is commutative, monotonic, bounded and idempotent.

In order to characterize the weighting vector of an OWA aggregation, Yager (1988) suggested the *degree of orness* and the *entropy of dispersion*. The degree of orness measures the tendency of the weights to the minimum or to the maximum. It is formulated as follows:

$$\alpha(W) = \sum_{j=1}^{n} w_j \left( \frac{n-j}{n-1} \right). \tag{2}$$

As we can see, if $\alpha(W) = 1$, the weighting vector uses the maximum and if $\alpha(W) = 0$, the minimum. The more of the weights located to the top, the higher is $\alpha$ and vice versa.

The entropy of dispersion is an extension of the Shannon entropy when dealing with OWA operators. It is expressed as:

$$H(W) = - \left( \sum_{j=1}^{n} w_j \ln(w_j) \right). \tag{3}$$

Observe that the highest entropy is found with the arithmetic mean ($H(W) = \ln(n)$) and the lowest one when selecting only one result such as the minimum or the maximum because in this case the entropy is 0.

The OWA operator includes the classical methods for decision making under uncertainty as particular cases. The optimistic criteria is found if $w_1 = 1$ and $w_j = 0$, for all $j \neq 1$. The pessimistic (or Wald) criteria is obtained when $w_n = 1$ and $w_j = 0$, for all $j \neq n$. The Laplace criteria is formed if $w_j = 1/n$, for all $a_i$. And the Hurwicz criteria is obtained when $w_1 = \alpha$, $w_n = 1 - \alpha$ and $w_j = 0$ for all $j \neq 1, n$. For further information on other cases of the OWA operator see Merigó and Gil-Lafuente (2009) and Yager (1993).

## 2.2 Portfolio Selection with Markowitz Approach

Consider a portfolio formed by $m$ individual assets, so that $r_i^k$ represents asset $k$'s return at state of nature $i$ for all $k = 1, 2, \ldots, m$, and $i = 1, 2, \ldots, n$. States of nature are distributed according to the probability vector $\pi = (\pi_1, \pi_2, \ldots, \pi_n)$. so that $\pi_i \in [0, 1]$ for all $i = 1, 2, \ldots, n$ and $\sum_{i=1}^{n} \pi_i = 1$.

Let $X = (x_1, x_2, \ldots, x_n)$ be the vector of wealth proportions invested in each individual asset of the portfolio so that $x_k \in [0, 1]$ for all $k = 1, 2, \ldots, m$, and $\sum_{k=1}^{m} x_k = 1$. The mean return of asset $k$ is then computed as

$$E(r^k) = \sum_{i=1}^{n} \pi_i r_i^k, \tag{4}$$

and the mean return portfolio is hence given by

$$E(r^p; x) = \sum_{k=1}^{m} x_k E(r^k), \tag{5}$$

as the expectation operator $E$ is linear.

Moreover, the covariance between assets $j$ and $k$ is given by

$$COV(r^j, r^k) = \sum_{i=1}^{n} E(r_i^j - E(r^j)) E(r_i^k - E(r^k))$$
$$= \sum_{i=1}^{n} \pi_i (r_i^j - E(r^j))(r_i^k - E(r^k)), \tag{6}$$

and the variance of the portfolio can then be computed as follows:

$$V(r^p; x) = \sum_{j=1}^{m} \sum_{k=1}^{n} x_j x_k COV(r^j, r^k), \tag{7}$$

given the linearity of the operator $E$.

A key aspect in Markowitz methodology is to characterize the *efficient frontier*, which collects all the pairs that yield the maximum mean return portfolio for a given level of risk (the portfolio variance or standard deviation), or alternatively, all the pairs representing the minimum portfolio risk for a given level of mean return portfolio. From this duality, two practical methodologies emerge to construct the efficient frontier: a quadratic or a parametric programming model. Both of them are summarized in Table 1.

**Table 1** Quadratic and parametric programming in Markowitz portfolio selection approach. $V^*$ and $E^*$ represent a given level of portfolio variance and portfolio mean return, respectively

|                        | Program 1                  | Program 2                  |
|------------------------|----------------------------|----------------------------|
| Objective function     | $\max_x E(r^p; x)$         | $\min_x V(r^p; x)$         |
| Parametric constraints | $V(r^p; x) = V^*$          | $E(r^p; x) = E^*$          |
| Budget constraints     | $\sum_{k=1}^{m} x_k = 1$   | $\sum_{k=1}^{m} x_k = 1$   |
| Non negativity         | $\forall\, x_k \in [0, 1]$ | $\forall\, x_k \in [0, 1]$ |

An important result coming from portfolio choice analysis is the investor's wealth allocation that allows him to bear the minimum level of risk, i.e., the so-called *minimum-variance portfolio*. This is a relevant result, especially for a too risk-averse investor who wants to minimize the variability of his position irrespective if this portfolio yields a quite low expected return. In formal terms, let us define $\bar{x}$ the minimum-variance portfolio, as follows:

$$\bar{x} = \arg\min_{x} V(r^p; x), \qquad (8)$$

where $\bar{X} = (\bar{x}_1, \bar{x}_2, \ldots, \bar{x}_m)$ is so that $\sum_{k=1}^{m} \bar{x}_k = 1$ for all $\bar{x}_k \in [0, 1]$.

## 3 The OWA Operator in Portfolio Selection

Markowitz's approach is based on the use of the mean and the variance. These techniques are usually studied with an arithmetic mean or a weighted average (or probability). However, many times the degree of uncertainty is more complex and it is necessary to represent the information in a deeper way. In this case, more general aggregation operators are needed in order to assess the information properly. A practical technique for doing so is the OWA operator because it provides a parameterized group of aggregation operators between the minimum and the maximum. Moreover, it is able to represent the *attitudinal* character of the decision maker in the specific problem considered. Therefore, the main advantage of this approach is that it represents the problem in a more complete way because it can consider any scenario from the most pessimistic to the most optimistic one and select the situation that is in closest accordance with the inverstor's interests.

In order to revise Markowitz's approach with the OWA operator, it is necessary to change the formulas used to compute the portfolio's mean return and risk, that is, Eqs. (5) and (7) of Sect. 2. Note that this is suggested for situations with high levels of uncertainty where probabilistic information is not available. In particular, when:

(1) The return of any asset is uncertain and cannot be assessed with probabilities. Therefore, the expected value cannot be used. In these uncertain environments, the Mean-OWA operator may become an alternative method for aggregating the information of the assets.
(2) The risk of any asset cannot be measured with the usual variance because probabilities are unknown. However, it is possible to represent it with the Variance-OWA (Merigó 2012; Yager 1996).

### 3.1 Asset Return and Risk Using OWA

One of the key ideas of this approach is to introduce a model that can adapt better to uncertain environments because Markowitz's approach is usually focused on risky environments. Note that with the expected value it is possible to consider subjective and objective probabilities but it is not possible to assess situations without any type of probabilities. An alternative for dealing with these situations is the OWA operator that aggregates the information according to the attitudinal character of the investor.

**Definition 2.** Let $r_i$ be an asset's return at state of nature $i$ for $i = 1, 2, ..., n$. The Mean-OWA operator can then be represented as follows:

$$E_{OWA} = OWA(r_1, r_2, ..., r_n) = \sum_{j=1}^{n} w_j q_j, \qquad (9)$$

where $q_j$ is the $j$th largest of the $r_i$.

Next, let us look into the other main perspective when dealing with portfolio selection. The analysis of risk in Markowitz's approach is based on the use of the variance measure. In this paper, we have suggested to use the Variance-OWA as a measure of risk. The main advantage is that this formulation is more general than the classical variance because it provides a parameterized family of variances between the minimum and the maximum one. Thus, it gives a better representation of the problem and selects the specific result that is in closest accordance with the attitude of the decision maker. Following Yager (1996) and Merigó (2012), an asset variance when using the OWA operator can be formulated as follows.

**Definition 3.** Given an asset with expected returns $(r_1, r_2, ..., r_n)$, let us define $s_i$ as:

$$s_i = (r_i - E_{OWA})^2$$

For $i = 1, 2, ..., n$, where $E_{OWA}$ is defined according to (7). The Variance-OWA can then be defined as follows:

$$V_{OWA} = OWA(s_1, s_2, ..., s_n) = \sum_{j=1}^{n} w_j t_j, \qquad (10)$$

where $t_j$ is the $j$th smallest of the $s_i$.

Observe that here we use an ascending order because usually it is assumed that a lower risk represents a better result and thus it should appear first in the aggregation. Similarly to the case of expected returns, with the OWA approach we can consider any risk from the minimum to the maximum one by using $w_1 = 1$ and $w_j = 0$ for all $j \neq 1$ (minimum variance) and $w_n = 1$ and $w_j = 0$ for all $j \neq n$ (maximum variance).

## 3.2 Portfolio Mean Return and Risk Using OWA

Consider now a portfolio in which it can be combined $m$ individual assets, so that $r_i^k$ represents asset $k$'s return at state of nature $i$ for all $k = 1, 2, ..., m$, and $i = 1, 2, ..., n$. Let $X = (x_1, x_2, ..., x_m)$ be the vector of wealth proportions invested in each individual asset of the portfolio so that $x_k \in [0, 1]$ for all $k = 1, 2, \ldots, m$, and $\sum_{k=1}^{m} x_k = 1$. Moreover, let us define $r_i^p$, the portfolio's return at state of nature $i$, as:

$$r_i^p = \sum_{k=1}^{m} x_k r_i^k$$

for all $i = 1, 2, \ldots, n$. The portfolio mean-OWA is then given by

$$E_{OWA}(r^p; x) = E_{OWA}(r_1^p, r_2^p, \ldots, r_n^p), \tag{11}$$

or equivalently, using definition given by (9), it becomes

$$E_{OWA}(r^p; x) = E_{OWA}(r_1^p, r_2^p, \ldots, r_n^p) = \sum_{j=1}^{n} w_j q_j^p, \tag{12}$$

where $q_j^p$ is the $j$th largest of the $r_i^p$.

Also, by applying (11) or (12), we can define $s_i^p$ as:

$$s_i^p = (r_i^p - E_{OWA}(r^p; x))^2, \tag{13}$$

for $i = 1, 2, \ldots, n$. The portfolio variance-OWA can then be formulated as follows:

$$V_{OWA}(r^p; x) = V_{OWA}(r_1^p, r_2^p, \ldots, r_n^p), \tag{14}$$

and it can be calculated as

$$V_{OWA}(r^p; x) = OWA(s_1^p, s_2^p, \ldots, s_n^p) = \sum_{j=1}^{n} w_j t_j^p,$$

where $t_j^p$ is the $j$th smallest of the $s_i^p$.

Once this initial information is calculated, the rest of the approach follows Markowitz methodology where a quadratic or parametric programming model is used (see Table 2).

**Table 2** Quadratic and parametric programming in portfolio selection using OWA. $V^*$ and $E^*$ represent a given level of portfolio variance OWA and portfolio mean-OWA return, respectively

|  | Program 1 | Program 2 |
|---|---|---|
| Objective function | $\max_x E(r^p; x)$ | $\min_x V(r^p; x)$ |
| Parametric constraints | $V_{OWA}(r^p; x) = V^*$ | $E_{OWA}(r^p; x) = E^*$ |
| Budget constraints | $\sum_{k=1}^{m} x_k = 1$ | $\sum_{k=1}^{m} x_k = 1$ |
| Non negativity | $\forall x_k \in [0, 1]$ | $\forall x_k \in [0, 1]$ |

As in the Markowitz approach, we define $\bar{x}_{OWA}$, the minimum-variance-OWA portfolio, as follows

$$\bar{x}_{OWA} = \arg\min_x V_{OWA}(r^p; x), \qquad (16)$$

where $\bar{x}_{OWA} = (\bar{x}_1, \bar{x}_2, \ldots, \bar{x}_m)$ is so that $\sum_{k=1}^{m} \bar{x}_k = 1$ for all $\bar{x}_k \in [0, 1]$.

## 3.3 Investor's Criteria for Mean-OWA

Observe that with the OWA operator, the expected returns are studied considering any scenario from the minimum to the maximum one. Thus, the decision maker does not lose any information in this initial stage. Once he selects a specific attitude, he opts for a specific result and decision although he still knows any extreme situation that can occur in the problem. This can be proved analyzing some key particular cases of the OWA aggregation including the minimum, the maximum and the arithmetic mean:

- If $w_1 = 1$ and $w_j = 0$ for all $j \neq 1$, the OWA becomes the maximum (or optimistic criteria).
- If $w_n = 1$ and $w_j = 0$ for all $j \neq n$, it is formed the minimum (or pessimistic criteria).
- If $w_j = 1/n$ for all $j$, it becomes the arithmetic mean (Laplace criteria).

By looking to the maximum and the minimum, it is proved that the OWA operator accomplishes the boundary condition:

$$\text{Min}\{a_i\} \leq f(a_1, a_2, \ldots, a_n) \leq \text{Max}\{a_i\}. \qquad (17)$$

Some other interesting particular cases of the OWA operator are the following:

- Hurwicz criteria: If $w_1 = \alpha$, $w_n = 1 - \alpha$ and $w_j = 0$, for all $j \neq 1, n$.
- Step-OWA: $w_k = 1$ and $w_j = 0$ for all $j \neq k$. Note that if $k = 1$ we get the maximum and if $k = n$, the minimum.
- Median-OWA: If $n$ is odd we assign $w_{(n+1)/2} = 1$ and $w_j = 0$ for all others. If $n$ is even we assign for example, $w_{n/2} = w_{(n/2)+1} = 0.5$ and $w_j = 0$ for all others.
- Olympic-OWA: When $w_1 = w_n = 0$ and for all others $w_j = 1/(n - 2)$.
- S-OWA: If $w_1 = (1/n)(1 - (\varepsilon + \delta)) + \varepsilon$, $w_n = (1/n)(1 - (\varepsilon + \delta)) + \delta$, $w_j = (1/n)(1 - (\varepsilon + \delta))$ for $j = 2$ to $n - 1$, where $\varepsilon, \delta \in [0, 1]$ and $\varepsilon + \delta \leq 1$.
- Centered-OWA: If it is symmetric, strongly decaying and inclusive. It is symmetric if $w_j = w_{j+n-1}$. Strongly decaying when $i < j \leq (n + 1)/2$ then $w_i < w_j$ and when $i > j \geq (n + 1)/2$ then $w_i < w_j$. It is inclusive if $w_j > 0$.

For further reading on other particular types of OWA operators see Merigó and Gil-Lafuente (2009) and Yager (1993).

# 4   Conclusions

The Markowitz mean-variance portfolio selection approach can be extended by using the OWA operator. The main advantage of this new formulation is that it can represent uncertain environments by using the attitudinal character of the decision maker. Thus, it aggregates the information considering the degree of optimism or pessimism that an individual has in a specific problem. This method assumes that probabilistic data is not available because the environment shows a high degree of uncertainty. The mean return and the risk of any asset have been studied with OWA operators instead of using the classical weighted and arithmetic mean. By doing so, the investor gets a more complete view of the problem because he can consider any scenario from the minimum to the maximum and select the one closest to his interests. Some particular cases have been studied in order to see how the aggregation process might be seen under different perspectives. Among others, the classical methods for decision making under uncertainty have been considered since the OWA operator includes them as special cases. That is, the optimistic, pessimistic, Laplace and Hurwicz criterion.

The degree of optimism and pessimism has been studied under a new formulation that takes into account the specific values of the arguments. Thus, rather than only considering the degree of orness through weights, it is also possible to consider the position of the arguments which also conditions the optimism or pessimism of the aggregation. Some numerical examples have also been presented in order to understand numerically the new approach. These examples have been developed considering different types of OWA positions that can be used in the decision and aggregation process including the optimistic, pessimistic, Olympic, Laplace and Hurwicz criterion. Various results emerge from this numerical exercise. First, the isolated ordering effect of OWA suggests that this operator induces more optimism *only* on the inefficient portfolio region, and thus, both efficient frontiers (Markowitz and OWA) are identical. Second, the combined ordering and weighting effect of OWA suggests that the minimum-risk portfolio is the *same* in both types of frontiers. This result is robust to different attitudinal characters of investors. Third, an optimistic investor faces better return-risk profile, so he expects, for a given level of either return or risk, a higher utility than a Markowitzian investor. This dominance gets exacerbated as the optimism degree increases, leading to an increasing polarization of portfolio choice in favor of the individual asset with the best return profile. Fourth, a pessimistic investor faces a worse return-risk trade-off, so he expects a lower utility than a Markowitzian investor. As a consequence, he always selects the minimum risk portfolio irrespective of his risk-aversion degree. Fifth, in the case of moderate, Laplace-type, and extremist investor, the results in general are more ambiguous as they depend on the specific probabilistic and weighting vectors under consideration.

In future research, further developments can be developed by using other extensions and generalizations of the OWA operator in Markowitz mean-variance portfolio selection model including the probabilistic OWA operator (Merigó 2012) and induced aggregation operators (Merigó and Gil-Lafuente 2009; Yager and Filev 1999). Moreover, some other portfolio selection methods can be considered with the OWA operator including the Sharpe model, the CAPM and the APT. Lastly, the approach here proposed can be seen as a starting point to provide alternative explanations to some controversial results in financial economics such as the two-fund separation puzzle.

# References

Beliakov, G., Calvo, T., Pradera, A.: Aggregation Functions: A Guide for Practitioners. Springer, Berlin (2007)

Emrouznejad, A., Marra, M.: Ordered weighted averaging operators 1988–2014: a citation based literature survey. Int. J. Intell. Syst. **29**, 994–1014 (2014)

Engemann, K.J., Filev, D.P., Yager, R.R.: Modelling decision making using immediate probabilities. Int. J. Gen. Syst. **24**, 281–294 (1996)

Fodor, J., Marichal, J.L., Roubens, M.: Characterization of the ordered weighted averaging operators. IEEE Trans. Fuzzy Syst. **3**, 236–240 (1995)

Garlappi, L., Uppal, R., Wang, T.: Portfolio selection with parameter and model uncertainty: a multi-prior approach. Rev. Financ. Stud. **20**, 41–81 (2007)

Grabisch, M., Marichal, J.L., Mesiar, R., Pap, E.: Aggregation functions: means. Inf. Sci. **181**, 1–22 (2011)

Kolm, P.N., Tütüncü, R., Fabozzi, F.J.: 60 years of portfolio optimization: practical challenges and current trends. Eur. J. Oper. Res. **234**, 356–371 (2014)

Maccheroni, F., Marinacci, M., Ruffino, D.: Alpha as ambiguity: robust mean-variance portfolio analysis. Econometrica **81**, 1075–1113 (2013)

Markowitz, H.M.: Portfolio selection. J. Financ. **7**, 77–91 (1952)

Markowitz, H.M.: Mean-variance approximations to expected utility. Eur. J. Oper. Res. **234**, 346–355 (2014)

Merigó, J.M.: Probabilities with OWA operators. Expert Syst. Appl. **39**, 11456–11467 (2012)

Merigó, J.M., Engemann, K.J., Palacios-Marqués, D.: Decision making with Dempster-Shafer theory and the OWAWA operator. Technol. Econ. Dev. Econ. **19**, S194–S212 (2013)

Merigó, J.M., Gil-Lafuente, A.M.: The induced generalized OWA operator. Inf. Sci. **179**, 729–741 (2009)

Merigó, J.M., Lobato-Carral, C., Carrilero-Castillo, A.: Decision making in the European Union under risk and uncertainty. Eur. J. Int. Manage. **6**, 590–609 (2012)

Sharpe, W.F.: A simplified model for portfolio analysis. Manage. Sci. **9**, 277–293 (1963)

Torra, V.: The weighted OWA operator. Int. J. Intell. Syst. **12**, 153–166 (1997)

Wu, H.L., Zeng, Y., Yao, H.X.: Multi-period Markowitz's mean-variance portfolio selection with state-dependent exit probability. Econ. Model. **36**, 69–78 (2014)

Xu, Z.S., Da, Q.L.: An overview of operators for aggregating the information. Int. J. Intell. Syst. **18**, 953–969 (2003)

Yager, R.R.: On ordered weighted averaging aggregation operators in multi-criteria decision making. IEEE Trans. Syst. Man Cybern. B **18**, 183–190 (1988)

Yager, R.R.: Families of OWA operators. Fuzzy Sets Syst. **59**, 125–148 (1993)

Yager, R.R.: On the inclusion of variance in decision making under uncertainty. Int. J. Uncertainty Fuzziness Knowl. Based Syst. **4**, 401–419 (1996)

Yager, R.R.: Heavy OWA operators. Fuzzy Optim. Decis. Making **1**, 379–397 (2002)

Yager, R.R., Filev, D.P.: Induced ordered weighted averaging operators. IEEE Trans. Syst. Man Cybern. **29**, 141–150 (1999)

Yager, R.R., Kacprzyk, J., Beliakov, G.: Recent Developments on the Ordered Weighted Averaging Operators: Theory and Practice. Springer, Berlin (2011)

Zopounidis, C., Doumpos, M., Fabozzi, F.J.: Preface to the special issue: 60 years following Harry Markowitz's contributions in portfolio theory and operations research. Eur. J. Oper. Res. **234**, 343–345 (2014)

# Part II
# Expert Systems and Forgotten Effects Theory

# Application of the Forgotten Effects Model to the Agency Theory

**Elena Arroyo and Elvira Cassú**

**Abstract** During the financial crisis, interest problems between shareholders (principal) and managers (agent) have raised due to the evidence of the dishonest behaviour of the Chief Executive Officer (CEO). Based on the agency theory, we use the model of forgotten effects in order to identify different solutions for each kind of agency problems. The aim of this study is to reduce agency problems and facilitate the companies' success, providing useful information to improve the decision-making process in management.

**Keywords** Forgotten effects model · Agency theory · Shareholders · Fuzzy sets · Management

## 1 Introduction

Principal-agent relationship may be described by any situation in which one or more people transfer authority to others to make decisions. The agency problem is one aspect of this relationship (Jensen and Meckling 1976). In that sense, agency problems exist when (1) the objectives of principal (who transfer authority) and agent (who is in charge of making decisions) are different and (2) asymmetric information exists and makes it difficult for the principal to monitor the agent's actions. Even though this problem can be found in many situations, we are focusing our analysis on the owner-manager relationship (Mitnick 2006; Jensen 1986; Fama and Jensen 1983).

E. Arroyo (✉)
Research Group on Statistics, Econometrics and Health (GRECS),
University of Girona, Girona, Spain
e-mail: elena.arroyo@udg.edu

E. Cassú
Department of Business Administration, University of Girona,
Carrer Universitat de Girona 10, 17071 Girona, Spain
e-mail: elvira.cassu@udg.edu

© Springer International Publishing Switzerland 2015
J. Gil-Aluja et al. (eds.), *Scientific Methods for the Treatment of Uncertainty in Social Sciences*, Advances in Intelligent Systems and Computing 377,
DOI 10.1007/978-3-319-19704-3_6

Agency theory affirms that when the ownership and control are separated and the structure of the firm is not defined in order to safeguard the interests of shareholders, the managers will not act to maximise the benefits of the shareholders and the value of the firm will decline (Boland et al. 2008; Donaldson and Davis 1991; Jensen and Meckling 1976).

It is important to describe the scenario and the characteristics of the participants. In that sense, one of the agent's characteristic is defined by "opportunistic behaviour". The main causes are the asymmetric information, divergence of interests and the attitude towards risk. Opportunism implies to act in favour of your own individual objective and maximise the own personal economic gain, even if it means to divert from shareholders' interests. The second characteristic that must be explained is the aversion to work which implies to invest less effort in tasks which are not important enough for the agent. Thus, the design of efficient incentives will improve performance and will reduce this problem. On the other hand, the characteristics of the principal include being a risk seeker and the desire of maximising the corporation's benefits. Therefore, the main elements in principal-agent relationship are the actors (principal and agent), but also the contract which acts as a mechanism to control and unify the relationship in order to minimise the different interests and potential disagreements (Alvarez 2008).

Not only does the separation of ownership and management suggest different interests, but also different information available for each part. This is what is called asymmetric information and it involves less information for the principal who is not able to know the effort and the decisions made by the agent, neither the factors that are affecting the daily activity of the company. This asymmetric environment benefits the agent who can focus the work on his own interest, and it does lead to reduce profits for the principal (Salas 2002). Information asymmetry deals with two main problems: adverse selection and moral hazard. The former is about immoral behaviour before signing the contract so that the principal does not know if the agent is good enough and the principal finds it difficult to forecast if the agent will do his best. The latter is about immoral behaviour after signing the contract so that the agent makes decisions only taking into consideration his own interests.

These divergence of interests lead to three types of agency costs: monitoring costs, bonding costs and residual losses. Monitoring costs include all actions taken in order to control and monitor the agent (stock options or the board of directors). Instead, bonding costs are incurred by the agent due to contractual obligations that restrict the own agent's activity. Finally, residual losses are those costs incurred due to the divergence of interests (Wang 2010).

The empirical literature focuses on some relationships between agency costs and mechanisms to solve the agency problem using statistical approaches, but few studies have used logic fuzzy. The aim of this article is to take advantage of the model for the detection of forgotten effects so that we may know the relationship between some agency problems and the mechanisms to solve them in order to reduce agency costs. The paper provides useful information to improve the decision-making process in firms where the ownership and management are separated.

# 2 The Recuperation of Forgotten Effects in the Resolution of the Agency Problem

The forgotten effects theory is broadly detailed by Kaufmann and Gil Aluja (1898). This methodology will let us identify forgotten effects of first and second generation.

This study pretends to be an application of the use of the forgotten effects model to solve some problems of the agency theory. The forgotten effects methodology will allow us to obtain useful and innovative information for the owner (principal) to reduce agency costs caused by the divergence of interests between agent and principal. According to Lozano et al. (2004), we will enumerate some solutions of the agency theory (set of causes) to solve the different sources of manager-owner conflict (set of effects).

## 2.1 The Main Agency Problems (Effects)

The objective of this section is to describe some sources of the agency problem that will be used as effects in the following analysis.

**b1**. Shirking: This problem is based on the asymmetric information and the agents' opportunistic behaviour. Shirking is about not putting the best effort to help to increase shareholders' benefits. Managers may spend time in activities which they enjoy more and it creates opportunity costs, costly for the principal and difficult to detect. This is a moral hazard problem caused by the lack of control by owners (Mascareñas 2007).

**b2**. Free cash flow (FCF): Free cash flow is cash flow in excess of that required to fund all projects that have positive net present values when discounted at the relevant cost of capital (Jensen 1986). Earnings retained in disposition of the manager. Free cash flow can be seen as financial resources at the executive's discretion to allocate and can be a result of waste of the corporate resources (Wang 2010).

**b3**. Managerial myopia: The time horizon can be another principal-agent conflict. In that case, owners' aim is to maximise profits in a long term. However, managers' decisions are more related to the short term. Thus, the executives would prefer projects at a low cost and benefits that can be gain in a short term as to enhance their reputation quickly. Then, managers tend to invest in projects with negative net present value (NPV) but with cash flows in a short term, instead of investing in profitable long-term projects (Álvarez 1999).

**b4**. Attitude towards risk: Principal and agent have different attitudes towards risk so that the two parts may prefer different strategies due to the different risk preferences. The manager assumes a higher risk than the principal because all decisions he makes are affecting his job. Managers may prefer less risky investments and lower leverage in order to reduce the probability of bankruptcy

(Salas 2004). Instead, shareholders are supposed to assume risky projects because they can diversify their portfolio.

**b5**. Job security: The security job may also be an agency problem. If the manager acts in an inefficient manner, his job is at risk. However, the agent knows that any change in management is quite expensive, so that the manager attempts to know the threshold, letting him manage the job with some tranquillity. Despite that fact, the most common behaviour in management is known as herd behaviour, which means that managers tend to make similar decisions to their competitors. Moreover, the job security also leads to maximise the size of the firm in order to reduce the likelihood of being taken over. Lastly, managers also search for diversification at the expense of the shareholders (who can diversify their individual portfolios more easily).

**b6**. Self-interests: The separation of ownership and control gives CEO an opportunity to work in his own self-interest. The manager may take advantage of perks such as use of the company jet, expense accounts or company cars. It represents a real cost to shareholders (Salas 2004).

## 2.2 Mechanisms to Solve the Agency Problems (Causes)

The goal of this section is to identify and describe the mechanisms to deal with the agency theory problems (Cuervo 1999; Jensen 1994). These are going to be the causes of the model.

**a1**. The board of directors: The board of directors is the most powerful mechanism inside the corporations. The board carries out the direction, administration and control. It also monitors opportunistic behaviours to improve the efficiency of the company. Moreover, the board guarantees shareholders' interest and makes the last decision to CEO's strategies. Therefore, it has an important paper in terms of control, supervision and evaluation in order to protect shareholders' interests. The efficiency depends on the size and composition of the board. First, the size is negatively related to the value of the firm. Second, the major proportion of external members on the board will lead to better results for the company (Fama and Jensen 1983; Baysinger and Hoskisson 1990).

**a2**. Dividend taxation: Dividend retention is one of the problems of the agency theory we have seen with the free cash flow. A decision to retain earnings instead of paying dividends would encourage executives to not maximize the shareholders' value. Thus, dividend taxation is a valuable tool to reduce managers' discretion. Also, it is a signal for the capital market which acts as a monitor. In short, large dividend taxation should influence negatively to agency costs (López and Saona 2007).

**a3**. The debt: The debt helps to mitigate agency problems. An increase in debt also increases the risk of bankruptcy; therefore it limits the executives' consumptions. Besides, debt forces the managers to pay cash out, reducing the free cash flow managers can waste. It also increases the control of the capital markets.

Therefore, there is a negative relationship between agency costs and the level of debt (López and Saona 2007).

**a4**. Executive compensation: To solve the agency problems shareholders also may use executive compensation. These incentives for managers can reward them financially for maximizing shareholders' interests. For instance, plans in which managers obtain shares, thus aligning financial interests with shareholders. Also, incentives related to future maximization of the firm in order to omit short-term executive actions.

**a5**. Stock options: Related to executive compensation, it is important that the executive owns part of the corporation in order to align interests with shareholders. However, it is important to take into account that if the managers' ownership increases too much, it can produce the executive retrenchment: the positive influence it may create at first might be cancelled as the participation of the executive in the ownership increases. However, once this threshold is achieved, exists again a convergence effect between interests (López and Saona 2007).

**a6**. Ownership concentration: The concentration of the ownership is also an important aspect in order to reduce agency costs. A larger concentration of shareholders might result in greater monitoring of CEO. It suggests a stronger monitoring power from investors over the managerial decisions because the incentives are greater. Instead, a low level of ownership concentration (diffuse ownership) might lead to an increase of the managers' power. However, if the concentration is too high it can produce a negative impact to the minority shareholders. Then, a new conflict will arise between large shareholders and minority shareholders. Also the composition of the ownership may be important. For instance, financial intermediaries are able to reduce agency problems because it creates an important relationship between the corporation and the financial providers.

**a7**. Goods and services market and the executive job market: Finally, the goods and services market is a competitive market and the CEO has to increase the efficiency of the company in order to guarantee the survival of the corporation (Alvarez 2008). In that sense, the goods and services market acts as a mechanisms to monitor the executives and penalize all members of the firm. Then, the competitiveness of the executive job market leads the manager to act properly because their strategies might be compared to the others. This market also influences positively to the agency problem and increases the value of the firm (Cuervo 1999).

## 2.3 Process Work Out for the Detection of Possible Forgotten Effects

To make this study possible we relied on four experts from the agency theory: one academic professor, one executive, one executive-owner and one owner. This number may result too little to pretend a result that could serve as a sufficient base to obtain valuations completely certain but it will serve to show the application of forgotten effects methodology in the field of agency theory. We asked each of the four experts, independent between them, to fill three matrixes of qualitative effects on incidence, one of cause-effect, another of cause-cause and finally one of effect-effect, giving them the freedom of expressing their valuations on each incidence from the confidence intervals comprised in the segment [0,1]. The question that each expert had to consider was: to what extent each solution of the principal-agent conflict (causes) has an impact to the sources of the agency problem (effects)? Also the semantic correspondence for the chosen values was given to them:

| | | | |
|---|---|---|---|
| 0: | no incidence | 0.6: | significant effect |
| 0.1: | virtually no effect | 0.7: | very significant effect |
| 0.2: | almost no effect | 0.8: | strong effect |
| 0.3: | very low effect | 0.9: | very strong effect |
| 0.4: | low effect | 1: | the highest effect |
| 0.5: | medium effect | | |

We present below the matrixes of the direct qualitative cause-effect incidences:

$M^{(1)}$

| | $b_1$ | $b_2$ | $b_3$ | $b_4$ | $b_5$ | $b_6$ |
|---|---|---|---|---|---|---|
| $a_1$ | [0.1,0.3] | [0.8,0.9] | [0.8,0.9] | [0.8,0.9] | [0.8,0.9] | [0.8,0.9] |
| $a_2$ | [0.3,0.5] | [0.8,0.9] | [0.2,0.3] | [0.7,0.8] | [0.5,0.6] | [0.3,0.4] |
| $a_3$ | 0.8 | [0.7,0.8] | [0.5,0.6] | [0.7,0.8] | [0.5,0.6] | [0.3,0.4] |
| $a_4$ | 0.9 | [0.1,0.3] | 0.8 | [0.2,0.3] | [0.7,0.8] | [0.7,0.8] |
| $a_5$ | 0.8 | 0.4 | 0.8 | [0.5,0.6] | [0.7,0.8] | [0.7,0.8] |
| $a_6$ | 0.7 | [0.6,0.7] | [0.3,0.4] | [0.6,0.7] | [0.4,0.5] | [0.3,0.4] |
| $a_7$ | 0.5 | 0.4 | [0.5,0.6] | 0.5 | [0.2,0.3] | [0.1,0.2] |

$M^{(2)}$

| | $b_1$ | $b_2$ | $b_3$ | $b_4$ | $b_5$ | $b_6$ |
|---|---|---|---|---|---|---|
| $a_1$ | [0.6,0.8] | 0.8 | 0.7 | [0.6,0.7] | 0.8 | 0.8 |
| $a_2$ | 0.4 | 0.8 | 0.4 | 0.6 | 0.4 | 0.8 |
| $a_3$ | 0.1 | 0.1 | 0.1 | 0.5 | 0.2 | 0.8 |
| $a_4$ | [0.8,0.9] | [0.6,0.7] | 0.8 | 0.8 | 0.6 | [0.5,0.6] |
| $a_5$ | 0.7 | 0.4 | 0.7 | 0.3 | 0.7 | 0.5 |
| $a_6$ | [0.6,0.8] | 0.2 | 0.6 | 0.2 | 0.5 | 0.7 |
| $a_7$ | [0.8,0.9] | 0.1 | 0.4 | 0.3 | 0.8 | 0.6 |

$M^{(3)}$

| | $b_1$ | $b_2$ | $b_3$ | $b_4$ | $b_5$ | $b_6$ |
|---|---|---|---|---|---|---|
| $a_1$ | 0.5 | 0.6 | 0.6 | 0.7 | 0.6 | 0.4 |
| $a_2$ | 0.7 | 0.9 | 0.6 | 0.8 | 0.7 | 0.6 |
| $a_3$ | 0.8 | 0.9 | 0.8 | 0.7 | 0.8 | 0.7 |
| $a_4$ | 0.9 | 0.6 | 0.8 | 0.5 | 0.7 | 0.6 |
| $a_5$ | 0.9 | 0.7 | 0.7 | 0.4 | 0.8 | 0.5 |
| $a_6$ | 0.8 | 0.6 | 0.6 | 0.8 | 0.5 | 0.5 |
| $a_7$ | 0.6 | 0.5 | 0.4 | 0.3 | 0.6 | 0.3 |

$M^{(4)}$

| | $b_1$ | $b_2$ | $b_3$ | $b_4$ | $b_5$ | $b_6$ |
|---|---|---|---|---|---|---|
| $a_1$ | 0.4 | 0.5 | 0.6 | 0.7 | 0.5 | 0.7 |
| $a_2$ | 0.1 | 0.7 | 0.1 | 0.1 | 0.1 | 0.1 |
| $a_3$ | 0.1 | 0.1 | 0.3 | 0.7 | 0.2 | 0 |
| $a_4$ | 0.8 | 0 | 0.7 | 0.7 | 0.3 | 0.8 |
| $a_5$ | 0.6 | 0.8 | 0.7 | 0.5 | 0.3 | 0.3 |
| $a_6$ | 0.8 | 0.8 | 0.8 | 0.8 | 0.4 | 0.8 |
| $a_7$ | 0.1 | 0.1 | 0.1 | 0.1 | 0.3 | 0.5 |

Next, we present the matrixes of direct qualitative cause-cause incidences resulting from each one of the four experts:

$\underset{\sim}{A}^{(1)}$

|  | $a_1$ | $a_2$ | $a_3$ | $a_4$ | $a_5$ | $a_6$ | $a_7$ |
|---|---|---|---|---|---|---|---|
| $a_1$ | 1 | 0.9 | [0.7,0.8] | [0.8,0.9] | 0.8 | 0.9 | [0.5,0.6] |
| $a_2$ | 0.9 | 1 | [0.4,0.5] | [0.8,0.9] | [0.8,0.9] | [0.8,0.9] | [0.2,0.3] |
| $a_3$ | [0.7,0.8] | [0.7,0.8] | 1 | [0.5,0.6] | [0.5,0.6] | [0.4,0.5] | [0.7,0.8] |
| $a_4$ | [0.8,0.9] |  | [0.2,0.3] | 1 | [0.6,0.7] | [0.4,0.5] | [0.2,0.3] |
| $a_5$ | [0.8,0.9] | [0.5,0.6] | [0.3,0.4] | [0.8,0.9] | 1 | [0.8,0.9] | [0.4,0.5] |
| $a_6$ | [0.8,0.9] | [0.7,0.8] | [0.4,0.5] | [0.7,0.8] | [0.8,0.9] | 1 | [0.2,0.3] |
| $a_7$ | [0.5,0.6] | [0.4,0.5] | [0.6,0.7] | [0.2,0.3] | [0.4,0.5] | [0.2,0.3] | 1 |

$\underset{\sim}{A}^{(2)}$

|  | $a_1$ | $a_2$ | $a_3$ | $a_4$ | $a_5$ | $a_6$ | $a_7$ |
|---|---|---|---|---|---|---|---|
| $a_1$ | 1 | [0.9,1] | 0.9 | [0.7,0.9] | 0.8 | 1 | [0.6,0.7] |
| $a_2$ | 1 | 1 | 0.9 | 0.8 | 0.5 | 1 | 0.5 |
| $a_3$ | [0.8,0.9] | [0.9,1] | 1 | [0.5,0.6] | 0.3 | [0.7,0.9] | 0.5 |
| $a_4$ | [0.8,0.9] | 0.8 | [0.7,0.8] | 1 | 0.3 | [0.6,0.7] | [0.8,0.9] |
| $a_5$ | 1 | [0.5,0.6] | 0.5 | [0.9,1] | 1 | [0.9,1] | [0.8,0.9] |
| $a_6$ | 1 | 1 | 0.9 | [0.5,0.6] | 0.3 | 1 | 0.3 |
| $a_7$ | 0.2 | 0.4 | 0.2 | [0.8,0.9] | 0.8 | [0.5,0.6] | 1 |

$\underset{\sim}{A}^{(3)}$

|  | $a_1$ | $a_2$ | $a_3$ | $a_4$ | $a_5$ | $a_6$ | $a_7$ |
|---|---|---|---|---|---|---|---|
| $a_1$ | 1 | 0.9 | 0.8 | 0.9 | 0.9 | 0.8 | 0.5 |
| $a_2$ | 0.9 | 1 | 0.9 | 0.8 | 0.8 | 0.8 | 0.7 |
| $a_3$ | 0.8 | 0.9 | 1 | 0.8 | 0.2 | 0.7 | 0.3 |
| $a_4$ | 0.9 | 0.8 | 0.8 | 1 | 0.9 | 0.3 | 0.7 |
| $a_5$ | 0.9 | 0.8 | 0.2 | 0.9 | 1 | 0.1 | 0.6 |
| $a_6$ | 0.8 | 0.8 | 0.7 | 0.3 | 0.1 | 1 | 0.3 |
| $a_7$ | 0.5 | 0.7 | 0.3 | 0.7 | 0.6 | 0.3 | 1 |

$\underset{\sim}{A}^{(4)}$

|  | $a_1$ | $a_2$ | $a_3$ | $a_4$ | $a_5$ | $a_6$ | $a_7$ |
|---|---|---|---|---|---|---|---|
| $a_1$ | 1 | 0.8 | 0.8 | 0.8 | 0 | 0 | 0 |
| $a_2$ | 0 | 1 | 0.9 | 0 | 0 | 0 | 0 |
| $a_3$ | 0 | 0.8 | 1 | 0.7 | 0 | 0 | 0 |
| $a_4$ | 0 | 0.2 | 0.2 | 1 | 0.1 | 0 | 0 |
| $a_5$ | 0.8 | 0.9 | 0.5 | 0.6 | 1 | 0 | 0 |
| $a_6$ | 1 | 1 | 1 | 1 | 0.2 | 1 | 0 |
| $a_7$ | 0 | 0 | 0 | 0 | 0 | 0 | 1 |

Finally, we show the matrixes of direct qualitative effect-effect resulting from each one of the four experts:

$\underset{\sim}{B}^{(1)}$

|  | $b_1$ | $b_2$ | $b_3$ | $b_4$ | $b_5$ | $b_6$ |
|---|---|---|---|---|---|---|
| $b_1$ | 1 | [0.2,0.3] | [0.7,0.8] | [0.4,0.5] | [0.2,0.3] | [0.4,0.5] |
| $b_2$ | [0.3,0.4] | 1 | [0.8,0.9] | [0.7,0.8] | [0.7,0.8] | [0.7,0.8] |
| $b_3$ | [0.5,0.6] | [0.7,0.8] | 1 | [0.6,0.7] | [0.8,0.9] | [0.7,0.8] |
| $b_4$ | [0.2,0.3] | [0.7,0.8] | [0.8,0.9] | 1 | [0.8,0.9] | [0.7,0.8] |
| $b_5$ | [0.2,0.3] | [0.8,0.9] | [0.8,0.9] | [0.8,0.9] | 1 | [0.7,0.8] |
| $b_6$ | [0.4,0.5] | [0.3,0.4] | [0.7,0.8] | [0.5,0.6] | [0.8,0.9] | 1 |

$\underset{\sim}{B}^{(2)}$

|  | $b_1$ | $b_2$ | $b_3$ | $b_4$ | $b_5$ | $b_6$ |
|---|---|---|---|---|---|---|
| $b_1$ | 1 | 0.1 | 0.8 | 0.8 | 0.8 | 0.8 |
| $b_2$ | 0.3 | 1 | [0.4,0.5] | [0.7,0.8] | 0.5 | 0.6 |
| $b_3$ | 0.8 | 0.5 | 1 | [0.7,0.8] | [0.7,0.8] | 0.6 |
| $b_4$ | [0.7,0.8] | [0.8,0.9] | 0.5 | 1 | 0.6 | 0.5 |
| $b_5$ | 0.8 | 0.5 | 0.5 | [0.6,0.7] | 1 | 0.7 |
| $b_6$ | 0.9 | [0.7,0.8] | 0.6 | [0.7,0.8] | [0.7,0.8] | 1 |

| $B^{(3)}$ | | | | | | | $B^{(4)}$ | | | | | |
|---|---|---|---|---|---|---|---|---|---|---|---|---|
| | $b_1$ | $b_2$ | $b_3$ | $b_4$ | $b_5$ | $b_6$ | | $b_1$ | $b_2$ | $b_3$ | $b_4$ | $b_5$ | $b_6$ |
| $b_1$ | 1 | 0.9 | 0.9 | 0.9 | 0.5 | 0.6 | $b_1$ | 1 | 0.1 | 0.1 | 0.6 | 0 | 0 |
| $b_2$ | 0.9 | 1 | 0.8 | 0.7 | 0.8 | 0.4 | $b_2$ | 0 | 1 | 0.5 | 0.7 | 0.5 | 0 |
| $b_3$ | 0.9 | 0.8 | 1 | 0.9 | 0.7 | 0.8 | $b_3$ | 0.3 | 0.7 | 1 | 0.8 | 0.7 | 0 |
| $b_4$ | 0.9 | 0.7 | 0.9 | 1 | 0.9 | 0.9 | $b_4$ | 0 | 0.8 | 0.9 | 1 | 0 | 0 |
| $b_5$ | 0.5 | 0.8 | 0.7 | 0.9 | 1 | 0.8 | $b_5$ | 0 | 0 | 0 | 0 | 1 | 0 |
| $b_6$ | 0.6 | 0.4 | 0.8 | 0.9 | 0.8 | 1 | $b_6$ | 0 | 0.8 | 0.8 | 0.8 | 0.8 | 1 |

From $M_{\sim j}$, $A_{\sim j}$ and $B_{\sim j}$, we can build three frequency tables for each one where it will be gathered, for each relation of incidence, the number of times the experts have assigned the same valuation as for the inferior extreme as well as for the superior extreme. Then, it could be considered that these statistical tables are the compilation of the four experts' opinions in each incidence relationship. It will be operated through the averages obtained from the statistical tables multiplying them by its correspondent level of confidence. Once calculated we will be able to build up three new matrixes $M(M)$, $M(A)$, $M(B)$ which will constitute the aggregate opinion of all four experts and, consequently, we will be able to operate as if it were an only expert.

We make the maximin convolution to obtain the matrix $M(M^*)$ that gathers the accumulated effects from first and second generation. It is necessary to compare it with the matrix of direct cause-effect incidence which will give us the possibility of revealing the forgotten effects. First, we must reduce the confidence intervals to ordinal numbers. This restriction is necessary since not always exists a strict order between two or more confidence intervals due to that the usual order relationship between confidence intervals is not a relationship of total order.

To make this comparison we will need to obtain the median value of each interval from matrixes $M(M)$ and $M(M^*)$. Thus,

| $\overline{M}(M^*) - \overline{M}(M)$ | | | | | | |
|---|---|---|---|---|---|---|
| | $b_1$ | $b_2$ | $b_3$ | $b_4$ | $b_5$ | $b_6$ |
| $a_1$ | 0.4 | 0.1 | 0.1 | 0.2 | 0 | 0 |
| $a_2$ | 0.3 | 0 | 0.4 | 0.1 | 0.3 | 0.2 |
| $a_3$ | 0.2 | 0.3 | 0.3 | 0 | 0.2 | 0.1 |
| $a_4$ | 0 | 0.3 | 0 | 0.2 | 0.1 | 0 |
| $a_5$ | 0.1 | 0.1 | 0.1 | 0.4 | 0.1 | 0.2 |
| $a_6$ | 0 | 0.2 | 0.1 | 0.1 | 0.2 | 0.1 |
| $a_7$ | 0 | 0.2 | 0.1 | 0.2 | 0 | 0.1 |

The last table only shows forgotten effects. Even though the current study does not seem to have very high level of forgotten effects, we must highlight the following ones (shown in bold):

$a_1 \rightarrow b_1$:  Incidence of the board of directors on shirking
$a_2 \rightarrow b_3$:  Incidence of the dividend taxation on managerial myopia
$a_5 \rightarrow b_4$:  Incidence of the stock options on the attitude towards risk

Previously to the analysis of the results, we will explain the methodology in one example in order to analyse the next cases.

For example, in the first case ($a_1 \rightarrow b_1$) exists a forgotten effect of degree of 0.4 in the incidence of the board of directors on the shirking. In a previous analysis, this relationship cannot be seen but the model allows us to describe this relationship through intermediary elements. Thus, the aim is to find intermediary elements that have provoked this forgotten relationship, searching the minimum between each element of the row 1 of the matrix $\overline{M}(A)$, and the minimum of each column of the matrix $\overline{M}(M)$. Then, the maximum of these minimums is chosen. Once the maximum values are taken, we proceed the maxmin composition again, but now with the column $b_1$ of the $\overline{M}(B)$.

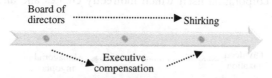

One can observe that the maximum of the minimum is 0.8, which corresponds to $a_4$, 'executive compensation' and this is the intermediary element.

To conclude, the board of directors has an incidence on shirking through the executive compensation scheme. The same analysis has been done for all significant incidences.

In the second case, $a_2 \rightarrow b_3$, with a degree of 0.4 there is a forgotten effect in the incident of dividend taxation to managerial myopia. In order to discover the intermediary elements, we will follow the same process we have done above. Thus, it appears to have 5 intermediary elements:

In this case 2A, there is an intermediary element with a degree level of 0.7. It is important to know that dividend taxation affects in the managerial myopia through the board of directors and this has an incidence on the free cash flow of the firm, which has an implication to the managerial myopia.

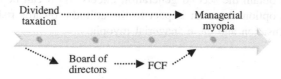

In that case 2B, the dividend taxation has an incidence on the managerial myopia through the board of directors.

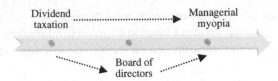

In the case 2C, the path shows that the dividend taxation affects the managerial myopia because the board of directors has an impact on self-interests which incidence on the managerial myopia.

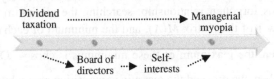

The case 2D also shows that the dividend taxation has an incidence on the free cash flow of the corporation itself which indirectly creates and affect to the managerial myopia.

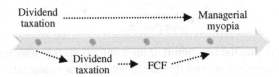

Finally (2E), the dividend taxation has an incidence on the managerial myopia because it has an effect to the debt and this has an incidence on the attitude towards risk, which acts against the managerial myopia.

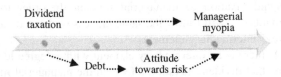

The last case, $a_5 \rightarrow b_4$, has a impact degree about 0.4. Again, the same analysis is done in order to obtain the second generation effects. As it can be seen from the graph, the stock options have an impact to the attitude towards risk through the executive compensation and the managerial myopia.

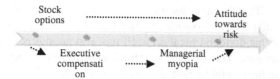

# 3 Results

The results show that the main intermediary elements are executive compensation, the board of directors and the free cash flow. As a first conclusion, we must suggest the owners that these are the three most important mechanisms. Thanks to their interactions, they have a high impact to reduce the agency costs. Now, the most important conclusions will be described for each case.

As for case 1, at first there is not a direct relationship between the board of directors and shirking. Even though the board of directors reduces the opportunism, they cannot monitor if the CEO is doing the properly effort. But, through an executive compensation they can have an incidence on shirking. In that way, the board of directors cannot directly detect the CEO effort but if they design an efficient executive compensation they are having an incidence on reducing shirking.

Regarding case 2, we cannot detect a causality relationship between dividend taxation and managerial myopia. But this relationship can be explained through the debt, the free cash flow and the board of directors, as follows:

As for case 2A, the dividend taxation policy reduces the CEO power. Hence, it implies a direct modification of the free cash flow of the corporation, which is an element very sensitive for the agent. From this point of view, a better dividend scheme might reduce the free cash flow, it reduces the discretionary, minimise the future overinvestment behaviour and forces the manager to improve the efficiency of the resources (López and Saona 2007). If the free cash flow is reduced, then the board of directors may look for external founds, which probably would not like to the executive because his monitoring is increasing. On the other hand, provided the executive has more monetary resources, he will make more expansionary decisions and he will take more growing strategies, this is to say more long-term decision-making.

Concerning case 2B, the dividend taxation affects in an indirect way the managerial myopia through the board of directors because they are who approve the dividend policies and make decisions about the funding, which might have an impact to the managerial myopia. Besides, the dividend refund gives the capital market reliable information about the future of the organization.

As for case 2C, the dividends have an impact on the managerial myopia through the board of directors which has an incidence on the self-interests of the CEO. Besides, CEO can assign tasks as the monitoring of the free cash flow and the salaries. Therefore, the executive is more monitored and the interests would converge with owners.

Regarding case 2D, the same discussion made in the case 2A can be used in order to analyse both dividend taxation and free cash flow as intermediate elements.

Case 2E shows that dividend taxation itself has also an indirect interaction with the managerial myopia. The dividends refund contributes to better control the executives because the capital market also monitors the strategies. As the dividends refund increases, the corporation has to look for external funding in order to finance profitable investment projects. In that sense, as the debt increases, the position of the corporation is more risky so that the CEO would not take some projects as to avoid the capital market control (due to the averse attitude towards risk).

Finally, referring to case 3, a forgotten effect exists between the stock options and the attitude towards risk. Provided the executives diversify their value participating in the ownership, the interests between executives and owners will converge. As a result, the managerial myopia will decrease and it will involve accepting long-term projects in spite of being more risky. Thus, the executive compensation will be the same as the owners (stock options).

## 4 Conclusions

This paper highlights some important aspects that will be summarized as follows:

1. The separation of the ownership and management may lead to agency problems due to the self-interest of managers. The agency theory is concerned with resolving the conflict between principals (shareholders) and agents (companies' executives).
2. Owners are interested in monitoring agency problems due to the agency costs. The agency costs include monitoring costs, bonding costs and residual loss.
3. The expertise concept of Kaufmann and Gil Aluja is presented as a suitable tool for the analysis and mathematical treatment under uncertainty of the experts' opinion.
4. Research into forgotten effects is a sort of mechanism to put on the alert and to allow bringing light to severe problem of the oblivion. For the owner it is important to know the external factors that may affect the sources of the agency problems in order to keep them present in the decision-making process.
5. On one hand, the effects considered in the study were the different sources of the agency problem: shirking, managerial myopia, free cash flow, different attitudes towards risk, the job security and the executives' self-interests. On the other hand, the solutions of the agency problem were taken as causes of the model: the board of directors, the debt, the dividend taxation, the executive compensation, the stock options, the ownership concentration and goods and services market and the executive job market. In order to assess the mechanisms to monitor the executives, four experts were asked for their interpretation of the effect of each mechanism on each source of the problem. It allows us to detect the forgotten effects.

6. From the forgotten effects model, we can conclude that the owner must be aware of the board of directors, dividend taxation and the stock options since they have an indirect effect upon the shirking, the managerial myopia and the attitude towards risk. We also have seen some forgotten effects such as the free cash flow, the board of directors and the executive compensation.

7. Finally, we must point out the contribution of this study to the agency theory since we provide some useful tools to reduce agency problems. It may facilitate the companies' success, providing useful information to improve the decision-making process in management.

# References

Alvarez, M.B.: Problemas de agencia y de elección contable derivados de la regulación legal de las cooperativas agrarias. Doctoral thesis, University of Oviedo, Spain (2008)

Baysinger, B., Hoskisson, R.: The composition of the boards of directors and strategic control: effects on corporate strategy. Acad. Manage. Rev. **15**(1), 72 (1990)

Boland, M., Golden, B., Tsoodle, L.: Agency theory issues in the food processing industry. J. Agric. Appl. Econ. **40**(2), 623–634 (2008)

Cuervo, A.: El Gobierno de la Empresa. AFDUAM **3**, 95–108 (1999)

Donaldson, L., Davis, J.: Stewardship theory or agency theory: CEO Governance and Shareholder Returns. Aust. J. Manag. **16**(1), 49–65 (1991)

Fama, E., Jensen, M.C.: Separation of ownership and control. J. Law Econ. XXVI (1983)

Ferrer, J.C.: Un estudi de la teoria dels subconjunts borrosos amb aplicacions a models econòmics i problemes empresarials. Doctoral thesis, University of Girona, Spain (1997)

Jensen, M.C.: The modern industrial revolution, exit, and the failure of internal control systems. J. Appl. Corp. Finan. **6**(4), 4–23 (1994)

Jensen, M.C.: Agency cost of free cash flow, corporate finance and takeovers. Am. Econ. Rev. **76**, 2 (1986)

Jensen, M.C., Meckling, W.: Theory of the firm: managerial behavior, agency costs and ownership structure. J. Financ. Econ. **3**(4), 305–360 (1976)

Kauffmann, A., Gil-Aluja, J.: Modelos para la investigación de efectos olvidados. Ed. Milladoiro. Spain (1989)

López, F., Saona, P.: Endeudamiento, dividendos y estructura de propiedad como determinantes de los problemas de agencia en la gran empresa española. Cuadernos de Economía y Dirección de la Empresa **21**, 119–146 (2007)

Lozano, M.B., De Miguel, A., Pindado, J.: El conflicto accionista-directivo: problemas y propuestas de solución. Tribuna de Economía, ICE. Febrero No. 813 (2004)

Mascareñas, J.: Contratos financieros Principal-Agente. Monografías de Juan Mascareñas sobre Finanzas Corporativas, Julio (2007)

Mitnick, B.: Origin of the Theory of Agency. An account by one of the Theory's Originators. Rev. University of Pittsburgh. January (2006)

Salas, V.: El gobierno de la empresa. Colección Estudios Económicos "la Caixa" 29, pp, 5–214. Barcelona (2002)

Wang, G.: The impacts of free cash flows and agency costs on firm performance. J. Serv. Sci. Manag. **3**(4), 408–418 (2010)

# Determining the Influence Variables in the Pork Price, Based on Expert Systems

Josep M. Jaile-Benitez, Joan Carles Ferrer-Comalat
and Salvador Linares-Mustarós

**Abstract** The meat industry is the most important economic activity in the agro-food sector. Globally, the pork meat is the most consumed, beating the poultry. The price of raw materials of natural type is based on reference quotes made from wholesale source products, and those are general prices. The contract wholesales publishes a price every week of the year. In this paper we studied the elements or circumstances, which we have named variables, that influence the price of pork and the quantification of each one's weight (degree of influence) has on the price quoted. This study is based in fuzzy logic through the use of the theory of *expertons* developed by Kaufmann (1987) and Kaufmann and Gil Aluja (1993).

**Keywords** Decision making · Expert systems · Expertons · Uncertainty · Pork price

## 1 Introduction

In the same way as many other biological raw materials, the price of pork is established in the markets of origin at the moment that buyers and producers reach an agreement. This agreement rarely remains stable because it is often reached after unwanted pressures, creating situations of dissatisfaction that involve one of the two parties. The consequences of this disharmony are an additional factor to the typical uncertainty in the markets for raw materials from biological character, therefore it is

J.M. Jaile-Benitez (✉) · J.C. Ferrer-Comalat · S. Linares-Mustarós
Department of Business Administration, University of Girona, C/Universitat de Girona, 10, 17071 Girona, Catalonia, Spain
e-mail: josepm.jaile@casademont.com

J.C. Ferrer-Comalat
e-mail: joancarles.ferrer@udg.edu

S. Linares-Mustarós
e-mail: salvador.linares@udg.edu

© Springer International Publishing Switzerland 2015                                         81
J. Gil-Aluja et al. (eds.), *Scientific Methods for the Treatment of Uncertainty in Social Sciences*, Advances in Intelligent Systems and Computing 377,
DOI 10.1007/978-3-319-19704-3_7

an unstable agreement with respect to the time. Uncertainty is a tangible element in this type of market.

In this work we use the theory of *Expertons* developed by Kaufmann and Gil Aluja with the objective of knowing the variables that influence the price of pork and how relevant the influence of each variable is. Our goal is to have control of this uncertainty. Thus, in another phase of our study, this control will allow us to create a prediction model based on neural networks.

## 2 The Experts

The expert person is different from a simple specialist in the sense that he has knowledge of the subject, but this subject begins in an unexplored field, without clear rules established. The experts selected to develop our work have been chosen based on the fulfilment of the following requirements:

(i) Years of experience in the pig alimentary sector: *reliable knowledge.*
(ii) Prestige in the alimentary sector: social *recognition* about their criteria, which are generally accepted in their field of action.
(iii) Securities acquired identifiable within the sector: demonstration of *ability*.
(iv) The expert needs the approval of the researcher to create the right environment for accuracy and access to the knowledge of the expert *to demonstrate the certainty* of judgment.

After performing the process to identify the experts, we have selected eighteen professionals to conduct the construction of an expert system that will indicate us the variables influencing the price of pork. This is the sector distribution of the experts:

7 experts from the production sector (farmers).
5 experts from the subsector of the slaughterhouses and cutting plants.
4 experts from processing industry or meat industry.
2 experts from other types.

The latter experts from other types are a manager of a business association and a *freelance*. These experts will give us a counterbalance of impartiality, and they will help us to reduce the noises that may arise due to the competition between subsectors.

All the experts meet the above requirements, and they are in the staff of major companies in Europe. We have identified them from the number 1 to 18, and we have guaranteed confidentiality in exchange for their complete and loyal co-operation in our study.

# 3   The Questionnaire

We must not confuse the concept of a questionnaire proposed to an expert with the idea of a general survey. A survey can be answered by any person with respect to a general question that affects our everyday life, but not everyone has some knowledge about a specific field of knowledge.

The questions about this specific field form a questionnaire. Following the concept of membership degree, we ask to each expert the intensity (weight) with a variable is present in determining the price of pork. This means the level of influence there are on the concrete results in the formation of a fact. The expert answers with a *valuation*, so, the expert must valuate as many variables as we need to analyse. The valuation could be performed considering several values with a bottom and a top (e.g. [0.4-0.6]). In this case, the expert considers that this issue is really between those two points; then we'll have a confidence interval. But if the expert only answers with a unique value, we'll have a singlet for this case.

In our study, we have proposed to use the named *endecanary* scale, this is, a scale with eleven levels ranging from 0 to 1. The linguistic interpretation of this scale is indicated in Table 1.

When we have obtained the valuations from the different experts, then we can build the corresponding *expertons*, obtaining an *experton* for each variable analyzed.

**Table 1** Linguistic interpretation of the endecanary scale

| | |
|---|---|
| 0 | No influence |
| 0.1 | Practically no influence |
| 0.2 | Very weak influence |
| 0.3 | Weak influence |
| 0.4 | Slight less than intermediate influence |
| 0.5 | Intermediate influence |
| 0.6 | Slight more than intermediate influence |
| 0.7 | Strong influence |
| 0.8 | Very strong influence |
| 0.9 | Almost total influence |
| 1 | Total influence |

# 4   The Variables

The first approach that we presented to the experts was based on a few variables proposed by the authors of this study. During the process, after some conversations with the different experts, some variables were being consolidated, some disappeared and others were incorporated in the list. This operation was long and laborious, but finally we established a list of 24 variables. In this paper, for the sake of brevity, we will only mention them. However, it must be stated that each variable was sufficiently defined before being put under consideration by the experts. Finally, this is the list of the variables:

$A_1$:    Cost of feed (for pigs).
$A_2$:    Supply and Demand.
$A_3$:    Climatology.
$A_4$:    Seasonality.
$A_5$:    Other Economic Factors.
$A_6$:    The influence of supranational external factors.
$A_7$:    Internal crisis in the pork subsector.
$A_8$:    Crisis of the Others Meat Subsector.
$A_9$:    Food crises outside de meat industry.
$A_{10}$:    Substitutes products.
$A_{11}$:    News in the media.
$A_{12}$:    Pressure of public opinion.
$A_{13}$:    Production methods.
$A_{14}$:    Pollution and the environment.
$A_{15}$:    Entry of new countries to the European Union.
$A_{16}$:    The Exports.
$A_{17}$:    Skilled Labor.
$A_{18}$:    Export restrictions in major importing countries.
$A_{19}$:    Public intervention: private storage and export refunds.
$A_{20}$:    Quote Euro/US Dollar, referring to the export competitiveness in foreign countries.
$A_{21}$:    Price Quote in European markets.
$A_{22}$:    Sector restructuring.
$A_{23}$:    Average carcass weights.
$A_{24}$:    Piglet price quote.

## 5    The Process of Valuation from Experts and Making *Expertons*

With the questionnaire completed by the experts, we have got their valuation where they showed their criteria and they indicated the degree of influence of the mentioned variable on the price of pork.

In Tables 1 and 2, we have summarized these valuations. The rows represent the 18 experts and the columns represent the variables listed in the previous section. A cell without value indicates that the expert has expressed no opinion.

For each variable is necessary to make an *experton*.

An *experton* is a fuzzy random subset containing the aggregated opinions of the different experts about every variable considered; therefore, there will be one *experton* for each variable analyzed.

When the valuation has been completed by the experts, for each variable we have made the corresponding *experton* using the algorithms and techniques described by Kaufmann (1987: The expertons) and Kaufmann and Gil Aluja (1993: Special Techniques for the management of Experts). As we have said, when this process is finished, each *experton* contains the global information about the opinion expressed by 18 experts consulted.

Then, for each variable we make the corresponding *experton* using the opinions expressed by the 18 experts about the degree of influence of the variable in the price indicated in the Tables 2 and 3.

In this way we make the *expertons* relating to 24 variables analyzed, and we represent by $\tilde{A}_i$ the corresponding *experton* where the subscript "$i$" indicates the number of the variable valued by the experts. These *expertons* are expressed in the Tables 4 and 5.

**Table 2** Table of valuation performed by experts ($E_1$ to $E_9$)

| | $E_1$ | $E_2$ | $E_3$ | $E_4$ | $E_5$ | $E_6$ | $E_7$ | $E_8$ | $E_9$ |
|---|---|---|---|---|---|---|---|---|---|
| $A_1$ | 0.8 | 0.7 | 0.6 | 0.6 | 0.3 | 0.7 | 0.6 | 0.5 | 0.3 |
| $A_2$ | 1 | 0.8 | 0.9 | 0.9 | 0.8 | 0.9 | 0.6 | 1 | 0.9 |
| $A_3$ | 0.9 | 0.8 | 0.7 | 0.6 | 0.5 | 0.5 | 0.6 | 1 | 0.8 |
| $A_4$ | 0.3 | 0.1 | 0.4 | 0.6 | 0.5 | 0.6 | 0.3 | 1 | 0.7 |
| $A_5$ | 0 | 0 | 0.3 | 0.3 | 0.3 | 0.4 | 0.1 | 0.2 | 0.8 |
| $A_6$ | 0 | 0.4 | 0.6 | 0.7 | 0.2 | 0.3 | 0.4 | 0.8 | 0.5 |
| $A_7$ | 1 | 1 | 0.8 | 0.8 | 0.7 | 0.8 | 0.8 | 1 | 0.8 |
| $A_8$ | 0.8 | 0.9 | 0.6 | 0.6 | 0.4 | 0.7 | 0.5 | 0.8 | 0.6 |
| $A_9$ | 0.3 | 0.8 | 0.7 | 0.5 | 0.3 | 0.4 | 0.2 | 0.6 | 0.5 |
| $A_{10}$ | 0.2 | 0.6 | 0.4 | 0.5 | 0.6 | 0.2 | 0.2 | 0.5 | 0.4 |
| $A_{11}$ | 0 | 0.4 | 0.5 | 0.5 | 0.5 | 0.6 | 0.3 | 0.8 | 0.6 |
| $A_{12}$ | 0 | 0.4 | 0.4 | 0 | 0.4 | 0.3 | 0.2 | 0.5 | 0.8 |
| $A_{13}$ | 0.5 | 0.4 | 0.3 | 0.3 | 0.3 | 0.2 | 0.2 | 0.5 | 0.2 |
| $A_{14}$ | 0.6 | 0.8 | 0.2 | 0.3 | 0.3 | 0.2 | 0.1 | 0.3 | 0 |
| $A_{15}$ | 0.7 | 0.5 | 0.8 | 0.5 | 0.5 | 0.5 | 0.3 | 0.8 | 0.7 |
| $A_{16}$ | 0.4 | 0.8 | 1 | 0.7 | 0.6 | 0.7 | 0.4 | 0.8 | 0.5 |
| $A_{17}$ | 0.4 | 0.2 | 0.3 | 0.6 | 0 | 0.1 | 0.1 | 0.3 | 0.1 |
| $A_{18}$ | 0.3 | 0.2 | 0.8 | 0.8 | 0.7 | 0.7 | 0.6 | 0.8 | 0.7 |
| $A_{19}$ | 0.7 | 0.7 | 0.8 | 0.8 | 0.9 | 0.7 | 0.6 | 0.6 | 0.6 |
| $A_{20}$ | 0.6 | 0.7 | 0.8 | 0.8 | 0.6 | 0.7 | 0.6 | 0.5 | 0.8 |
| $A_{21}$ | | | | | | | 0.6 | | 0.9 |
| $A_{22}$ | | | | | | | 0.4 | | 0.9 |
| $A_{23}$ | | | | | | | | | |
| $A_{24}$ | | | | | | | | | |

**Table 3** Table of valuation performed by experts ($E_{10}$ to $E_{18}$)

|        | $E_{10}$ | $E_{11}$ | $E_{12}$ | $E_{13}$ | $E_{14}$ | $E_{15}$ | $E_{16}$ | $E_{17}$ | $E_{18}$ |
|--------|------|------|------|------|------|------|------|------|------|
| $A_1$    | 0.5  | 0.9  | 0.8  | 0.7  | 0.8  | 0.4  | 0.6  | 0.7  | 0.5  |
| $A_2$    | 1    | 0.9  | 0.8  | 1    | 0.9  | 1    | 1    | 0.9  | 1    |
| $A_3$    | 0.7  | 0.6  | 0.5  | 0.7  | 0.4  | 0.6  | 0.7  | 0.6  | 0.7  |
| $A_4$    | 0.7  | 0.6  | 0.4  | 0.6  | 0.6  | 0.7  | 0.6  | 0.4  | 0.5  |
| $A_5$    | 0.6  | 0.2  | 0.5  | 0.3  | 0.3  | 0.5  | 0.5  | 0.5  | 0.2  |
| $A_6$    | 0.6  | 0.3  | 0.4  | 0.7  | 0.6  | 0.6  | 0.5  | 0.7  | 0.3  |
| $A_7$    | 0.8  | 0.8  | 0.8  | 0.6  | 0.7  | 0.9  | 0.8  | 0.8  | 0.5  |
| $A_8$    | 0.5  | 0.7  | 0.6  | 0.6  | 0.6  | 0.6  | 0.7  | 0.7  | 0.3  |
| $A_9$    | 0.4  | 0.3  | 0.4  | 0.4  | 0.5  | 0.4  | 0.6  | 0.6  | 0.3  |
| $A_{10}$   | 0.4  | 0.2  | 0.4  | 0.4  | 0.6  | 0.3  | 0.6  | 0.6  | 0.3  |
| $A_{11}$   | 0.5  | 0.4  | 0.5  | 0.7  | 0.7  | 0.5  | 0.8  | 0.8  | 0.5  |
| $A_{12}$   | 0.5  | 0.3  | 0.4  | 0.7  | 0.7  | 0.4  | 0.7  | 0.7  | 0.5  |
| $A_{13}$   | 0.2  | 0.1  | 0.3  | 0.4  | 0.6  | 0.1  | 0.7  | 0.7  | 0.3  |
| $A_{14}$   | 0.3  | 0.3  | 0.3  | 0.3  | 0.6  | 0.2  | 0.5  | 0.6  | 0.3  |
| $A_{15}$   | 0.5  | 0.6  | 0.3  | 0.7  | 0.6  | 0.5  | 0.7  | 0.8  | 0.3  |
| $A_{16}$   | 0.4  | 0.7  | 0.5  | 0.8  | 0.7  | 0.6  | 0.7  | 0.7  | 0.3  |
| $A_{17}$   | 0.2  | 0.2  | 0.3  | 0.6  | 0.6  | 0.1  | 0.2  | 0.7  | 0.1  |
| $A_{18}$   | 0.7  | 0.6  | 0.6  | 0.8  | 0.7  | 0.8  | 0.8  | 0.8  | 0.5  |
| $A_{19}$   | 0.8  | 0.4  | 0.6  | 0.6  | 0.6  | 0.8  | 0.6  | 0.7  | 0.4  |
| $A_{20}$   | 0.8  | 0.7  | 0.6  | 0.7  | 0.7  | 0.7  | 0.7  | 0.8  | 0.3  |
| $A_{21}$   | 0.9  | 0.9  | 0.5  | 0.8  | 0.8  | 0.9  | 0.7  | 0.8  | 0.5  |
| $A_{22}$   | 0.5  | 0.3  | 0.5  | 0.7  | 0.6  | 0.4  | 0.8  | 0.6  | 0.3  |
| $A_{23}$   | 1    | 0.8  | 0.7  | 0.7  | 0.7  | 0.9  | 0.6  | 0.6  | 0.5  |
| $A_{24}$   | 0.8  | 0.3  | 0.4  | 0.6  | 0.7  |      | 0.2  | 0.7  | 0.2  |

After that, for each *experton* we will compute the mathematical expectation that will help us to make comparisons and decisions. These comparisons will be used to order all the variables according to their significance to the formulated hypothesis of study. The mathematical expectation is useful because it gives us only one value obtained through full information collected from the experts.

The mathematical expectation is the indicator through which we can make decisions and we can make comparisons between different variables. It helps us in our process of decision making when the topic of study is uncertain and risky.

To calculate the mathematical expectation of an *experton* we must do the addition of the contents of all levels except the first, corresponding to 0 (where the degree of agreement is unanimous), and divide by 10 which is the number of levels that remain after underestimating the first level, this is:

$$E\left(\tilde{\tilde{A}}\right) = \frac{1}{10} \cdot \sum_{\alpha \in I^*} a(\alpha)$$

where $I^* = \{0.1, 0.2, 0.3, 0.4, 0.5, 0.6, 0.7, 0.8, 0.9, 1\} = I - \{0\}$.

**Table 4** Expertons corresponding to 24 variables analyzed (variables from $A_1$ to $A_{12}$)

| | α | a(α) | | α | a(α) | | α | a(α) | | α | a(α) |
|---|---|---|---|---|---|---|---|---|---|---|---|
| $A_1 =$ | 0 | 1 | $A_2 =$ | 0 | 1 | $A_3 =$ | 0 | 1 | $A_4 =$ | 0 | 1 |
| | 0.1 | 1 | | 0.1 | 1 | | 0.1 | 1 | | 0.1 | 1 |
| | 0.2 | 1 | | 0.2 | 1 | | 0.2 | 1 | | 0.2 | 0.9 |
| | 0.3 | 1 | | 0.3 | 1 | | 0.3 | 1 | | 0.3 | 0.9 |
| | 0.4 | 0.89 | | 0.4 | 1 | | 0.4 | 1 | | 0.4 | 0.8 |
| | 0.5 | 0.83 | | 0.5 | 1 | | 0.5 | 0.9 | | 0.5 | 0.7 |
| | 0.6 | 0.67 | | 0.6 | 1 | | 0.6 | 0.8 | | 0.6 | 0.6 |
| | 0.7 | 0.44 | | 0.7 | 0.9 | | 0.7 | 0.5 | | 0.7 | 0.2 |
| | 0.8 | 0.22 | | 0.8 | 0.9 | | 0.8 | 0.2 | | 0.8 | 0.1 |
| | 0.9 | 0.06 | | 0.9 | 0.7 | | 0.9 | 0.1 | | 0.9 | 0.1 |
| | 1 | 0 | | 1 | 0.1 | | 1 | 0.1 | | 1 | 0.1 |
| $A_5 =$ | 0 | 1 | $A_6 =$ | 0 | 1 | $A_7 =$ | 0 | 1 | $A_8 =$ | 0 | 1 |
| | 0.1 | 0.89 | | 0.1 | 0.94 | | 0.1 | 1 | | 0.1 | 1 |
| | 0.2 | 0.83 | | 0.2 | 0.94 | | 0.2 | 1 | | 0.2 | 1 |
| | 0.3 | 0.67 | | 0.3 | 0.89 | | 0.3 | 1 | | 0.3 | 1 |
| | 0.4 | 0.39 | | 0.4 | 0.72 | | 0.4 | 1 | | 0.4 | 0.9 |
| | 0.5 | 0.33 | | 0.5 | 0.56 | | 0.5 | 1 | | 0.5 | 0.9 |
| | 0.6 | 0.11 | | 0.6 | 0.44 | | 0.6 | 0.9 | | 0.6 | 0.8 |
| | 0.7 | 0.06 | | 0.7 | 0.22 | | 0.7 | 0.9 | | 0.7 | 0.4 |
| | 0.8 | 0.06 | | 0.8 | 0.06 | | 0.8 | 0.8 | | 0.8 | 0.2 |
| | 0.9 | 0 | | 0.9 | 0 | | 0.9 | 0.2 | | 0.9 | 0.1 |
| | 1 | 0 | | 1 | 0 | | 1 | 0.2 | | 1 | 0 |
| $A_9 =$ | 0 | 1 | $A_{10} =$ | 0 | 1 | $A_{11} =$ | 0 | 1 | $A_{12} =$ | 0 | 1 |
| | 0.1 | 1 | | 0.1 | 1 | | 0.1 | 0.9 | | 0.1 | 0.9 |
| | 0.2 | 1 | | 0.2 | 1 | | 0.2 | 0.9 | | 0.2 | 0.9 |
| | 0.3 | 0.94 | | 0.3 | 0.78 | | 0.3 | 0.9 | | 0.3 | 0.8 |
| | 0.4 | 0.72 | | 0.4 | 0.67 | | 0.4 | 0.9 | | 0.4 | 0.7 |
| | 0.5 | 0.44 | | 0.5 | 0.39 | | 0.5 | 0.8 | | 0.5 | 0.4 |
| | 0.6 | 0.28 | | 0.6 | 0.28 | | 0.6 | 0.4 | | 0.6 | 0.3 |
| | 0.7 | 0.11 | | 0.7 | 0 | | 0.7 | 0.3 | | 0.7 | 0.3 |
| | 0.8 | 0.06 | | 0.8 | 0 | | 0.8 | 0.2 | | 0.8 | 0.1 |
| | 0.9 | 0 | | 0.9 | 0 | | 0.9 | 0 | | 0.9 | 0 |
| | 1 | 0 | | 1 | 0 | | 1 | 0 | | 1 | 0 |

In Table 6 we indicate the results of the mathematical expectations of the twenty-four variables analyzed:

**Table 5** Expertons corresponding to 24 variables analyzed. (variables from $A_{13}$ to $A_{24}$)

| | $\alpha$ | $a(\alpha)$ | | $\alpha$ | $a(\alpha)$ | | $\alpha$ | $a(\alpha)$ | | $\alpha$ | $a(\alpha)$ |
|---|---|---|---|---|---|---|---|---|---|---|---|
| $A_{13}=$ | 0 | 1 | $A_{14}=$ | 0 | 1 | $A_{15}=$ | 0 | 1 | $A_{16}=$ | 0 | 1 |
| | 0.1 | 1 | | 0.1 | 0.9 | | 0.1 | 1 | | 0.1 | 1 |
| | 0.2 | 0.9 | | 0.2 | 0.9 | | 0.2 | 1 | | 0.2 | 1 |
| | 0.3 | 0.7 | | 0.3 | 0.7 | | 0.3 | 1 | | 0.3 | 1 |
| | 0.4 | 0.4 | | 0.4 | 0.3 | | 0.4 | 0.8 | | 0.4 | 0.9 |
| | 0.5 | 0.3 | | 0.5 | 0.3 | | 0.5 | 0.8 | | 0.5 | 0.8 |
| | 0.6 | 0.2 | | 0.6 | 0.2 | | 0.6 | 0.5 | | 0.6 | 0.7 |
| | 0.7 | 0.1 | | 0.7 | 0.1 | | 0.7 | 0.4 | | 0.7 | 0.6 |
| | 0.8 | 0 | | 0.8 | 0.1 | | 0.8 | 0.2 | | 0.8 | 0.2 |
| | 0.9 | 0 | | 0.9 | 0 | | 0.9 | 0 | | 0.9 | 0.1 |
| | 1 | 0 | | 1 | 0 | | 1 | 0 | | 1 | 0.1 |
| $A_{17}=$ | 0 | 1 | $A_{18}=$ | 0 | 1 | $A_{19}=$ | 0 | 1 | $A_{20}=$ | 0 | 1 |
| | 0.1 | 0.9 | | 0.1 | 1 | | 0.1 | 1 | | 0.1 | 1 |
| | 0.2 | 0.7 | | 0.2 | 1 | | 0.2 | 1 | | 0.2 | 1 |
| | 0.3 | 0.4 | | 0.3 | 0.9 | | 0.3 | 1 | | 0.3 | 1 |
| | 0.4 | 0.3 | | 0.4 | 0.9 | | 0.4 | 1 | | 0.4 | 0.9 |
| | 0.5 | 0.2 | | 0.5 | 0.9 | | 0.5 | 0.9 | | 0.5 | 0.9 |
| | 0.6 | 0.2 | | 0.6 | 0.8 | | 0.6 | 0.9 | | 0.6 | 0.9 |
| | 0.7 | 0.1 | | 0.7 | 0.7 | | 0.7 | 0.5 | | 0.7 | 0.7 |
| | 0.8 | 0 | | 0.8 | 0.4 | | 0.8 | 0.3 | | 0.8 | 0.3 |
| | 0.9 | 0 | | 0.9 | 0 | | 0.9 | 0.1 | | 0.9 | 0 |
| | 1 | 0 | | 1 | 0 | | 1 | 0 | | 1 | 0 |
| $A_{21}=$ | 0 | 1 | $A_{22}=$ | 0 | 1 | $A_{23}=$ | 0 | 1 | $A_{24}=$ | 0 | 1 |
| | 0.1 | 1 | | 0.1 | 1 | | 0.1 | 1 | | 0.1 | 1 |
| | 0.2 | 1 | | 0.2 | 1 | | 0.2 | 1 | | 0.2 | 1 |
| | 0.3 | 1 | | 0.3 | 1 | | 0.3 | 1 | | 0.3 | 0.8 |
| | 0.4 | 1 | | 0.4 | 0.8 | | 0.4 | 1 | | 0.4 | 0.6 |
| | 0.5 | 1 | | 0.5 | 0.6 | | 0.5 | 1 | | 0.5 | 0.5 |
| | 0.6 | 0.8 | | 0.6 | 0.5 | | 0.6 | 0.9 | | 0.6 | 0.5 |
| | 0.7 | 0.7 | | 0.7 | 0.3 | | 0.7 | 0.7 | | 0.7 | 0.4 |
| | 0.8 | 0.6 | | 0.8 | 0.2 | | 0.8 | 0.3 | | 0.8 | 0.1 |
| | 0.9 | 0.4 | | 0.9 | 0.1 | | 0.9 | 0.2 | | 0.9 | 0 |
| | 1 | 0 | | 1 | 0 | | 1 | 0.1 | | 1 | 0 |

## 6 Analyzing Results

The mathematical expectation of an *experton* expresses the average level of incidence of the variable represented by the corresponding *experton* on the influence that the determination of the pork price at the time of its listing. To determine the

**Table 6** Mathematical expectations corresponding to 24 expertons

| Variables | Mathematical expectation |
|---|---|
| $A_1$: *Cost of feed (for pigs)* | 0.61 |
| $A_2$: *Supply and Demand* | 0.86 |
| $A_3$: *Climatology* | 0.66 |
| $A_4$: *Seasonality* | 0.53 |
| $A_5$: *Other Economic Factors* | 0.33 |
| $A_6$: *The influence of supranational external factors* | 0.48 |
| $A_7$: *Internal crisis in the pork subsector* | 0.80 |
| $A_8$: *Crisis of the Others Meat Subsector* | 0.62 |
| $A_9$: *Food crises outside the meat industry* | 0.46 |
| $A_{10}$: *Substitutes products* | 0.41 |
| $A_{11}$: *News in the media* | 0.41 |
| $A_{12}$: *Pressure of public opinion* | 0.44 |
| $A_{13}$: *Production methods* | 0.35 |
| $A_{14}$: *Pollution and the environment* | 0.34 |
| $A_{15}$: *Entry of new countries to the European Union* | 0.57 |
| $A_{16}$: *The Exports* | 0.63 |
| $A_{17}$: *Skilled Labor* | 0.28 |
| $A_{18}$: *Export restrictions in major importing countries* | 0.66 |
| $A_{19}$: *Public intervention* | 0.66 |
| $A_{20}$: *Quote Euro/US Dollar* | 0.67 |
| $A_{21}$: *Price Quote in European markets* | 0.75 |
| $A_{22}$: *Sector restructuring* | 0.55 |
| $A_{23}$: *Average carcass weights* | 0.72 |
| $A_{24}$: *Piglet price quote* | 0.49 |

variables that must be considered influential, certain requirements must be established. As a general rule, we consider that the variable Ai has a higher incidence than the variable Aj, and we express that idea writing it as a $A_i \gg A_j$ if, and only if, the *mathematical expectation* of the *experton* $\tilde{A}_j$ is bigger than *mathematical expectation* of *experton* $\tilde{A}_j$. This is expressed as follows:

$$A_i \gg A_j \Leftrightarrow E\left(\tilde{A}_i\right) > E\left(\tilde{A}_j\right)$$

The degree of agreement that the experts reach at a valuation of a particular variable expresses quantitatively the level of agreement between the experts that accept to have the same opinion on the same subject. Certainly, one of the weaknesses of the *mathematical expectation* is that it doesn't show the dispersion of the various opinions.

After that, we interpret the results of the study following the definition of a boundary and using the linguistic interpretation of Table 1. The boundary that it

**Table 7** Influential variables in the pork price

| Influent variables | $E\left(\dot{\tilde{A}}\right)$ | Lingüístic interpretation |
|---|---|---|
| $A_2$: Supply and Demand | 0.86 | Very strong influence |
| $A_7$: Internal crisis in the subsector | 0.80 | Very strong influence |
| $A_{21}$: Price Quote in European markets | 0.75 | Strong influence |
| $A_{23}$: Average carcass weights | 0.72 | Strong influence |
| $A_{20}$: Quote Euro/US Dollar | 0.67 | Slight more than intermediate influence |
| $A_3$: Climatology | 0.66 | Slight more than intermediate influence |
| $A_{18}$: Export restrictions | 0.66 | Slight more than intermediate influence |
| $A_{19}$: Public intervention | 0.66 | Slight more than intermediate influence |
| $A_{16}$: The Exports | 0.63 | Slight more than intermediate influence |
| $A_8$: Crisis of the Others Meat Subsector | 0.62 | Slight more than intermediate influence |
| $A_1$: Cost of feed | 0.61 | Slight more than intermediate influence |
| $A_{15}$: Entry of new countries to the EU | 0.57 | Intermediate influence |
| $A_{22}$: Sector restructuring | 0.55 | Intermediate influence |
| $A_4$: Seasonality | 0.53 | Intermediate influence |

indicates us which variable will be considered as influential and which not, is set at 0.50 level (intermediate influence). In this way, we have excluded ten of the twenty-four variables that we had studied, and finally we will consider only the fourteen variables which have a *mathematical expectation* higher than 0.50.

In the Table 7 we list the **influential variables** in an orderly manner from most to least important to justify its *mathematical expectation*:

The variables that have been chosen and that exceed the boundary established are considered to be influential. They range from intermediate influence to very strong influence, according to the established linguistic interpretation given by Table 1.

# 7 Conclusions

1. The theory of *Expertons* that we have used in our analysis has been very efficient as it has allowed us to identify which variables affect the price finally established in the market, and what is the degree of influence of each variable.
2. The theory of *Expertons* has been adapted comfortably in this field, that among others, is characterized by uncertainty where it is not possible a mere projection of past data to predict a future situation. It has also been a very competent methodology due to its collection of all the information obtained in the questionnaire, with a minimal fraction of entropy (loss of information in the process).
3. The cooperation of experts is absolutely essential. They provide the material that allows us to work efficiently with this method. Its location and the contribution of their wisdom is absolutely irreplaceable.

4. The variables that the model has been rejected as influential are:
   $A_5$: Other Economic Factors.
   $A_6$: The influence of supranational external factors.
   $A_9$: Food crises outside the meat industry.
   $A_{10}$: Substitutes products.
   $A_{11}$: News in the media.
   $A_{12}$: Pressure of public opinion.
   $A_{13}$: Production methods.
   $A_{14}$: Pollution and the environment.
   $A_{17}$: Skilled Labor.
   $A_{24}$: Piglet price quote.

5. The variables that this methodology considers really influential are the variables with a mathematical expectation higher than 0.50. We list these variables from more to less influential.
   With a very strong influences level:
   $A_2$: Supply and Demand.
   $A_7$: Internal crisis in the subsector.

   With a strong influence level:
   $A_{21}$: Price Quote in European markets.
   $A_{23}$: Average carcass weights.

   With a slight more than intermediate influence level:
   $A_{20}$: Quote Euro/US Dollar, referring to the export competitiveness in foreign countries.
   $A_3$: Climatology.
   $A_{18}$: Export restrictions in major importing countries.
   $A_{19}$: Public intervention: private storage and export refunds.
   $A_{16}$: The Exports.
   $A_8$: Crisis of the Others Meat Subsector.
   $A_1$: Cost of feed.

   With an intermediate influence level:
   $A_{15}$: Entry of new countries to the European Union.
   $A_{22}$: Sector restructuring.
   $A_4$: Seasonality.

6. We show that the variables influential, as well as its weight, in the pork price determination is cyclical and consequently they suffer variations in the course of events. Thus a variable can be influential during a contingency and lose influence in another. The same occurs with their weight. A variable can have much influence at one given point but not in another context. For this reason, the permanent contact with the experts is essential at the moment of implementing a forecasting model based in this study.

# References

Kaufmann, A.: Les expertons. Ed. Hermes, París (1987)

Kaufmann, A., Gil Aluja, J.: Introducción de la teoría de los subconjuntos borrosos a la gestión de las empresas. Ed.Milladoiro, Santiago de Compostela (1992)

Kaufmann, A., Gil Aluja, J.: Técnicas especiales para la gestión de expertos. Ed.Milladoiro. Santiago de Compostela (1993)

Kaufmann, A., Gil Aluja, J., Terceño, A.: Matemática para la economía y la gestión de empresas. Aritmética de la Incertidumbre. Ed. Foro Científico, Barcelona (1994)

Kaufmann, A., Gupta, M.M.: Introduction to fuzzy arithmetic. Theory and Applications. International Thomson Computer Press, USA (1991)

MAPA, Ministerio de Agricultura, Pesca y Alimentación: Sector cárnico español. Ed. Dirección General de Alimentación. Subdirección general de fomento y desarrollo agroindustrial, Madrid (2001)

Monroe, K.: Política de precios para hacer más rentables las decisiones. Ed. McGraw-Hill, Madrid (1995)

Ordinas-Koenig, M.: L'ampliació de la Unió Europea: efectes a Catalunya. Patronat català pro-Europa, Barcelona (2006)

Sambug, R.: Fonctions Φ-flous. Application au diagnostic en pathologie thiroïdienne. Ph. Thesis. Université de Marseille (1975)

Tugores, J.: La economía por sectores. Fira de Barcelona y Universitat de Barcelona, Barcelona (2001)

Zadeh, L.A.: The concept of a linguistic variable and its applications to approximate reasoning. Parts 1, 2, 3. Information Sciences 8, 199–249 (part 1), 301–357 (part 2) (1975) Information Sciences 9, 43–80 (part 3) (1976)

# Forgotten Effects Analysis Between the Regional Economic Activity of Michoacan and Welfare of Its Inhabitants

Anna M. Gil-Lafuente, Jaume Balvey, Víctor G. Alfaro-García
and Gerardo G. Alfaro-Calderón

**Abstract** The creation of effective policies for the acceleration of economic, social and environmental development of regions is a demanding issue for government's agenda. Applying the Forgotten Effects Theory, this research aims to quantify the incidence of economic activity on the welfare level of citizens. In this paper we analyze 10 regions located in the Latin-American State of Michoacán, México, each with different characteristics and specificities. Results show the relevant cumulative indirect effect that the primary sector has on family income and occupation. Manufacturing activities display high indirect cumulative effect on higher education and health. Construction and service sectors play a dominant role in environmental quality. This research presents a first step in order to release the nature of the problem and to show to what extent, in quantitative terms, the economic activity types influence welfare. This analysis could help decision makers for the effective allocation of resources and the creation of sustainable policies.

**Keywords** Economic sectors · Welfare indicators · Fuzzy sets · Regional development · Forgotten effects theory

## 1 Introduction

In the last decades the economic openness and globalization has simplified trade barriers making economies more dependent on each other and so affecting the competitiveness of organizations and regions. In this context, governments have been reflecting about competitiveness and sustainable growth, focusing their efforts

A.M. Gil-Lafuente (✉) · J. Balvey · V.G. Alfaro-García · G.G. Alfaro-Calderón
Department of Business of Business Administration, Universitat de Barcelona,
Av. Diagonal 690, 08034 Barcelona, Spain
e-mail: amgil@ub.edu

V.G. Alfaro-García
e-mail: valfarga7@alumnes.ub.edu

© Springer International Publishing Switzerland 2015
J. Gil-Aluja et al. (eds.), *Scientific Methods for the Treatment of Uncertainty in Social Sciences*, Advances in Intelligent Systems and Computing 377,
DOI 10.1007/978-3-319-19704-3_8

on the promotion of economic policies as dominating key element to boost the regional de-velopment within a worldwide economic context.

One of the main problems for governments and institutions is the imbalance between its different regions. These institutions have to carry out some actions in order to manage well the imbalances between regions and rebalance the whole territory, which will allow its all citizens to enjoy a similar standard of living.

The objective pursued in this work is especially relevant to emergent economies such as the Mexican market, which since the 1980's has been transiting from a closed economy to an open market strategy. Such external and internal economic liberation has affected in a certain degree to all the enterprises within the territory. The increasing commercial openness and free market treaties signed by the nation have increased the competitive environment in a way that productivity and higher quality of manufactured goods needs to be continuously pursed in order to retain market share (Chauca 1999).

In order to assess the task of identifying the role that economic indicators play in the welfare level of citizens, we present a methodology that relates in general terms how the economy of a region determines the quality of life of the population in relation to all others (Zadeh 1965; Bellman and Zadeh 1970). At the end of the study, we present the direct and indirect connections established between the main economic activity of regions and the quality of life of its inhabitants, within the whole State of Michoacán, México (Kaufmann and Gil – Aluja 1988).

The study is based on data obtained mainly from the National Institute of Statistics and Geography of Mexico (INEGI 2014), but we also consulted some information retrieved from the National Council of Population (CONAPO 2015), specifically the human development indexes.

# 2 Theoretical Background

A total of 32 States integrate the United Mexican States, 31 of them are free and sovereign entities, which have the right to establish an own constitution and legislative powers. The last State is the Federal District, territory under the share dominance of the Mexican Federation and the local government entities.

The State of Michoacán de Ocampo occupies the 3.0 % of the National territory with a surface of 58,599 km$^2$, it has a total population of 4,351,037 habitants from which 2,248928 are women, and 2,102,109 are men. The State presents a distribution of 69 % of urban locations and 31 % of rural locations. In terms of schooling it is observed that from every 100 people: 10.7 do not have access to education, 61.8 have the primary education, 0.4 have a technical or commercial primary school, 4.8 have finalized the high school, 11.8 have concluded university, 0.5 is not specified. In the State 92 % of the population is considered catholic, less than 3 % of the population speaks a native language. In 2010 there were 1,066,061 private homes, from which: 87.7 %, have access to potable water, 88.6 %, have access to drainage 98.0 % counts with electricity. The main economic indicators present that the State provides 2.5 % to the GDP being the activity of commerce the most relevant one, the primary sector provides 11.27 % of the State GDP, the secondary activities provide the 19.97 % to the State GDP and the third sector of economic activities provide 68.76 % to the State GDP, (INEGI 2014).

The State of Michoacán is divided into 113 municipalities, these municipalities are geographical and economically grouped in 10 main regions, see Table 1.

Table 1 Main regions of Michoacán

| Region | | Number of municipalities |
|--------|----------------|--------------------------|
| 1 | Morelia | 12 |
| 2 | Zacapu | 8 |
| 3 | Pátzcuaro | 8 |
| 4 | Bajío | 8 |
| 5 | Oriente | 16 |
| 6 | Tierra Caliente | 10 |
| 7 | Costa | 7 |
| 8 | Meseta | 13 |
| 9 | Apatzingán | 10 |
| 10 | Chapala | 21 |

Source: Valerio and Chávez (2002)

In our research we discuss, first, the main economic activities in each of the regions of Michoacán. We consider three main economic activities grouped together by the similarity of different tasks and jobs, as same as several subsectors. We will work on the following economic activities (INEGI 2014):

**Primary sector:** In this sector, jobs are grouped that involve primary sector activities such as: farming, extractive industries, fishing and forestry. It is important to mention that in order to facilitate the collection of statistical data, we decided to treat the whole of the primary sector, regardless of the region that may have, taking into account different weights in its sub-activities.

**Manufacture:** This sector includes all economic activities related to the process and treatment of both raw materials and semi-finished products at different stages.

**Construction and Services:** In this section we grouped the entire service sector and the construction industry information. As stated before this is the predominant sector of the entity and by far the one that most contributes to the GDP, incorporating all activities related to services, such as: education, health, commerce, transportation, public administration, food and beverage, tourism.

The size of the State, along to the different environmental characteristics shape the economic activities that each region displays (Valerio and Chávez 2002). For instance, statistics show a predominance of manufacturing activity in the coast region, in fact the port of Lázaro Cárdenas, one of the most important connections of the country to the Pacific Ocean; incentivize the growth of transformation enterprises. The Region Meseta is characterized for its fertile lands portraying Michoacán as world leader of avocado production; statistics reveal the predominance of the primary sector in this region. The Region Morelia is the center of the state and holds the capital of the State, therefore displaying strong commerce and construction indexes.

# 3 Methodology

All events and activities around us, including ourselves, are a part, one way or another, some kind of a system or subsystem, that is, we could say that almost all activities are a consequence of a cause-effect impact. Despite the establishment of good control systems, it is always possible to set aside or ignore some causal relationships which are not always explicit, evident or visible, and usually not perceived di-rectly. It is common that those relations remain hidden because they are effects over effects, and thus the accumulation of causes that are provoked by them. Human Intelligence need to rely on models and tools which are able to create a technical foundation that can enable the creation of all comparisons from the information obtained from the environment and bring out all the possible direct and indirect causal relationships.

The underlying idea behind the model we apply is based on the so-called "butterfly effect" that we have heard often. Popularly it is said that when a butterfly beats its wings in Mexico it causes a typhoon in Philippines. The idea of direct and indirect causal relationships is an accumulated evident. The direct relationship can be exampled such as: the beat of butterfly wing would cause the typhoon. The indirect relationship would be the accumulation of the elements that boost the movement of the air, from the beat of butterfly wings to typhoon.

Kaufmann and Gil – Aluja established the "Theory of the Forgotten Effects" (Kaufmann and Aluja 1988). This theory allows obtaining all direct and indirect relations, with no possibility of errors, recovering the effects as it is called: "Forgotten Effects". According to the authors, all happenings that surround us are part of a system or subsystem. It means we could almost ensure that any activity is subject to a problem is a result of "causes" and "effects". Despite a good system control, there is always the possibility of leaving voluntarily or involuntarily some causal relationships that are not always explicit, obvious or visible, and usually they are not directly perceived. It is common that there are some hidden reasons of the problems that we encounter due to effects of incidence effects on outcomes. The forgotten effects theory is an innovative and efficient approach taking into account many aspects of the relations, and which enables minimizing the errors that may occur in many processes (Gil-Lafuente 2005). For this study, we have two sets of elements (causes and effects):

$$A = \{a_i / i = 1, 2, \ldots, n\} : \text{Economic activities.}$$
$$B = \{b_j / j = 1, 2, \ldots, m\} : \text{Welfare variables.}$$

We assume that there is an occurrence of ai over bj if the value of the membership functions of the feature pair (ai, bj) is estimated at (the value of each cell in the array cannot be smaller than 0 or greater than 1 as if we had valued from 0 to 10 but in decimals):

$$\forall (a_i, b_j) \Rightarrow \mu(a_i, b_j) \in [0, 1]$$

The set of rated pair elements define the "direct relationship matrix", which shows the cause relationships - effect on different levels that occur between the joint elements of the A set (causes-economic activities) and the joint elements of the B set (effects - welfare variables):

$$M = \begin{array}{c|ccccc|c}
\vec{} & b_1 & b_2 & b_3 & b_4 & L & b_j \\
\hline
a_1 & \mu_{a_1 b_1} & \mu_{a_1 b_2} & \mu_{a_1 b_3} & \mu_{a_1 b_4} & L & \mu_{a_1 b_j} \\
a_2 & \mu_{a_2 b_1} & \mu_{a_2 b_2} & \mu_{a_2 b_3} & \mu_{a_2 b_4} & L & \mu_{a_2 b_j} \\
a_3 & \mu_{a_3 b_1} & \mu_{a_3 b_2} & \mu_{a_3 b_3} & \mu_{a_3 b_4} & L & \mu_{a_3 b_j} \\
a_4 & \mu_{a_4 b_1} & \mu_{a_4 b_2} & \mu_{a_4 b_3} & \mu_{a_4 b_4} & L & \mu_{a_4 b_j} \\
a_5 & \mu_{a_5 b_1} & \mu_{a_5 b_2} & \mu_{a_5 b_3} & \mu_{a_5 b_4} & L & \mu_{a_5 b_j} \\
\hline
M & M & M & M & M & M & \vdots \\
a_i & \mu_{a_i b_1} & \mu_{a_i b_2} & \mu_{a_i b_3} & \mu_{a_i b_4} & L & \mu_{a_i b_j} \\
\end{array}$$

This matrix can also be represented by its associated relationship graph. In the case where the characteristic function of the property is null, the hoop that connects the elements of the sets A and B would be deleted:

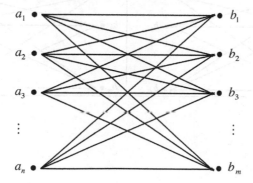

The accumulation process:

To continue with the explanation of the model, below we show how the effects can accumulate. If we suppose now the appearance of a third set of elements, different from the previous two:

$$C = \{c_k / k = 1, 2, \ldots, z\}$$

This consists of the elements that are the effects of the B set:

$$N = \begin{array}{c} \rightarrow \\ b_2 \\ b_2 \\ \vdots \\ b_m \end{array} \begin{array}{cccc} c_1 & c_2 & L & c_z \\ \boxed{\mu_{b_1 c_1}} & \boxed{\mu_{b_1 c_2}} & L & \boxed{\mu_{b_1 c_z}} \\ \boxed{\mu_{b_2 c_1}} & \boxed{\mu_{b_2 c_2}} & L & \boxed{\mu_{b_2 c_z}} \\ L & L & L & L \\ \boxed{\mu_{b_m c_1}} & \boxed{\mu_{b_m c_2}} & L & \boxed{\mu_{b_m c_z}} \end{array}$$

and having the common elements of the B set:

$$M = \begin{array}{c} \rightarrow \\ a_1 \\ a_2 \\ \vdots \\ a_n \end{array} \begin{array}{cccc} b_1 & b_2 & L & b_m \\ \boxed{\mu_{a_1 b_1}} & \boxed{\mu_{a_1 b_2}} & L & \boxed{\mu_{a_1 b_m}} \\ \boxed{\mu_{a_2 b_1}} & \boxed{\mu_{a_2 b_2}} & L & \boxed{\mu_{a_2 b_m}} \\ L & L & L & L \\ \boxed{\mu_{a_n b_1}} & \boxed{\mu_{b_m c_2}} & L & \boxed{\mu_{b_m c_z}} \end{array} \qquad N = \begin{array}{c} \rightarrow \\ b_2 \\ b_2 \\ \vdots \\ b_m \end{array} \begin{array}{cccc} c_1 & c_2 & L & c_z \\ \boxed{\mu_{b_1 c_1}} & \boxed{\mu_{b_1 c_2}} & L & \boxed{\mu_{b_1 c_z}} \\ \boxed{\mu_{b_2 c_1}} & \boxed{\mu_{b_2 c_2}} & L & \boxed{\mu_{b_2 c_z}} \\ L & L & L & L \\ \boxed{\mu_{b_m c_1}} & \boxed{\mu_{b_m c_2}} & L & \boxed{\mu_{b_m c_z}} \end{array}$$

The incident graphs associated to each of the two matrices would be as following:

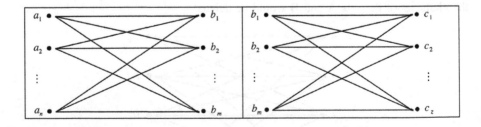

We can therefore say that the matrix P defines the causal relationships between the first of elements A set and the elements of the third C set, with the intensity or degree that leads to consider the elements belonging to set B.

$$M \subset A \times B \quad \text{i} \quad N \subset B \times C.$$

The mathematical operator to find out the effect of A on C is the max-min composition (least of all, we are left with the most). In fact, when there are three impact relationships we would have:

$$M \subset A \times B, \quad N \subset B \times C, \quad P \subset A \times C$$

And the Kaufmann (1975) equation is induced to:

$$M \circ N = P$$

Where the ∘ symbol represents precisely the max-min composition. The composition of two relations is uncertain such that:

$$\forall (a_i, c_z) \in A \times C:$$
$$\mu(a_i, c_z)_{M \circ N} = \underset{bj}{V}\left(\mu_M(a_i, b_j) \wedge \mu_N(b_j, c_z)\right)$$

We, therefore, affirm the relationship of impact $P$ defines the causal relationships between the elements of the $A$ set and the $C$ set in the intensity or degree implying the elements belonging to the $B$ set.

## 4 Application of the Model

Below we present two tables that include all the obtained statistical information, and have been normal-ized which means we divided all values of the column by the largest value. The aim is that all column values refer to the highest value, which takes the value 1. We need normalized values on base 1 because, as you can see, Tables 1 and 2 include, respectively, the variables that meet for each of the regions of Michoacán, values between 0 and 1. Table 2 shows the importance of each welfare indicator for each region compared to the best position.

Following, Table 3 shows the relative weight of economic sectors in each region, including in relation to the best position. In this context the sum of the values of the three main economic sectors is not equal to 1, since the normalization has been done on the basis of one regional indicator (per columns and not per rows):

These two tables provide us the extent of fulfillment level for analyzed variables of each region in relation with the region that has the highest indicator. But this information is not useful to make a global study. A part from that, we should take

**Table 2** Welfare indicators per regions of Michoacán

| Region | Occupation | Family income | Higher education | GDP per capita |
|---|---|---|---|---|
| Morelia | 1.000 | 1.000 | 1.000 | 0.891 |
| Zacapu | 0.166 | 0.079 | 0.093 | 0.857 |
| Pátzcuaro | 0.252 | 0.060 | 0.112 | 0.906 |
| Bajío | 0.260 | 0.130 | 0.142 | 0.921 |
| Oriente | 0.218 | 0.079 | 0.143 | 0.896 |
| Tierra Caliente | 0.128 | 0.045 | 0.073 | 0.855 |
| Costa | 0.397 | 0.301 | 0.203 | 0.904 |
| Meseta | 0.533 | 0.273 | 0.262 | 0.997 |
| Apatzingán | 0.254 | 0.110 | 0.102 | 1.000 |
| Chapala | 0.290 | 0.158 | 0.142 | 0.995 |

| Region | Human capital | Social capital | HDI | Environmental quality | Health |
|---|---|---|---|---|---|
| Morelia | 0.932 | 1.000 | 0.965 | 0.467 | 1.000 |
| Zacapu | 0.842 | 0.503 | 0.966 | 0.511 | 0.667 |
| Pátzcuaro | 0.939 | 0.360 | 0.971 | 0.567 | 0.511 |
| Bajío | 0.878 | 0.523 | 0.986 | 0.319 | 0.556 |
| Oriente | 0.899 | 0.300 | 0.951 | 0.605 | 0.722 |
| Tierra Caliente | 0.807 | 0.287 | 0.904 | 0.858 | 0.333 |
| Costa | 0.856 | 0.680 | 0.945 | 1.000 | 0.222 |
| Meseta | 1.000 | 0.536 | 0.989 | 0.634 | 1.000 |
| Apatzingán | 0.876 | 0.389 | 0.974 | 0.655 | 0.444 |
| Chapala | 0.874 | 0.570 | 1.000 | 0.440 | 0.489 |

Source: INEGI (2015); CONAPO (2010)

**Table 3** Weight of economic sectors, per region

| Sector | Morelia | Zacapu | Pátzcuaro | Bajío | Oriente |
|---|---|---|---|---|---|
| Primary Sector | 0.197 | 0.133 | 0.225 | 0.251 | 0.200 |
| Manufacture | 0.109 | 0.059 | 0.009 | 0.014 | 0.017 |
| Construction and Services | 1.000 | 0.037 | 0.006 | 0.070 | 0.030 |

| Sector | Tierra Caliente | Costa | Meseta | Apatzingán | Chapala |
|---|---|---|---|---|---|
| Primary Sector | 0.543 | 0.301 | 1.000 | 0.338 | 0.333 |
| Manufacture | 0.005 | 1.000 | 0.033 | 0.011 | 0.038 |
| Construction and Services | 0.000 | 0.337 | 0.140 | 0.085 | 0.057 |

Source: INEGI (2015)

into account the impact of economic activities over regions, which will help us to determine the welfare of the population in relation to the rest of indicators of the other regions. Thus, we perform a max-min composition between Tables 1 and 2, in order to perform a combination of indicators based on the regions.

Below we present the max-min composition of the three studied economic sectors (primary, manufactur-ing and construction and services). The detailed operation consist on identifying, first, the minimum val-ues between the economic sectors (rows), and the welfare indicators (columns): occupation, family income, Higher education, GDP per capita, etc., then displaying the maximum number of the minimum values identified, see Table 4.

**Table 4** Max-min matrix Economic Sectors and Welfare Indicators

| Sector | Occupation | Family income | Higher education | GDP per capita |
|---|---|---|---|---|
| Primary sector | 0.533 | 0.301 | 0.262 | 0.997 |
| Manufacture | 0.397 | 0.301 | 0.203 | 0.904 |
| Construction and services | 1.000 | 1.000 | 1.000 | 0.891 |
| Sector | Human capital | Social capital | HDI | Environmental quality | Health |
| Primary sector | 1.000 | 0.536 | 0.989 | 0.634 | 1.000 |
| Manufacture | 0.856 | 0.680 | 0.945 | 1.000 | 0.222 |
| Construction and services | 0.932 | 1.000 | 0.965 | 0.467 | 1.000 |

Source: Self – elaborated

Since this matrix is not normalized in basis 1, we proceed to divide each column by the largest value, which will serve us as a reference to evaluate other levels, obtaining the normalized matrix in Table 5. This matrix is the direct relationship matrix, as the economy sectors affect the indicators of economic welfare, taking into

**Table 5** Normalized relationship matrix

| Sector | Occupation | Family income | Higher education | GDP per capita |
|---|---|---|---|---|
| Primary sector | 0.533 | 0.301 | 0.262 | 1.000 |
| Manufacture | 0.397 | 0.301 | 0.203 | 0.907 |
| Construction and services | 1.000 | 1.000 | 1.000 | 0.894 |
| Sector | Human capital | Social capital | HDI | Environmental quality | Health |
| Primary sector | 1.000 | 0.536 | 1.000 | 0.634 | 1.000 |
| Manufacture | 0.856 | 0.680 | 0.956 | 1.000 | 0.222 |
| Construction and services | 0.932 | 1.000 | 0.976 | 0.467 | 1.000 |

Source: Self – elaborated

account all the information related to the regions which has been absorbed in the max-min composition calculations above.

As it has been justified in previous pages, the direct impacts are not enough to make an overall analysis, given that causes (economic activity) are conditioned by other causes, as well as the effects (welfare variables of the population) are affected not only by the direct causes, but also by other cross effects. Therefore it is

**Table 6** Relationship matrix between sectors

| Sector | Primary sector | Manufacture | Construction and services |
|---|---|---|---|
| Primary sector | 1.000 | 0.500 | 0.100 |
| Manufacture | 0.000 | 1.000 | 0.800 |
| Construction and services | 0.300 | 0.200 | 1.000 |

Source: Self – elaborated

necessary to construct two additional matrices. Table 6 shows the relations between sectors. You can see that each sector is 100 % related to itself, but in relation to other sectors the impact is not symmetric, that is, for instance, the construction

**Table 7** Relationship matrix between welfare indicators

| Region | Occupation | Family income | Higher education | GDP per capita |
|---|---|---|---|---|
| Occupation | 1.000 | 1.000 | 0.350 | 1.000 |
| Family income | 0.350 | 1.000 | 0.950 | 0.600 |
| Higher education | 1.000 | 0.450 | 1.000 | 0.850 |
| GDP per capita | 0.800 | 1.000 | 0.750 | 1.000 |
| Human capital | 0.950 | 0.100 | 0.400 | 0.200 |
| Social capital | 0.350 | 0.100 | 0.200 | 0.100 |
| HDI | 0.850 | 0.000 | 0.500 | 0.600 |
| Environmental quality | 0.200 | 0.150 | 0.350 | 0.100 |
| Health | 0.350 | 0.850 | 0.400 | 0.100 |

| Region | Human capital | Social capital | HDI | Environmental quality | Health |
|---|---|---|---|---|---|
| Occupation | 0.900 | 0.750 | 0.950 | 0.250 | 0.700 |
| Family income | 0.400 | 0.300 | 0.750 | 0.150 | 0.850 |
| Higher education | 0.800 | 0.600 | 0.500 | 0.750 | 1.000 |
| GDP per capita | 0.750 | 0.500 | 0.900 | 0.250 | 0.800 |
| Human capital | 1.000 | 0.150 | 0.400 | 0.450 | 0.150 |
| Social capital | 0.000 | 1.000 | 0.000 | 0.200 | 0.350 |
| HDI | 0.900 | 0.450 | 1.000 | 0.100 | 0.800 |
| Environmental quality | 0.350 | 0.000 | 0.000 | 1.000 | 0.750 |
| Health | 0.100 | 0.400 | 0.800 | 0.150 | 1.000 |

Source: Self – elaborated

sector is not equally affected by the industry sector that on the service and construction industry.

On the other hand, we build another matrix that relates welfare economic indicators between each other. In Table 7, we show an asymmetric matrix that containing a maximum value (100 %) when an indicator is related by itself:

**Table 8** Final result matrix

| Sector | Occupation | Family income | Higher education | GDP per capita |
|---|---|---|---|---|
| Primary sector | 0.950 | 1.000 | 0.750 | 1.000 |
| Manufacture | 0.856 | 0.907 | 0.800 | 0.907 |
| Construction and services | 1.000 | 1.000 | 1.000 | 1.000 |
| Sector | Human capital | Social capital | HDI | Environmental Quality | Health |
| Primary sector | 1.000 | 0.536 | 1.000 | 0.634 | 1.000 |
| Manufacture | 0.900 | 0.800 | 0.956 | 1.000 | 0.800 |
| Construction and services | 0.932 | 1.000 | 0.976 | 0.750 | 1.000 |

Source: Self – elaborated

Finally, from matrices 5, 6, and 7 we calculate the final max-min composition matrix. This cause-effect matrix includes the impact of economic activities on the quality of life of the population; taking into ac-count the possible effects of any variable that may have a direct effect. The results of the cumulative ef-fects matrix are shown in Table 8.

# 5 Results and Discussion

Once the calculations and relationship matrices arc analyzed, we observe that on the cumulative basis any kind of economic activity affects the welfare of the population of the regions. We can see that, either directly or indirectly, the activity that has the minimum relation is 0.536 over 1 (in case of the primary sector over the Social Capital).

In the results we can see that the primary sector is the sector that has the least influence over analyzed welfare indicators. Regarding to the influence of manufacture, services and construction sectors we ob-serve that their impact on the quality of life is positive. Finally, we see that the service sector is crucial to determine the quality of life of the people of the regions.

The Forgotten Effects Theory permits the analysis of the direct and indirect incidence that each of the economic sectors has on the welfare level of the population. Table 9 shows the absolute difference be-tween the direct effect of the economic sectors and the indirect effect calculated by the composition of the cumulative effect, i.e. the final result matrix.

**Table 9** Indirect Incidence

| Sector | Occupation | Family income | Higher education | GDP per capita |
|---|---|---|---|---|
| Primary Sector | 0.417 | 0.699 | 0.488 | 0.000 |
| Manufacture | 0.459 | 0.606 | 0.597 | 0.000 |
| Construction and services | 0.000 | 0.000 | 0.000 | 0.106 |

| Sector | Human capital | Social capital | HDI | Environmental quality | Health |
|---|---|---|---|---|---|
| Primary sector | 0.000 | 0.000 | 0.000 | 0.000 | 0.000 |
| Manufacture | 0.044 | 0.120 | 0.000 | 0.000 | 0.578 |
| Construction and services | 0.000 | 0.000 | 0.000 | 0.283 | 0.000 |

Source: Self-elaborated

For example, if we analyze the effect of the primary sector on family income of a region, and if we only consider the direct effect (see Table 5) we observe that the impact is 0.301 over 1. When we look at the direct and indirect cumulative effect the result increases to 1. The difference between the two values 0.699 (1-0.301) represents the isolated indirect effect. This means that the indirect effect is more im-portant than the direct effect. In this case it is possible to know which variables are inter-jected, making the indirect effect more important. To find it out it is necessary to follow the max-min composition process in the calculations. In this case we have:

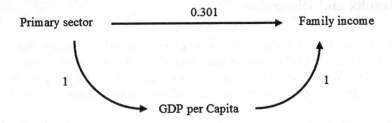

The primary sector is fully (1) linked to the GDP per capita variable on a maximum level (1) with the family income variable. In this way, it is possible to find out the more important indirect effects for each relationship. This result is coherent to the historic importance that farming, extractive industries and forestry, has had in the economy of the State.

# 6 Conclusions

The economic activities of the regions have a significant impact on the quality of life of its inhabitants. Data comparison and normalized max-min composition processes have been realized in order to quantify the direct and indirect effect that variables display between each other. This operation allowed us to establish a benchmark and presenting bounded data (between 0 and 1), which facilitates the consequent analysis of the results.

It is shown that, despite all economic activities influence the welfare of the people, the impact of each sector is singular with the following order from high to less: service and construction, manufacture and primary sector. The regions with higher service sector presence have higher quality of life indicators and they are more sensitive to the developed economic activity. On the other hand it is possible to identify the elements or variables that increase the direct and indirect causal relations that will permit performing certain actions at the institutional level with a multiplicative effect.

Lastly, the application of the Forgotten Effects Theory allows the prevision of economic actions as it gives a first insight about the relationships between different economic sectors and their effect on the quality of life of citizens. That information could aid decision makers to manage resources in an efficient way, hence building sustainable and effective public policies.

# References

Bellman, R., Zadeh, L.: Decision-making in a fuzzy environment. Manag. Sci. **17**(4), B-141 (1970)

Chauca, P.: La micro, pequeña y mediana empresa manufacturera en Michoacán. México (1999)

CONAPO: Desarrollo Humano. http://www.conapo.gob.mx/es/CONAPO/Desarrollo_Humano (2000)

Gil-Lafuente, A.M.: Fuzzy Logic in Financial Analysis. Springer, Berlin (2005)

INEGI: Banco de Información de Michoacán. http://www3.inegi.org.mx/sistemas/descarga/ (2014)

Kaufmann, A.: Introduction to the Theory of Fuzzy Subsets. Academic Press, New York (1975)

Kaufmann, A., Gil-Aluja, J.: Modelos para la investigación de efectos olvidados. Milladoiro, Spain (1988)

Valerio, V.A., Chávez, J.N.: Economía y desarrollo regional en México. Universidad Mi-choacana de San Nicolás de Hidalgo, México (2002)

Zadeh, L.: Fuzzy Sets. Inf. Control **8**(3), 338–353 (1965)

# Interval Numbers Versus Correlational Approaches Analyzing Corporate Social Responsibility

Sefa Boria-Reverter, Montserrat Yepes-Baldó, Marina Romeo and Luis Torres

**Abstract** Our purpose is to explore the relationship between Corporate Social Responsibility (CSR), work-life balance (WLB) and effectiveness comparing (a) a correlational approach, (b) expertons method, and (c) uncertain averaging operators (UA, UWA, UPA, and UPWA). These methodologies are common in the field of economics and engineering, but very innovative in the human resources research, allowing more accurate analyses of workers' perceptions. The Survey Work-Home Interaction – NijmeGen (SWING -SSC) and the Balanced Scorecard (BSC) were used. Results showed differences between companies with different levels of CSR development on individual effectiveness, and relations between WLB and individual effectiveness. Expertons methodology and uncertain averaging operators allows more accurate results than correlational statistics.

**Keywords** Expertons · Uncertain averaging operators · Corporate social responsibility · Effectiveness

**JEL code** D81 · Criteria for Decision-Making under risk and uncertainty · M14 · Corporate culture · Diversity · Social responsibility · C18 · Methodological issues · General

S. Boria-Reverter (✉)
Department of Economy and Business Organization, University of Barcelona Campus Diagonal Sud, Labour Relations, 1st. Floor Baldiri Reixac, 13, 08028 Barcelona, Spain
e-mail: jboriar@ub.edu

M. Yepes-Baldó · M. Romeo
Department of Social Psychology, University of Barcelona, Campus Mundet, Faculty of Psychology, Passeig de La Vall D'Hebron, 171, 08035 Barcelona, Spain
e-mail: myepes@ub.edu

M. Romeo
e-mail: mromeo@ub.edu

L. Torres
Centre for Organizational Health and Development, University of Nottingham, Jubilee Campus, Wollaton Road, Nottingham NG8 1BB, UK
e-mail: Luis.Torres@nottingham.ac.uk

© Springer International Publishing Switzerland 2015
J. Gil-Aluja et al. (eds.), *Scientific Methods for the Treatment of Uncertainty in Social Sciences*, Advances in Intelligent Systems and Computing 377,
DOI 10.1007/978-3-319-19704-3_9

# 1 Introduction

In organizations, when dealing with management decision making processes, as those related to the relationship between Corporate Social Responsibility (CSR) and effectiveness, we can assume that all the information we have is clear and can be assessed with aggregation operators based on exact numbers. Therefore, there exists a wide range of aggregation operators, as weighted average, the probabilistic aggregation, the OWA operator, the Choquet integral, distance measures, norms, logarithm aggregations, heavy aggregations or induced aggregator operators.

The most used aggregation operators based on exact numbers are weighted average and the probabilistic aggregation. The first operator aggregates the information by giving different levels of importance to each argument in the problem. On the other hand, the probabilistic aggregation uses probabilities to aggregate the data. Merigó (2009) proposed a combination of both previous operators, the probabilistic weighted average (PWA) that considers the degree of importance of each concept in the analysis including the objective and the subjective information of the environment.

However, organizations are complex realities where information is not always so clear. To deal with this reality, Moore (1966) proposed interval numbers as a useful technique for representing uncertainty because it considers the minimum and the maximum results that may occur. When using interval numbers to aggregate the available information we form uncertain aggregation operators. One operator based on interval numbers is the experton methodology. It enhances scientific rigor by measuring subjectivity and pooling the opinion of several experts to establish a single valuation. It is regarded as a major step forward in the development of fuzzy subsets, given that it seeks objectivity, thanks to the use of aggregation as a solution. This methodology has shown its suitability as an analysis tool in such different fields as quality assessment, monitoring of stocks management, marketing and business ethics.

Other operators based on interval numbers are the uncertain average (UA), the uncertain weighted average (UWA) and the uncertain probabilistic aggregation (UPA). Merigó (2011) proposed a new decision making approach based on an operator that is the result of the union between the uncertain weighted average (UWA) and the uncertain probabilistic aggregation (UPA) operators. This result considers not only the subjective part of the assessment of the individual but also the objective of the subject being treated, and it is called the uncertain probabilistic weighted averaging (UPWA) or the interval probabilistic weighted average (IPWA).

The purpose of this study is to explore the relationship between CSR, work-life balance (WLB) and effectiveness comparing (a) the correlational approach, (b) expertons method, and (c) uncertain averaging operators (UA, UWA, UPA, and UPWA).

# 2 Theoretical Background: CSR and Effectiveness

The Corporate Social Responsibility (CSR) approach has grown in importance and has become a possible answer to the necessity of developing a world class culture of responsible businesses, accompanied by more government regulation and the claim that every company must take into account all its stakeholders. It seems that the era of world economic crisis and recessions may be a favourable scene to develop a business concept in which not only financial improvement is important, but also sustainable growth.

The use of CSR and its focus inside the organization generates competitive advantages in order to improve the organizational effectiveness. There exists theoretical and empirical evidence to accept that this affirmation is valid; however, there is little evidence regarding the internal impact of CSR.

The empirical research has been related with external aspects such as social licence and tax advantages, customer satisfaction, customer loyalty, purchase intention, recruitment, financial performance, company reputation, and brand value and performance.

In contrast, academic interest in the internal impact of CSR is recent and has been related with variables such as employee satisfaction, organizational commitment, organizational identification, turnover, and job motivation.

The CSR addresses the strategic level of the organization, which in turn includes the development of different policies. In this paper, we will focus on the analysis of family-responsible policies and their effect on perceived balance between the personal and work areas (WLB). The analysis of this relationship has scarcely been studied from the scientific literature and some of the research has limitations in measuring the perception of WLB from the negative and one-dimensional side (Yuile et al. 2012). In order to overcome this weakness, in this paper we propose the Geurts and Demerouti model of WLB (2003, in Geurts et al. 2005). The authors defined WLB as a process in which an employee is affected positively or negatively by the interaction of both labour and non-labour ambits. In addition, this definition may be complete because it considers the interactions of both the labour and non-labour aspects in four different directions: negative work-home and home-work interaction, positive work-home and positive home-work interaction.

Related to effectiveness, Matthews (2011) indicates that there are several ways to understand it, however, this can be understood in terms of accomplishment of objectives, while performance is the measure by which organizations know if they are or not effective. In this way, a measure of effectiveness assesses the capacity of an organization to achieve its planned goals.

One of the most influential frameworks to assess effectiveness is the Balanced Scorecard (BSC) (Evans 2004). The BSC is designed to provide companies with a measureable roadmap with so that they may know if they are achieving their goals. Thus, with the use of the BSC companies define their strategic objectives and arrange them on a map according to the four perspectives: financial, customers, internal processes, and growth and learning (Kaplan and Norton 1992, 2004).

## 3   OWA Operators: UPA, UWA and UPWA

Considered the UA (uncertain average), as the basis for using the UPA (uncertain probabilistic average), UWA (uncertain weighted average), and the UPWA (uncertain probabilistic weighted average) operators, because they can analyse situations with subjective and objective information in the same formulation.

UPWA operator is an aggregation operator, which unifies the probability and the weighted average, then joins the UPA (uncertain probabilistic average) and UWA (uncertain weighted average).

An UPA operator of dimension n is a mapping UPA: $\Omega n \rightarrow \Omega$ ($\Omega$ be the set of interval number) that has an associated probabilistic vector P, with $\tilde{p}_i \in [0, 1]$ and $\sum_{i=1}^{n} \tilde{p}_i = 1$ thus:

$$UPA(\tilde{a}_1, \tilde{a}_2, \ldots, \tilde{a}_n) = \sum_{i=1}^{n} \tilde{p}_i \tilde{a}_i$$

An UWA operator of dimension n is a mapping UPA: $\Omega n \rightarrow \Omega$ ($\Omega$ be the set of interval number) that has an associated weighting vector W, with $\tilde{p}_i \in [0, 1]$ and $\sum_{i=1}^{n} p_i = 1$ thus:

$$UWA(\tilde{a}_1, \tilde{a}_2, \ldots, \tilde{a}_n) = \sum_{i=1}^{n} \tilde{w}_i \tilde{a}_i$$

An UPWA operator link UPA operator with UWA operator, this union combines objective and subjective probabilities. The sum of the two operators must be equal to 1, and for this we use ß and (1-ß).

An UPWA operator of dimension n is a mapping UPWA: $\Omega n \rightarrow \Omega$ ($\Omega$ be the set of interval number), it has an associated probabilistic vector P, with $\tilde{p}_i \in [0, 1]$ and $\sum_{i=1}^{n} \tilde{p}_i = 1$ and a weighting vector W, with $\tilde{w}_i \in [0, 1]$ and $\sum_{i=1}^{n} \tilde{w}_i = 1$, thus

$$UPWA(\tilde{a}_1, \tilde{a}_2, \ldots, \tilde{a}_n) = ß \sum_{i=1}^{n} \tilde{p}_i \tilde{a}_i + (1 - ß) \sum_{i=1}^{n} \tilde{w}_i \tilde{a}_i$$

## 4   Procedure and Sample

Two Chilean companies were chosen according to their external CSR reputation and family friendly policies using the following criteria: (1) National CSR rankings, (2) National CSR awards, (3) International CSR awards, and (4) IFREI level. As a result, companies were classified according to CSR policy development.

Company 1 is a company in the lumber industry with 1,712 employees in total. It has a high-level CSR policies development because it appears in the first 25 positions in a CSR national ranking and has several national and international awards and a B level on IFREI (Company has some family friendly policies and practices). Company 2 is involved in resorts and casinos industry with 6,284 employees in total. It has a low-level CSR policy development because it does not appear in any CSR national ranking and does not have any national or international awards. This company obtained a C level on IFREI (Company has some policies but their use is limited).

A non-probability for convenience sampling method was used. Instruments were sent to employees from both companies by the online platform QuestionPro in June and July 2012.

In company 1, 114 employees answered the survey of which 21.1 % were women ($n = 24$) and 78.9 % men ($n = 90$) with 43.8 average age ($SD = 7.8$). Of them 99.1 % were permanent employees ($n = 113$) and 0.9 % temporary ($n = 1$); all full-time workers (100 %, $n = 114$) with an average of 13.1 years working for the same company ($SD = 8.5$). 84.2 % participants worked extra time the last 3 months and 4.4 % live alone ($n = 5$), 87.7 % life with a partner ($n = 100$), 4.4 % live with parents ($n = 5$), and 87.7 % have children ($n = 100$).

In company 2, 700 employees answered the survey of which 44.4 % were women ($n = 290$) and 58.6 % men ($n = 410$) with 33.8 average age ($SD = 8.2$). Of them 95.7 % were permanent employees ($n = 670$) and 4.2 % temporary ($n = 30$); working full-time (92 %, $n = 644$) or part-time (8.0 %, $n = 56$) with an average of 4.9 years working for the same company ($SD = 5.6$). 80.3 % participants worked extra time the last 3 months and 14.9 % live alone ($n = 104$), 64.1 % live with a partner ($n = 449$), 16.3 % live with parents ($n = 114$), and 61.3 % have children ($n = 429$).

**Instruments**

- Work-life Balance. Survey Work-Home Interaction - Nijmegen for Spanish Speaking Countries (SWING-SSC) (Romeo et al. 2014). A 27 item survey using a five points Likert's scale (1 = *Strongly disagree* to 5 = *Strongly agree*). This instrument evaluates perceived WLB through four possible interactions: Negative work-home interaction (NWHI), Negative home-work interaction (NHWI), Positive work-home interaction (PWHI), and Positive home-work interaction (PHWI) (Geurts et al. 2005).
- The reliability of the original version has reported a good internal consistency ($\alpha = 0.80$) (Geurts et al. 2005). The version used in this study obtained a good internal consistency with a general $\alpha$ coefficient of 0.84, and similar levels of internal consistency for all dimensions ($\alpha = 0.90$ for NWHI; $\alpha = 0.88$ for NHWI; $\alpha = 0.87$ for PWHI; $\alpha = 0.85$ for PHWI) (Romeo et al. 2014).
- Effectiveness. Balanced Scorecard (BSC) (Kaplan and Norton 1992, 2004; Becker et al. 2001). The Balanced Scorecard allows companies to design a performance appraisal based on their own key performance indicators to measure objectives accomplishment and strategic competencies development at the employee level (Becker et al. 2001). In this study, in order to know this indicator the following question was added: "Please, report the result of the performance appraisal you obtained in 2011". Participants answered according to a five points scale (1 = *Critical* to 5 = *Excellent*).

**Data analyses**

A three data analysis strategy was employed: (a) correlational approach by t-test and correlations, (b) data analysis based on experton method and (c) uncertain averaging operators (UA, UPA, UWA, and UPWA).

From experton methodology the WLB scale establishes confidence intervals based on five anchors. In this study, a group of ten academic and professional experts, with a minimum of 10 years' experience in the field of CSR, was asked to establish confidence intervals based on the five anchors. To do this, they used a 11-point scale (11 values between 0 - *Null* and 1- *Totally*) and considered confidence levels as [0,1]. They allocated the *Strongly agree* response the value of 1, the value of 0 to *Strongly disagree*. Likewise, the responses *Disagree, Neither agree nor disagree* and *Agree*, being an uncertain level, were ascribed the values of [0.1, 0.4], [0.5, 0.5] and [0.6, 0.9] respectively. In this way, the imprecision in the validation corresponds to a specific situation that is semantically acceptable respectively (Kaufmann et al. 1994, p. 45).

Finally, we considered the UA (uncertain average), as the basis for using the UPA, UWA and the UPWA operators, because they can analyze situations with subjective and objective information in the same formulation. For the UPA and UWA weights, experts give their own opinions. These opinions are agreed in order to form the collective weights to be used in the aggregation process. The UPA operator use de following vector $P = (0.2, 0.1, 0.4, 0.3)$ and for UWA $V = (0.1, 0.1, 0.4, 0.4)$. It is important to note that the experts agreed, on one hand, higher coefficients for the interaction Work-Home versus the interaction Home-Work. On the other hand, they agreed higher coefficients for the positive interactions than for the negative ones. The UPWA operator uses UPA vector with a 40 % of importance and the UWA vector with a 60 %.

## 5 Results and Conclusions

The present study had as a purpose to go deeper into the influence of CSR within the organization, particularly, on WLB and individual effectiveness.

It was expected to find differences in WLB and effectiveness between companies but this was not confirmed for WLB from the correlational approach. Knowing to what extent employees' WLB has an impact on their individual effectiveness, results showed a significant negative relationship, although small, for the negative home-work interaction (NHWI) which is consistent with other previous studies showing negative relationships between WLB and other measures of effectiveness such as productivity or intention to quit. Additionally, the low relation in this study could be explained because of the difficulties of giving a nominal and operational definition of this variable.

However, the methodologies based on the uncertain aggregation operators, namely expertons, UA, UPA, UWA and UPWA have allowed us a better approach to a complex reality from the multi-person evaluation. Following Merigó (2011) we can affirm that these operators allow us to "unify decision-making problems under objective risk and under subjective risk in the same formulation and considering the degree of importance that each concept has in the analysis".

For the UPA and UWA weights, each expert of our selected group gives their own opinions, and these had been agreed in order to form the collective weights to be used in the aggregation process. They decided higher coefficients for the interaction Work-Home versus the interaction Home-Work. On the other hand, they agreed higher coefficients for the positive interactions than for the negative ones. This is aligned with the perspective related to the optimal functioning of individuals and groups in organizations, as well as the effective management of wellbeing at work and the development of healthy organizations (Seligman 1999). In this sense, our work gives priority to management decisions oriented to the improvement of WLB against those directed towards reducing the negative effects of imbalance.

# References

Becker, B., Huselid, M., Ulrich, D.: The HR Scorecard: Linking people, strategy, and performance. Harvard Business School Press, Boston (2001)

Evans, J.: An exploratory study of performance measurement systems and relationship with performance results. J. Oper. Manage. **22**, 219–232 (2004)

Geurts, S., Taris, T., Kompier, M., Dikkers, J., Van Hooff, M., Kinnunen, U.: Work-home interaction from a work psychological perspective: development and validation of a new questionnaire, the SWING. Work Stress **19**, 319–339 (2005)

Greening, D., Turban, D.: Corporate social performance as a competitive advantage in attracting a quality workforce. Bus. Soc. **39**(3), 254–280 (2000)

Kaplan, R., Norton, D.: The balanced scorecard: measures that drive performance. Harvard Bus. Rev. **70**(1), 71–79 (1992)

Kaplan, R., Norton, D.: Strategy maps: converting intangible assets into tangible outcomes. Harvard Business School Press, Boston (2004)

Kaufmann, A., Gil-Aluja, J., Gil-Lafuente, A.M.: La creatividad en la gestión de las empresas. Pirámide, Madrid (1994)

Matthews, J.: Assessing organizational effectiveness: the role of performance measures. Libr. Q. **81**(1), 83–110 (2011)

Merigó, J.M.: On the use of the OWA operator in the weighted average. In: Proceedings of the World Congress on Engineering, pp.82–87, London (2009)

Merigó, J.M.: The uncertain probabilistic weighted average and its application in the theory of expertons. Afr. J. Bus. Manage. **5**(15), 6092–6102 (2011)

Moore, R.: Interval Analysis. Prentice Hall, Englewood Cliffs (1966)

Romeo, M., Berger, R., Yepes-Baldó, M., Ramos, B.: Adaptation and validation of the spanish version of the survey work-home interaction–nijmegen (swing) to spanish speaking countries. Anales de Psicología **30**(1), 287–293 (2014)

Seligman, M.E.P.: The president's address. Am. Psychol. **54**, 559–562 (1999)

Yuile, C., Chang, A., Gudmundsson, A., Sawang, S.: The role of life friendly policies on employees work-life balance. J. Manage. Organ. **18**(1), 53–63 (2012)

# Second-Order Changes on Personnel Assignment Under Uncertainty

Ruben Chavez, Federico González-Santoyo, Beatriz Flores
and Juan J. Flores

**Abstract** In this work we present some basics about the commitment in organizations and how it influences this heavily on the change to a new strategy in the organization. This task involves using the Hamming Distance (fuzzy logic) to measure the level of commitment to employees with the new strategy to establish. The process consists of four states to the changes in strategy called second-order, in such a process it includes the assessment of parameters through fuzzy numbers. In this regard, we propose estimating parameters a and b, using Expertons as a first approximation and to provide justification to the calculations of the second order change SOC (Sigismund and Oran 2002).

**Keywords** Second-order change · Expertons · Hamming distance

## 1 Introduction

The first part of this proposal exposes the strategy changes as an important element to generate a new order within the organization. According to Sigismund, this process is related to the quantitative evaluation of the inertia of strategies for a set of

R. Chavez (✉) · F. González-Santoyo · B. Flores
Facultad de Ciencias Contables Y Administrativas,
Universidad Michoacana, Morelia, Mexico
e-mail: pintachavez@gmail.com

F. González-Santoyo
e-mail: fsantoyo@umich.mx

B. Flores
e-mail: betyf@umich.mx

J.J. Flores
Division de Estudios de Posgrado, Facultad de Ingenieria Electrica,
Universidad Michoacana, Morelia, Mexico
e-mail: juanf@umich.mx

© Springer International Publishing Switzerland 2015                                             115
J. Gil-Aluja et al. (eds.), *Scientific Methods for the Treatment of Uncertainty
in Social Sciences*, Advances in Intelligent Systems and Computing 377,
DOI 10.1007/978-3-319-19704-3_10

workers within an organization. According to the changes in the market and competitiveness levels, organizations are forced to set new strategies to produce significant changes in them. This also has to do with the new information the company manages; this information produces an entropic state and reluctance to adopt the new strategy.

Second-order change (SOC) is related to these differences between inertia (backed up by the strategy's commitment) and the tension generated by the new strategy. SOC is evaluated according to Sigismund and Oran (2002), where accumulated tension must be greater than accumulated inertia. Generating SOC involves several variables, some of which are related or are a function of commitments, skills, and competences, in the presence of new market requirements and challenges arising with the new strategy.

This part proposes the evaluation of parameters $a$ and $b$ using expertons. This evaluation is performed using fuzzy numbers, so that they provide decision making with adequate support to reach the goals of the company. The model allows us to determine whether or not a SOC can be accomplished with the new strategy.

## 2 Methodology

### 2.1 Employees Commitment Level Evaluation

The workers commitment is fundamental to obtain good results in the company. According to Davenport (2000), commitment and dedication open up a path to human resources investment. Commitment reveals a strong positive association with personal motivation. There also exists a strong correlation with objective performance measures. Furthermore, commitment establishes a negative correlation with the desire to abandon the company and the rotation index. These findings indicate that in organizations that keep the right amount and quality of human resources, one can induce commitment to invest, with the hope to increase the profits. Loyalty-based commitment presents a similar relation, but less tied to profits. So, SOC cannot be achieved easily, since they are related to commitment levels and motivation, which not necessarily match with those of the employees simultaneously. A set of motivational mechanisms to transform the organization has been created. Those changes force directors to design a formal structure for the new strategy, including all elements properly justified.

Davenport and Bass (1999) established that commitment has three levels: attitude, loyalty-based, and pragmatic. Attitude commitment is based on the existing links with the organization and the employee's satisfaction to belong to the organization. Loyalty-based commitment relates to a chain of elements that provide employees with a sense of duty and respect to the organization. The greatest profit is on pragmatic commitment, related with the way an employee is deployed. This fact takes into account limitations due to lack of experience and the advantage to work there, with respect to other companies. This situation forces the employee to accept the options programmed by the organization.

According to Baldoni (2007), commitment involves several variables: the product of the unhappiness with status quo, a future vision and the need to act; those have to overcome the reluctance to change.

Reluctance to change is related to many aspects in the organization. Rejecting change is related to human costumes (O'Toole 1996). Individuals have a mean intellect and inclinations, so they do not have a strong enough desire to push forward and start something new.

It is up to experts and leaders to convince employees of acting and breaking old habits and paradigms. Therefore, it is necessary to know the level of unhappiness of employees. Training may have a strong influence on the acceptance of new strategies.

To measure those characteristics that are a function of commitment, motivation, unhappiness, etc., we propose to apply a Hamming Distance to compare the employees' performance level with respect to the ideal level, to determine SOC. According to Kaufmann et al. (2002) the Hamming Distance between two fuzzy sets $\underline{A}$ and $\underline{B}$ is defined by Eq. (1).

$$N(\underline{A}, \underline{B}) = \left[ \sum |\mu_i - \mu_j| \right] \tag{1}$$

## 2.2 Fuzzy SOC

In changing a company we distinguish four phases: (1) review elements that reduce results due to following old practices; (2) consolidation of mechanisms that maintain the organization in sync with its environment; (3) deterioration of control information; (4) growth of defensive routines. This makes necessary to reach a consensus in the conditions of the current strategy and the required changes to produce a SOC. The establishment of the right conditions for a strategic debate helps employees to accept new models.

Fuzzy SOC is also tied to the different phases of the process (Sigismund and Oran 2002). State I (normal business conditions): adaptation of the current strategy. The change process is related to fuzzy cumulative inertia as an element to measure strength or weakness of the company. Fuzzy cumulative inertia is defined by Eq. (2).

$$\tilde{I}_1(t+1) = I(t) + \tilde{b}I(t)[1 - I(t)] \tag{2}$$

$$0 \le \tilde{b} \le 1$$

$$0 \le I(t) \le 1$$

Equation (1) represents the level of commitment. Cumulative inertia on state I is $\tilde{I}_1(t+1)$. I(t) describes the increase in inertia, which can be interpreted as the percentage of committed employees with the current strategy. $[1 - I(t)]$ represents the employees not committed yet. Fuzzy parameter $\tilde{b}$, can be seen as the appeal of the current strategy. Parameter $b$ is an 11-valued fuzzy number used in inertia and tension equations. We reach consensus from a set of expert, to form an experton (Tables 1 and 2).

**Table 1** Confidence interval for $\tilde{a}$, given by a group of experts

| | |
|---|---|
| Expert 1 | [0.2, 0.4] |
| Expert 2 | [0.3, 0.5] |
| Expert 3 | [0.2, 0.6] |
| Expert 4 | [0.3, 0.6] |
| Expert 5 | [0.7] |

**Table 2** Confidence interval for $\tilde{b}$, given by a group of experts

| | |
|---|---|
| Expert 1 | [0.2, 0.5] |
| Expert 2 | [0.3, 0.6] |
| Expert 3 | [0.2, 0.7] |
| Expert 4 | [0.3, 0.8] |
| Expert 5 | [0.4, 0.8] |

The experton for fuzzy parameter $\tilde{a}$ is computed based on statistical information and absolute frequencies on lower and upper bounds, $E[\tilde{a}] = [0.4, 0.6]$. Similarly, $E[\tilde{b}] = [0.3, 0.7]$. We are now ready to start computing the SO. Equation (3) provides the means to determine the cumulative tension.

$$\tilde{S}_1(t+1) = S(t) + z(t) - \widetilde{H}_1(t) \tag{3}$$

$$\widetilde{H}_1(t) = \tilde{a}S(t) \tag{4}$$

$$0 \leq a \leq 1$$

Fuzzy parameter $\tilde{a}$ is a function of the tension reduction (internal adjustments on the organization) on the new strategy. We assume the organization stays on state I as long as the homeostatic mechanisms of the firs-order change maintain the tension level under a critical threshold $\bar{S}$.

When fuzzy tension accumulates quickly and the difference between $\tilde{S}(t)$ and $\bar{S}$ increases, the probability to consider a significant SOC increases. The probability to leave state I and transition to state II of the SOC process, $P_{12}(t)$, equals the positive difference between the tension on the organization, $\tilde{S}(t)$ and the critical threshold, $\bar{S}$.

$$P_{12}(t) = \begin{cases} \tilde{S}(t) - \bar{S}, & \text{if } \tilde{S}(t) > \bar{S} \\ 0, & \text{otherwise} \end{cases} \tag{5}$$

$\tilde{S}(t)$ and $\bar{S}$ are designed to guarantee that $0 < P_{12}(t) < 1$. Homeostatic first-order change, $H_1$ is driven by $\tilde{a}$. When a new tension is generated (a Poisson process with parameter $\lambda$ – expected frequency of tension events per time unit).

$$P_r = [z(t) = nz] = \frac{\lambda^n e^{-\lambda}}{n!} \tag{6}$$

where n is the number of tension generating events during time t, and z is a fixed amount on Poisson's equation. We observe how parameter $\lambda$ (see Fig. 1), is set by an experton.

**Fig. 1** Tension and Inertia States for SOC. Source: Sigismund and Oran (2002)

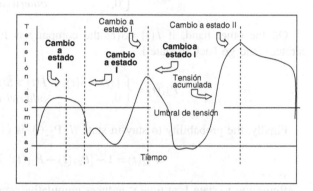

SOC is based on communication and trust. It is particularly important in a process of strategic decisions that require links of different hierarchical levels. Communication with people unrelated to specific situations motivate a SOC. We assume that efficiency of the fuzzy homeostatic answers to tension in state II, $\tilde{H}_2(t)$ is affected by the level of commitment with the current strategy $\tilde{I}(t)$, as shown by Eqs. (7) and (8).

$$\tilde{S}_2(t+1) = \tilde{S}(t) + z(t) - \tilde{H}_2(t) \tag{7}$$

$$\tilde{H}_2(t) = \tilde{a}I(t)\tilde{S}(t) \tag{8}$$

$$0 \leq \tilde{a} \leq 1$$

$$0 \leq I(t) \leq 1$$

Change is proportional to the difference between inertia and tension in the previous period (Eq. (9)).

$$\tilde{I}_2(t+1) = \tilde{I}(t) + \tilde{b}\tilde{I}(t)[\tilde{I}(t) - \tilde{S}(t)][1 - \tilde{I}(t)] \tag{9}$$

$$0 \le \tilde{b} \le 1$$

$$0 \le \tilde{I}(t) \le 1$$

If $\tilde{S} > \tilde{I}$ commitment with the current strategy decreases; in a fuzzy SOC, it means to transition to state III. The probability to transition from state II to state II at time t, $P_{23}(t)$ is given by Eq. (10).

$$P_{23}(t) = \begin{cases} \tilde{S}(t) - \tilde{I}(t), & \tilde{S}(t) > \tilde{I}(t) \\ 0, & \textit{otherwise} \end{cases} \tag{10}$$

On the other hand, if $\tilde{I}(t) > \tilde{S}(t)$ the company is forced to take the current strategy and go back to state I.

$$P_{21}(t) = \begin{cases} \tilde{I}(t) - \tilde{S}(t), & \tilde{I}(t) > \tilde{S}(t) \\ 0, & \textit{otherwise} \end{cases} \tag{11}$$

Finally, the probability to stay in state II, $P_{22}(t)$, is given by Eq. (12).

$$P_{22}(t) = 1 - [P_{23}(t) + P_{21}(t)] \tag{12}$$

Returning to state I, at time t, reduces cumulative tension, and inertia increases (see Eqs. (13) and (14)).

$$\tilde{S}_{23}(t) = \tilde{S}(t) + \tilde{a}\tilde{S}(t) \tag{13}$$

$$\tilde{I}_{21}(t) = \tilde{I}(t) + \tilde{b}\tilde{I}(t)[1 - \tilde{I}(t)] \tag{14}$$

Returning to state I separates tension and inertia, eliminating S(t) and state commitment. Nonetheless, if at least some components of a feasible option are set, interactive processes will decrease even more the commitment with the old strategy. Increasing the rationale contributes to explain the selection process of the orientation to follow with the SOC.

SOC alternatives start defining a new strategy, marked by *. Sigismund and Oran (2002) states that to maintain inertia to originate significant, permanent, and growing changes, the company needs to transition from state I to state II. The initial commitment level with the current strategy is given by Eq. (15).

$$I^*(t) = \bar{I} \tag{15}$$

where $\bar{I}$ corresponds to the commitment reduction on the current strategy. So, transition from state II to state III, as a function of inertia, is given by Eq. (16).

$$I_{23}(t) = I^*(t) - \bar{I} \tag{16}$$

The commitment level increases with the new strategy when people join to it. Therefore, fuzzy inertia, created around the new strategy in state III, $1_3^*(t+1)$ is given by Eq. (17).

$$\tilde{I}_3^*(t+1) = I^*(t) + \tilde{b}I^*(t)\left[1 - I^*(t)\right] \tag{17}$$

While in state III, the organization must withdraw resources and energy to maintain the old strategy. Old strategy in state III is given by Eq. (18).

$$\tilde{I}_3^*(t+1) = I(t) + \tilde{b}[I(t) - S(t)][1 - I(t)] \tag{18}$$

This implies a decrease in fuzzy inertia, making the fuzzy tension increase. An important consequence of moving from the old strategy is a reduction on the effectiveness of the homeostatic controls on the fuzzy tension at the now in danger strategy, since the internal mechanisms are tightly related to commitment. $S_3(t)$ increases rapidly as inertia decreases, as shown by Eqs. (19) and (20).

$$\tilde{S}_3(t+1) = \tilde{S}(t) + z(t) - \widetilde{H}_3(t) \tag{19}$$

$$\widetilde{H}_3(t) = \tilde{a}I(t)\tilde{S}(t) \tag{20}$$

The possibility to transition from state III to state IV, at time t, is given by Eq. (21).

$$P_{34}(t) = \begin{cases} \tilde{I}_3^*(t) - \tilde{I}_3(t), & \tilde{I}_3^*(t) > \tilde{I}_3(t) \\ 0, & otherwise \end{cases} \tag{21}$$

The company may keep the old strategy if cumulative tension is less than the inertia. In this situation, it returns to state I, which is proportional to the positive difference between inertia and tension. See Eq. (22).

$$P_{31}(t) = \tilde{I}_3(t) - \left[\tilde{S}_3(t)\right], \tilde{I}(t) > \tilde{S}(t) \tag{22}$$

The probability to continue searching for new alternatives is given by Eq. (23).

$$P_{33}(t) = 1 - [P_{31}(t) + P_{34}(t)] \tag{23}$$

where s(t) and I(t) are set to guarantee non-negative transition probabilities. Following its goals, the company may announce a sudden transition from state III to state IV. In the formal model, we assume that adopting the new strategy produces a reduction on fuzzy tension – see Eq. (24).

$$\tilde{S}_{34}^*(t) = \tilde{S}(t) - \tilde{a}\tilde{S}(t) \tag{24}$$

It is convenient to reduce the initial fuzzy tension, so it is not ahead of time and it does not decrease before the new strategy produces its effects. So, tension changes to the previous situation.

$$\tilde{S}_4^*(t+1) = \tilde{S}_4^*(t) + z(t) - \tilde{a}\tilde{I}_4^*(t)\tilde{S}_4^*(t) \tag{25}$$

As far as the fuzzy inertia is concerned, it assumes that initial commitment to the newly adopted strategy $\tilde{I}_{34}^*(t)$, is given by Eq. (26).

$$\tilde{I}_{34}^*(t) = I^*(t) + \tilde{b}I^*(t)\left[1 - I^*(t)\right] \tag{26}$$

So, inertia increases steadily:

$$\tilde{I}_4^*(t+1) = I^*(t) + \tilde{b}I^*(t)[1 - I^*(t)] \tag{27}$$

In this case, the organization evolves to state IV. If the commitment with the new strategy, $\tilde{I}^*(t) > \tilde{S}^*(t)$, the probability of transition to state I is given by Eq. (28).

$$P_{41}(t) = \begin{cases} \tilde{I}^*(t) - \tilde{S}^*(t), & \tilde{I}^*(t) > \tilde{S}^*(t) \\ 0, & otherwise \end{cases} \tag{28}$$

If $\tilde{S}^*(t) > \tilde{I}^*(t)$, the probability to search for another alternative (return to state III) is given by Eq. (29).

$$P_{43}(t) = \begin{cases} \tilde{S}^*(t) - \tilde{I}^*(t), & \tilde{S}^*(t) > \tilde{I}^*(t) \\ 0, & otherwise \end{cases} \tag{29}$$

The probability to continue with the new strategy is given by Eq. (30).

$$P_{44}(t) = 1 - [P_{41}(t) + P_{43}(t)] \tag{30}$$

# 3   Study Case

We have developed fuzzy quantitative tools to solve the SOC problem. The proposal works in two stages: the first one is personnel selection to perform leadership tasks and support the SOC; the second one is the application of the SOC model, to make efficient and optimize the potential of human resources to produce results and influence strategy changes as a result on market changes and permanent threats to the enterprise. The company's personnel has to respond and propose innovative strategies to continue being competitive. Under those circumstances, SOC have to be made quickly.

### Stage 1

The most important of the goals has to do with selecting the personnel involved with the new strategy. This personnel will influence and direct the rest of the people towards the changes the company needs. The Hamming Distance method is used to select the personnel selection through an ideal profile (designed by the experts). The Hungarian method is used to assign specific tasks to perform SOC (Table 3).

Considering the efficiency of the training program, we have considered a fuzzy assessment of the problem. To deploy the Hamming Distance, in the first column of Table 4, we set goals, followed by the expected results (experton), and finally the results of each operator p.

Applying the distance method, we have the distances to the expected results, and the similarity results, which represent the performance of the training program. See Table 5.

Table 5 shows that the program has a mean performance of 73.9 %. It also shows that p2 must be the group leader, since its profile is the closest one the ideal profile for SOC.

| **Table 3** Ideal values for the characteristics included in the new strategy | Characteristics | Value assigned by experts $\underline{B}$ |
|---|---|---|
| | Knowledge about the new strategy | 1 |
| | Availability | 0.9 |
| | Innovation and involvement about the new strategy | 0.9 |
| | Responsibility | 0.9 |
| | Leadership – proactivity | 0.8 |
| | Socialization level | 0.9 |
| | Decision capability | 0.9 |
| | Results orientation | 0.9 |
| | Resistance to stress | 1 |
| | Credibility | 0.8 |
| | Training | 1 |
| | Time to personal training | 10.8 |
| | Time with an instructor | 1 |
| | Bibliographic review time | 0.8 |
| | Problem analysis | 1 |
| | Decision making ability | 0.9 |

**Table 4** Goal, experton (column 2), and value of each operator p (columns 3, 4, 5 and 6)

| Objetivos | Expertos [0,1] | Operador 1 | Operador 2 | Operador 3 | Operador 4 |
|-----------|----------------|------------|------------|------------|------------|
| a | 1 | 0.6 | 0.7 | 0.6 | 0.6 |
| b | 0.9 | 0.7 | 0.5 | 0.5 | 0.6 |
| c | 0.9 | 0.7 | 0.4 | 0.5 | 0.6 |
| d | 0.9 | 0.6 | 0.7 | 0.6 | 0.8 |
| e | 0.8 | 0.5 | 0.6 | 0.4 | 0.8 |
| f | 0.9 | 0.6 | 0.6 | 0.7 | 0.7 |
| g | 0.9 | 0.8 | 0.6 | 0.7 | 0.4 |
| h | 0.9 | 0.6 | 0.7 | 0.8 | 0.6 |
| i | 1 | 0.7 | 0.7 | 0.6 | 0.6 |
| j | 0.8 | 0.8 | 0.7 | 0.7 | 0.8 |
| k | 1 | 0.7 | 0.8 | 0.7 | 0.6 |
| l | 0.8 | 0.5 | 0.8 | 0.6 | 0.7 |
| m | 1 | 0.5 | 0.7 | 0.5 | 0.6 |
| n | 0.8 | 0.6 | 0.6 | 0.8 | 0.8 |
| o | 1 | 0.8 | 0.7 | 0.5 | 0.6 |
| P | 0.9 | 0.6 | 0.8 | 0.8 | 0.6 |

**Table 5** Distance-similarity from the ideal fuzzy profile

| | p 1 | p 2 | p 3 | p 4 |
|---|-----|-----|-----|-----|
| Alejamiento Ã | 0.2625 | 0.24375 | 0.28125 | 0.25625 |
| 1- Ã (acercamiento) | 0.7375 | 0.75625 | 0.71875 | 0.74375 |

## Stage 2

Consider the tension and inertia values (Table 6) from personnel, after the training program. Applying formulae 1–12, assuming that parameters take on the values $E[\tilde{a}] = [0.4, 0.6]$ and $E[\tilde{b}] = [0.3, 0.7]$, assigning a value to z corresponding to market changes, an exogenous factor (see Fig. 3), and the probability to stay in state (see Fig. 2).

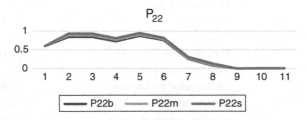

**Fig. 2** Probability to stay in state II, $P_{22}$

**Table 6** Tension S(t) and inertia I(t)

| Tension (S) | Inertia (I) |
|---|---|
| 0 | 1 |
| 0.4 | 0.8 |
| 0.4 | 0.8 |
| 0.4 | 0.7 |
| 0.2 | 0.6 |
| 0.2 | 0.5 |
| 0.4 | 0.4 |
| 0.6 | 0.4 |
| 0.7 | 0.3 |
| 0.7 | 0.3 |
| 0.7 | 0.3 |

**Fig. 3** Market volatilities z

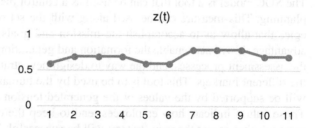

z(t)

where $P_{22b}$ = low probability, $P_{22m}$ = medium probability, and $P_{22s}$ = high probability.

The difference between inertia and tension is represented by the commitment (in state I if the difference is less than 1, and in state II if it is greater than 0) to the new strategy, according to $\tilde{S}(t+1) < \tilde{I}(t+1)$, as shown in Fig. 4.

Difference located under zero indicates that inertia dominates. That is, personnel are still attached to the old strategy, while a difference greater than zero show a level of involvement with the new strategy. Notice the decrease in the fuzzy tension in Fig. 4. This is due to the new strategy, which makes the fuzzy tension decrease to a level close to zero and then increasing again. Figure 5 shows the probability to stay in state II, where the SOC is settling.

**Fig. 4** State transition

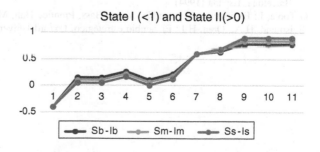

State I (<1) and State II(>0)

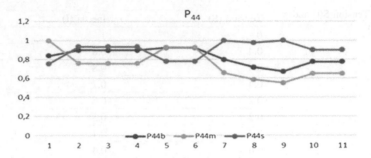

**Fig. 5** SOC settlement with fuzzy parameters $\tilde{a}$ and $\tilde{b}$

## 4 Conclusions

The SOC model is a tool that can be used as a control measure to perform strategic planning. This measure can be used along with the set of goals and tactical strategies that allow us to accomplish the mission and goals to achieve a competitive advantage. It will also enable the formation and generation of knowledge on each of the assessment processes, and the way to design new strategies to be competitive in the different markets. This tool is to be used by the human resources manager, who will be supported by the values of the generated tension and inertia of employees. Those values indicate how employees tend to keep the old strategy, and to determine whether or not the new strategy will be successful. This work is based on the evaluation of SOC, using the fuzzy parameters $\tilde{a}$ and $\tilde{b}$, to quantify the changes in tension and inertia of employees. In the future we will consider more fuzzy elements and tools, e.g., fuzzy linear regression. These tools allow us to justify strategic changes with respect to time more effectively.

## References

Baldoni, J.: Qué hacen los líderes para obtener los mejores resultados. Primera edición en español. McGraw-Hill Companies Inc., México (2007)

Davenport, O.T., Bass, J.: Capital humano, Ediciones Gestión 2000 S.A Barcelona, España (1999)

Gil Aluja, J., Kaufmann, A., Terceño, G.A.: Matemáticas para la economía. Foro científico Barcelona, España (1994)

O'Toole, J.: El liderazgo del cambio. Jossey-Bass, Prentice Hall, México (1995)

Sigismund, H.A., Oran, H.J.: El cambio estratégico. Oxford University, México (2002)

# Part III
# Forecasting Models

# Advanced Spectral Methods and Their Potential in Forecasting Fuzzy-Valued and Multivariate Financial Time Series

Vasile Georgescu and Sorin-Manuel Delureanu

**Abstract** In this paper we explore the effectiveness of two nonparametric methods, based upon a matrix spectral decomposition approach, namely the *Independent Component Analysis* (ICA) and the *Singular Spectrum Analysis* (SSA). The intended area of applications is that of forecasting fuzzy-valued and multivariate time series. Given a multivariate time series, ICA assumes that each of its components is a mixture of several independent underlying factors. Separating such distinct time-varying causal factors becomes crucial in multivariate financial time series analysis, when attempting to explain past co-movements and to predict future evolutions. The multivariate extension of SSA (MSSA) can be employed as a powerful prediction tool, either separately, or in conjunction with ICA. As a first application, we use MSSA to recurrently forecasting triangular-shaped fuzzy monthly exchange rates, thus aiming at capturing both the randomness and the fuzziness of the financial process. A hybrid ICA-SSA approach is also proposed. The primarily role of ICA is to reveal certain fundamental factors behind several parallel series of foreign exchange rates. More accurate predictions can be performed via these independent components, after their separation. MSSA is employed to compute forecasts of independent factors. Afterwards, these forecasts of underlying factors are remixed into the forecasts of observable foreign exchange rates.

**Keywords** Independent component analysis · Singular spectrum analysis · Forecasting fuzzy-valued · Multivariate time series

V. Georgescu (✉) · S.-M. Delureanu
Department of Statistics and Informatics, University of Craiova, Craiova, Romania
e-mail: v_geo@yahoo.com

S.-M. Delureanu
e-mail: s.m.delureanu@gmail.com

© Springer International Publishing Switzerland 2015
J. Gil-Aluja et al. (eds.), *Scientific Methods for the Treatment of Uncertainty in Social Sciences*, Advances in Intelligent Systems and Computing 377,
DOI 10.1007/978-3-319-19704-3_11

# 1 Introduction

*Independent Component Analysis* (ICA) and *Singular Spectrum Analysis* (SSA) are both nonparametric methods for time series analysis, based upon a matrix spectral decomposition approach.

ICA is used to separate a set of *signal mixtures* into a corresponding set of statistically *independent component (source)* signals. There are two alternative ways to accomplish such separation: either to maximize a measure of nongaussianity (Hyvärinen et al. 2001), or to maximize a measure of independence. The idea behind maximizing nongaussianity is that source signals are generally assumed to have non-Gaussian distributions. By contrast, any mixture of source signals tends to have a distribution that is more Gaussian than that of any of its constituent source signals, whatever their distribution is. Then, extracting independent signals with non-Gaussian distributions and low complexity structure from a set of signal mixtures should recover the original signals. The alternative way is finding independent signals by maximizing entropy (a quantity which is related in a useful way to independence), method known as *infomax* (Bell and Sejnowski 1995). SSA is also a nonparametric method for time series structure recognition. It has been first introduced in (Broomhead and King 1986), and further developed by many authors (Goljadina et al. 2001; Vautard et al. 1992). Both the fuzzy-valued and the multivariate time series may benefit from SSA via one of its straightforward extensions: the multi-channel SSA (MSSA). The forecasting potential of MSSA is explored in this paper either as a separate tool, or in conjunction with ICA.

# 2 Approaches to Extracting Independent Factors from Signal Mixtures: The FastICA Algorithm

Let us consider a sample $\{x_t\}_{t=1}^{T}$, drawn from a random vector $x \in \mathfrak{R}^N$. We assume preprocessing $x_t$. First, it is centered, being transformed into a zero mean vector $y_t$. The already centered vector $y_t$ is whitened, i.e., is linearly transformed into a new vector $z_t$, which is white (spherical), i.e., its covariance matrix equals the identity matrix: $\Sigma_z = E\{zz'\} = I$. Whitening uses the eigenvalue decomposition (EVD) of the covariance matrix of $y_t$, $\Sigma_{y_t} = E\{y_t y'_t\} = EDE'$, where $E$ is the orthogonal matrix of eigenvectors, and $D$ is the diagonal matrix of its eigenvalues. Whitening can now be done by $z_t = D^{-1/2}E' y_t$. Now, the preprocessed signal vector $z_t \in \mathfrak{R}^N$ is considered to be a linear mixture of some unknown source vector $s_t \in \mathfrak{R}^M$:

$$z_t = A s_t, \quad t = 1, 2, \ldots, T \tag{1}$$

where $A \in \mathfrak{R}^{N \times M}$ is the unknown *mixing* matrix and $M \leq N$. The objective is to recover the source vector realizations $\{s_t\}$ from the corresponding realizations of the observed vector $\{z_t\}$. This means to find a matrix $W \in \mathfrak{R}^{M \times N}$ (called a *separating* matrix) that makes the components of $z_t$ statistically independent. The process involves, explicitly or implicitly, the inversion of matrix $A$, which is thus assumed to have full column rank. The separator output is computed as $\hat{s}_t = W z_t$. A solution to the problem can be found based on the optimization of a cost function, also called *contrast* function.

FastICA is a computationally efficient algorithm for performing ICA estimations by optimizing orthogonal contrast functions of the form

$$Y_G(W) = \sum_{i=1}^{N} E\{G_i(w_i'z)\} \tag{2}$$

where the data $z$ are prewhitened and matrix $W$ is constrained to be orthogonal. The functions $G_i(\cdot)$ can fundamentally be chosen in two ways:

- In a maximum likelihood framework, $G_i(\hat{s}_i) = \log p_{s_i}(\hat{s}_i)$ where $p_{s_i}$ is the pdf of the $i$-th independent component. The functions $G_i(\cdot)$ can be fixed based on prior information of the independent components, or they can be estimated from the current estimates $\hat{s}_i$. In many applications, the independent components are known to be super-Gaussian (sparse), in which case all the functions $G_i(\cdot)$ are often set a priori as

$$G_i(\hat{s}) = -\log\cosh\hat{s} \tag{3}$$

- This corresponds to a symmetric super-Gaussian pdf and has the great benefit of being very smooth; a non-smooth $G_i(\cdot)$ can pose problems for any optimization method. Maximum likelihood estimation is equivalent, under some assumptions, to information-theoretic approaches such as minimum mutual information and minimum marginal entropy.
- A second framework consists in using cumulants. Consider the simple case of maximizing the kurtosis $kurt(\hat{s}) = E\{\hat{s}^4\} - 3(E\{\hat{s}^2\})^2$ of the estimated signals $\hat{s}$. Because we assume that the estimated signals have unit variance, due to prewhitening and the orthogonality of $w$, this reduces to maximizing the fourth moment $E\{\hat{s}^4\}$, thus $G_i(\hat{s}) = \hat{s}^4$. This means that the contrast is essentially the sum of kurtosis. If we know a priori that the independent components are super-Gaussian, the maximization of this contrast function is a valid estimation method. If the sources are all sub-Gaussian, we switch the sign and choose $G_i(\hat{s}) = -\hat{s}^4$, which becomes again a valid contrast.

Function $G_i(\cdot)$ is implicitly expressed in the algorithm via its derivative, a nonlinearity $g_i(\cdot)$; that corresponding to $G_i(\hat{s}) = -\log\cosh\hat{s}$ is the hyperbolic tangent function (tanh).

FastICA is an efficient fixed point based algorithm. For a given sample of the prewhitened vectors $z$, an iteration is defined as follows:

1. Take a random initial vector $w_0$ of norm 1. Let $k = 1$.
2. Let $w_k = E\{z(w'_{k-1}z)^3\} - 3w_{k-1}$. The expectation can be calculated using a sample of $z$ vectors.
3. Divide $w_k$ by its norm, i.e. $w_k = w_k / \|w_k\|$.
4. If $\|w_k - w_{k-1}\|$ is not small enough, let $k = k + 1$ and go back to step 2. Otherwise, output the vector $w_k$.

The final vector $w_k$ given by the algorithm separates one of the underlying factors in the sense that $w'_k z(t)$, $t = 1, 2, \ldots$ equals one of the factors. To separate $m$ independent factors, we run this algorithm $m$ times. To ensure that we separate each time a different factor, we use the deflation algorithm that adds a simple projection inside the loop. Recalling that the rows of the separating matrix $W$ are orthonormal because of the prewhitening (sphering), we can separate the factors one-by-one by projecting the current estimate $w_k$ on the space orthogonal to the rows of the separating matrix $W$ previously found.

## 3   Singular Spectrum Analysis, Its Multivariate Extension, and the Recurrent SSA Forecasting Procedure

The SSA algorithm has two complementary main stages: *decomposition* and *reconstruction*.

The *decomposition stage* is carried out in two steps:

**(D1) Embedding**. This step consists of mapping the original one-dimensional time series $\{x_1, x_2, \ldots, x_N\}$ into a sequence of $K = N - M + 1$ lagged vectors of dimension $M$ (where $M$ is called *window length*): $X_i = (x_i, \ldots, x_{i+M-1})'$, $i = 1, \ldots, K$, $1 < M < N$. These lagged vectors form the columns of the *trajectory matrix* $X$, where $X = [X_1, X_2, \ldots, X_K] = (x_{ij})_{i,j=1}^{M,K}$. Note that the trajectory matrix $X$ is actually a Hankel matrix, which means that all the elements along the $i + j - 1 = constant$ are equal.

**(D2) SVD**. This step is the singular value decomposition of the trajectory matrix $X$. Denote by $\lambda_1, \ldots, \lambda_M$ the eigenvalues of the lag-covarince matrix $R = X \cdot X'$, of size $M \times M$, taken in the decreasing order of magnitude ($\lambda_1 \geq \ldots \geq \lambda_M \geq 0$) and by $U_1, \ldots, U_M$ the orthonormal system of the eigenvectors of the matrix $X \cdot X'$ corresponding to these eigenvalues. Let $d = \max\{i \mid \lambda_i > 0\} = \text{rank}(X)$. If we denote $V_i = X' U_i / \sqrt{\lambda_i}$ ($i = 1, \ldots, d$), then the SVD of the trajectory matrix $X$ can be written as $X = X_1 + \ldots + X_d$, where $X_i = \sqrt{\lambda_i} U_i \cdot V'_i$. The matrices $X_i$ are elementary matrices (have rank one). According to the SVD terminology, the collection $(\sqrt{\lambda_i}, U_i, V_i)$, consisting of the singular values $\sqrt{\lambda_i}$ and the left and right singular vectors $U_i$ and $V_i$, respectively, is called the $i$th eigentriple of the matrix $X$.

The *reconstruction stage* is also carried out in two steps:

**(R1) Groupping**. The aim of grouping SVD components (eigentriples) is to obtain the representation of the trajectory matrix of initial time series as a sum of trajectory matrices, which uniquely determine the expansion of the time series into a sum of additive components. Actually, the *grouping* step consists of partitioning the set of indices $\{1, \ldots, d\}$ into $m$ disjoint subsets $I_1, \ldots, I_m$, and summing the matrices $X_i = \sqrt{\lambda_i}\, U_i \cdot V_i'$ within each group: $X = X_{I_1} + \ldots + X_{I_m}$, where $X_{I_k} = \sum_{i \in I_k} X_i$ and $k = 1, \ldots, m$. The grouping with $I_k = \{i\}$ is called elementary.

Hankelization can be made by means of the convolution operator.

**(R2) Diagonal averaging.** The last step transforms each matrix $X_{I_k}$, corresponding to each group in the decomposition (3), into a new time series of length $N$, by diagonal averaging. It consists of averaging over the diagonals $i + j - 1 = \mathrm{const.}$ $(i = 1, \ldots, M, j = 1, \ldots, K)$ of the matrices $X_{I_k}$. A typical SSA decomposition of a time series is the decomposition into slowly-varying trend, seasonal and residual components or the decomposition into some pattern, regular oscillations and noise.

MSSA is a natural extension of the standard SSA to the case of multivariate time series. The (joint) trajectory matrix of multi-channel time series consists of stacked trajectory matrices of the individual time series. It is block-Hankel rather than simply Hankel. Other stages of MSSA are identical to the stages of the univariate SSA. Note that the window also produces subseries of length $M$, but we apply this window to all time series, i.e., to the one-dimensional vectorization of the multivariate time series.

In recurrent SSA forecasting, diagonal averaging is first performed to obtain the reconstructed series and then the LRR is applied. The algorithm produces a new series which is expected to 'continue' the current series on the basis of the given SSA decomposition. It sequentially projects the incomplete embedding vectors (either original or from reconstructed series) onto the subspace spanned by the selected eigentriples of the decomposition to forecast new values of such vectors.

Denote by $h$ the desired length of the forecasted series (forecasting horizon), by $I$ the subset of eigentriples to be used in the forecast, by $U_i \in \Re^M$, $i \in I$, the corresponding eigenvectors, by $\underline{U}_i$ their first $M - 1$ coordinates, and by $\pi_i$ the last coordinate of $U_i$, with $v^2 = \sum_i \pi_i^2$. Define $R = (a_{M-1}, \ldots, a_1)'$ as $R = \frac{1}{1 - v^2} \sum_{i \in I} \pi_i \underline{U}_i$.

The recurrent forecasting algorithm can be formulated as follows.

1. Firstly, the time series $\{y_i\}_{i=1}^{N+h}$ is defined by

$$
y_i - \begin{cases} \tilde{x}_i & i = 1, \ldots, N \\ \sum_{j=1}^{M-1} a_j\, y_{i-j} & i = N+1, \ldots, N+h \end{cases} \tag{4}
$$

2. The numbers $y_{N+1}, \ldots, y_{N+h}$ form the $h$ terms of the recurrent forecast.

Thus, the recurrent forecasting is performed by the direct use of the LRR with coefficients $\{a_i; i = 1, \ldots, M-1\}$.

For multichannel SSA the column forecast is obtained by applying the LRR to each series separately. Assume the signal is related to the leading $r$ eigentriples $(\lambda_i, U_i, V_i)$. Denote by $Y$ the matrix consisting of the last $M-1$ values of the reconstructed signals, by $U_i^\nabla$ the vectors of the first $M-1$ coordinates of the eigenvectors $U_i$, by $\pi_i$ the last coordinates of the eigenvectors, with $v^2 = \sum_i \pi_i^2$. Denote by $R_N$ the vector of forecasted signal values for each channel of the multivariate time series. If $v < 1$, then the forecast exists and can be calculated by the formula

$$R_N = Y\mathcal{R}_M, \quad where \quad \mathcal{R}_M = \frac{1}{1-v^2} \sum_{i=1}^{r} \pi_i U_i^\nabla \in \mathfrak{R}^{M-1}. \tag{5}$$

**Fig. 1** Triangular-shaped fuzzy monthly exchange rates (minimum, average, maximum values)

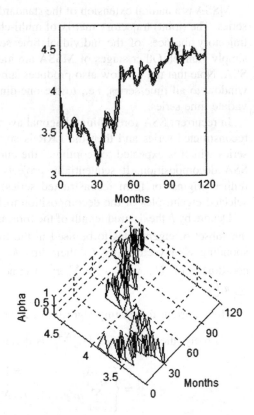

# 4 Applying MSSA to Forecasting Triangular-Shaped Fuzzy Monthly Exchange Rates

In what follows, MSSA will be used for smoothing and forecasting of a fuzzy time series. The later consists of minimum, average and maximum monthly exchange rates (Romanian New Leu to Euro) calculated from daily registrations over a period of 120 months. The fuzzy time series is shown in Fig. 1. The corresponding empirical fuzzy cumulative distribution function is represented in the left part of Fig. 2. The right part of Fig. 2 depicts the fraction and cumulated variance explained by the singular values of the trajectory matrix. An incomplete MSSA reconstruction of the fuzzy-valued time series, using the only 3 leading eigentriples, is shown in Fig. 3. With each new eigentriple added to produce a more complete reconstruction, the incorporated amount of fluctuations (interpreted as pure noise

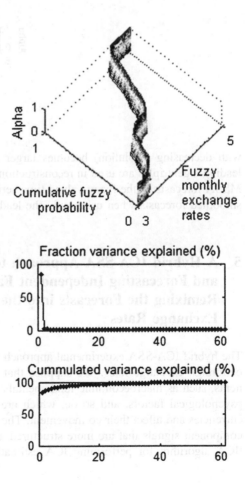

**Fig. 2** Empirical fuzzy cumulative distribution function (left); Singular values (right)

**Fig. 3** Incomplete MSSA
reconstruction of the
fuzzy-valued time series,
using 3 eigentriples

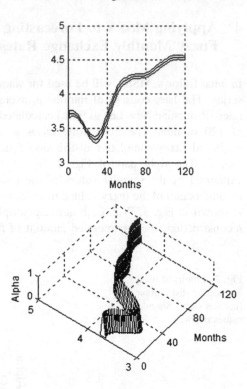

with decreasing resolution) becomes larger and larger (see Fig. 4, where the 5 leading eigentriples are used in reconstruction). Finally, 12 months ahead recurrent MSSA forecasts of the fuzzy-valued time series are computed. Figure 5 shows the smoothed forecasts when using only the leading 5 eigentriples.

# 5 A Hybrid ICA-SSA Approach to Separating and Forecasting Independent Factors and then Remixing the Forecasts into the Observable Foreign Exchange Rates

The hybrid ICA-SSA experimental approach starts by considering 6 parallel series of foreign exchange rates. We suppose that there are some independent components, such as macroeconomic fundamentals, extraneous influences, interest rates, psychological factors, and so on, which are closely tied to the evolution of the currencies and affect their co-movement. The ICA transformation tends to produce component signals that are more structured and regular. We actually use the FastICA algorithm for performing ICA estimations. Figure 6 shows the 6 original

**Fig. 4** Incomplete MSSA
reconstruction of the
fuzzy-valued time series,
using 5 eigentriples

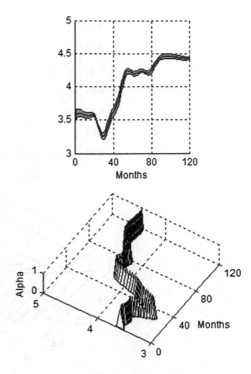

foreign exchange series (mixtures, before preprocessing). As we easily can see, the time series are clearly non-stationary and present some trends. Figure 7 (left) shows the 6 series, after centering and whitening. Clearly, after preprocessing, the time series become stationary. Whitening uses the eigenvalue decomposition (EVD) of the covariance matrix of centered data. The independent components and their 1 to 276 step-ahead forecasts are shown in Fig. 7 (right). By contrast with the original time series, which display similar shapes, due to common influences affecting their co-movement, the independent components have very distinctive shapes, corresponding to mutually different factors.

The goal of our hybrid ICA-SSA approach is to predict financial time series observed as some mixture signals by first going to the ICA space (spanned by the independent components), doing the prediction there, and then transforming back to the observable time series. The prediction for the vector of independent components can be done simultaneously using the multivariate extension of SSA, namely MSSA. The in- sample and out-of-sample estimates or forecasts of the mixtures can then be reconstructed starting from the independent component estimates or forecasts $\hat{s}$ and using the mixing matrix $A$, i.e., $\hat{z} = A\,\hat{s}$. Applying the mixing matrix $A$ to the *h-step-ahead* forecasts of fundamental factors $\hat{s}_{T+h|T}$ will give the corresponding forecasts of the mixtures $\hat{z}_{T+h|T} = \hat{A}\,\hat{s}_{T+h|T}$. Additionally, we can apply to $\hat{z}$ the de-whitening matrix $E\,D^{1/2}$ in order to reconstruct the centered data, i.e.,

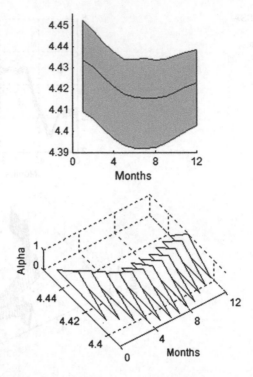

**Fig. 5** 12 months ahead recurrent MSSA forecasts of the fuzzy-valued time series, using the leading 5 eigentriples

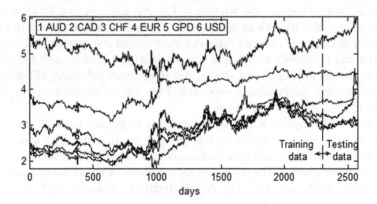

**Fig. 6** Original foreign exchange time series (mixtures)

$\hat{y} = E\,D^{1/2}\,\hat{z}$, and finally we can add the mean, i.e. $\hat{x} = \hat{y} + m_x$. Our experiments show that the in-sample reconstructions when using the estimates $\hat{s}$ is absolutely exact for all these time series, i.e., $\hat{z} = z$, $\hat{y} = y$ and finally $\hat{x} = x$.

**Fig. 7** Series after centering
and whitening (left);
Independent components and
their 1 to 276 step-ahead
recursive forecasts (right)

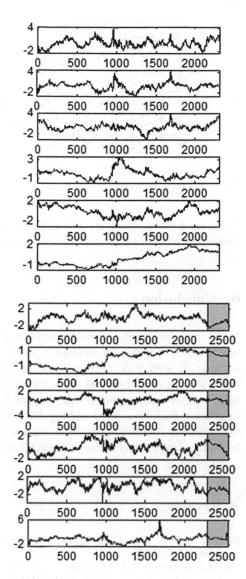

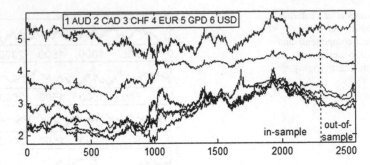

**Fig. 8** In-sample estimates and out-of-sample 276 step-ahead forecasts

Figure 8 shows the remixed in-sample estimates and the out-of-sample 1 to 276 step-ahead recursive forecasts of the observable foreign exchange rates.

## 6  Conclusion

In this paper we explored the extra-potential of MSSA for predicting multivariate objects, such as fuzzy-valued time series, as well as its effectiveness when using in conjunction with ICA. Such a hybrid approach allows interpreting financial time series as mixtures of several distinct underlying factors. MSSA clearly proved to have high forecasting performance, especially in long run prediction. In an attempt to produce more structured and regular representations and more accurate predictions, we also exploited the ICA's capability of revealing fundamental factors behind several parallel series of foreign exchange rates.

## References

Broomhead, D.S., King, G.P.: Extracting qualitative dynamics from experimental data. Phys. D **20**, 217–236 (1986)

Bell, A.J., Sejnowski, T.J.: An information-maximization approach to blind separation and blind deconvolution. Neural Comput. **7**, 1129–1159 (1995)

Goljadina, N., Nekrutkin,V., Zhigljavky, A.: Analysis of time series structure: SSA and related techniques. Chapman & Holl/CRC, New York, London (2001)

Vautard, R., Yiou, P., Ghil, M.: Singular-spectrum analysis: a toolkit for short, noisy chaotic signals. Phys. D **58**, 95–126 (1992)

Hyvärinen, A., Karhunen, J., Oja, E.: Independent component analysis. Wiley, New York (2001)

# Goodness of Aggregation Operators in a Diagnostic Fuzzy Model of Business Failure

Valeria Scherger, Hernán P. Vigier, Antonio Terceño-Gómez
and M. Glòria Barberà-Mariné

**Abstract** The aim of the following paper is proposed a mechanism of analysis useful to verify the capacity of the Vigier and Terceño (2008) diagnostic fuzzy model to predict diseases. The model is enriched by the inclusion of aggregation operators because this allows reducing the detected map of causes or diseases in strategic areas of continuous monitoring. And at the same time this causes can be disaggregated once some alert indicator is identified. The capacity of explanation and prediction of estimated diseases are measured through this mechanism; and also are detected the monitoring key areas that warning insolvency situations. In this approach are introduced aggregation operators of causes of business failure, and a goodness measure using approximate solutions. This index of goodness allows testing the degree of fit of the predictions of the model. Also, as an example, the empirical estimation and the verification of the improvement proposal to a set of small and medium- sized enterprises (SMEs) of the construction sector are presented.

**Keywords** Economic-financial · Diagnosis · Prediction · Symptoms and causes · Fuzzy relations

V. Scherger (✉) · H.P. Vigier
Economic Department, Universidad Nacional Del Sur – CEDETS UPSO,
12 de Octubre 1198 (7°), 8000 Bahía Blanca, Argentina
e-mail: valeria.scherger@uns.edu.ar

H.P. Vigier
e-mail: hvigier@uns.edu.ar

A. Terceño-Gómez · M.G. Barberà-Mariné
Department of Business Management, Universitat Rovira I Virgili,
Avinguda Universitat, 1, 43204 Reus, Spain
e-mail: antonio.terceno@urv.cat

M.G. Barberà-Mariné
e-mail: gloria.barbera@urv.cat

© Springer International Publishing Switzerland 2015
J. Gil-Aluja et al. (eds.), *Scientific Methods for the Treatment of Uncertainty
in Social Sciences*, Advances in Intelligent Systems and Computing 377,
DOI 10.1007/978-3-319-19704-3_12

# 1  Introduction

The Vigier and Terceño (2008) model of business diagnosis is based on fuzzy binary relations. This model advances over other prediction models, understanding the process of business failure through the relations between causes and symptoms, and the analyst's or expert's knowledge. This model has the capacity to predict insolvency situations and to diagnose problems and simulate (or complement) the analyst's task.[1] Also, it provides a different point of view from the traditional models (Altman 1968; Beaver 1966; Keasey and Watson 1987, etc.), incorporating elements of subjectivity and uncertainty formalized by fuzzy logic, and introduces improvements concerning most of the methodological problems discussed in the literature.[2]

Due to characteristics of the proposed problem, which uses a large number of qualitative variables or expert analysts' opinions, it is very difficult to find a comprehensive solution using a classical method. These considerations open the way to tools and methodology of fuzzy logic relations. As this model and others similar (Delcea et al. 2012; Behbood and Lu 2011; Scarlet et al. 2010; Arias-Aranda et al. 2010; Behbood et al. 2010; Delcea and Scarlat 2009; Mărăcine and Delcea 2009; Thapar et al. 2009; Gil Aluja 1990; Gil Lafuente 1996) don't have an evaluation system of diagnosis and prediction capacity, nor a mechanism of formal treatment of causes. In this work is proposed a method to group causes through aggregate operators and a goodness index to evaluate the estimation results. Furthermore, despite the model's originality in the literature, it has not been empirically proven. So in this approach is also done an empirical application that overcomes many of its constraints, described in Sect. 3. This supposes the identification and selection of causes and symptoms and the adjustment of theoretical hypothesis to simulate the model. These contributions enrich the Vigier and Terceño model (2008) and can be applied to any model of fuzzy diagnosis.

# 2  The Diagnosis Model

The model presented by Vigier y Terceño (2008), is based on the estimation of an economic-financial knowledge matrix (R) that starts in the estimation of the symptoms-causes matrices. The construction of R is determined by a group of

---

[1]This model is chosen because it has a comprehensive analysis of causes and symptoms over the pioneers models of Aluja Gil (1990) and Gil Lafuente (1996), and the recent proposals of Mărăcine and Delcea (2009), Scarlat et al. (2010), among others. These recents approaches taken as reference Vigier and Terceño model, derive the matrix of causes and symptoms by difference with the performance matrix, and therefore introduce a greater degree of subjectivity and a turn on in favor of experts point of views.

[2]Revisions by Zmijewski (1984); Jones (1987); Balcaen and Ooghe (2006); Tascón and Castaño (2012), etc., can be consulted. They comment the main methodological problems of the traditional models and the new applied techniques.

symptoms $S = \{Si\}$, where $i = 1, 2,..., n$, of causes $C = \{Cj\}$, where $j = 1, 2,..., p.$, of periods $T = \{Tk\}$, where $k = 1, 2,..., t.$, and of firms $E = \{Eh\}$, where $h = 1, 2, ..., m$, in which is possible to identify symptoms and causes. So, $R = Q^t \alpha P$, being, $Q^t = [q_{hi}]^t = [q_{ih}]$: the transposed membership matrix of the firms' symptoms; $P = [p_{hj}]$: the membership matrix of the firms' causes; and, $\alpha$: the fuzzy relations operator. That is, $R = Q^t \alpha P = [q_{ih}] \alpha [p_{hj}] = [r_{ij}]$, where, $[r_{ij}] = 1$ if $q_{ih} \leq p_{hj}$; and, $[r_{ij}] = p_{hj}$ if $q_{ih} > p_{hj}$. Each $r_{ij}$ shows the level of incidence between the symptom $S_i$ ($q_{ih}$) and the cause $C_j$ ($p_{hj}$).

The set of symptoms (S) is built from ratios taken from previous papers on solvency prediction. Once the cardinal matrix of symptoms has been obtained (S), the membership matrix of each symptom (Q) is determined by the cumulative relative frequency.

With regard to the causes, the model proposes the construction of a group of subjectively measured causes; and a group of objectively measured causes. The membership matrix of objective causes is obtained by the same mechanism used for building the membership matrix of symptoms. In the case of the subjective causes, the expert's opinion is formalized by linguistic labels in the interval [0,1] that reflect the incidence of each cause for the firm performance. The model logics indicate that the higher values correspond with causes that have a greater influence (Zimmermann 1987). The incidence level of each cause is determined by the cumulative frequencies by the labels in each scale and it is expressed in the causes' membership matrix (P).

To perform a study as homogeneous as possible, the set of selected companies has to be from a region and a specific productive sector (E) and also be composed of healthy and unhealthy companies to detect differences in indicators of both groups.

Also, to overcome potential inconsistency problems, Vigier and Terceño (2012) provide the use of a filtering method based on the decomposition and operation of each of the $r_{ij}$. The inconsistency occurs when there are high intensity levels for many companies, and low ones for a few, therefore the alpha indicator may incorrectly select the lowest intensity. The methodology consists of the removal of those companies that can cause inconsistent levels of incidence.[3] Furthermore, the model develops the distinctive features in the aggregation of matrices and the verification of trends that may distort the results of the model. These techniques allow obtaining a matrix with temporary validity and capacity to predict ($\Re$).

The matrix $\Re$ is used to make predictions (Terceño and Vigier 2011), that is, it is able to detect possible diseases in firms through the operation max- min between the membership matrix of symptoms (Q) and the aggregate matrix of financial-economic knowledge ($\Re$), $P' = Q \alpha \Re = [p'_{hj}]$; being $p'_{hj} = \max (\min (q_{hi}, r_{ij}))$, where $P'$ expresses the estimate causes (or diseases) that generate symptoms in the firms. Thus, it can be determined the degree of occurrence of different diseases (for similar firms in size and sector), and by monitoring the firm's causes, measures

---

[3]According to Vigier y Terceño (2012), this methodology is consistent with Sánchez (1982) contributions regarding the general conditions imposed on the solution of fuzzy binary equations.

can be taken to correct the situation. The empirical estimation and the analysis of diseases are available at Scherger et al. (2014).

# 3 Limitations of the Model

While the Vigier and Terceño model (2008) advances over others regarding the possibility to diagnose and predict causes of failure, some limitations can be pointed out because is a theoretical model:

1. Lack of application. Vigier and Terceño is a theoretical model that not defines specifically the set of causes and symptoms involved in the diagnosis. Added to its simulation requires a large amount of firms' performance information, till the moment it has not been empirically tested, although is demonstrated the internal consistency of the model.
2. Verification of prediction capacity. The model does not have any mechanism to test their capacity to diagnosis and prediction. It is not scheduled any guide for evaluate the degree of fit of the predictions to the responses provided by the experts.
3. Selection of symptoms. Regarding the symptoms, the model refers that can be financial economic ratios that arise from the financial statements of companies, without saying which ones should be used.
4. Detection of causes. The model only mentions the existence of qualitative and quantitative causes from the internal and the external environment of the firm and identifies the difference and relationship between symptoms (or effects) and diseases (or causes). The causes are not explicitly defined in a list or vademécum that should be considered by the analyst in the diagnosis. In this sense in Terceño et al. (2009, 2014) is proposed the integration of the fuzzy logic with the Balanced Scorecard to detect the causes of failure.

Therefore, the model fails to verify its capacity to simulate the expert's task, and judge the economic reasonableness of the relations between the variables. This is only possible with the specific definition of the set of causes and symptoms that is presented in Terceño et al. (2012). The exposed limitations have motivated this global proposal of analysis of causes through aggregation operators to facilitate the expert task in business diagnosis. Also is introduced a formal mechanism to test and verify the model's capacity of diagnosis and prediction.

# 4 Grouping the Causes

Given P′, estimated by the inverse operation between Q and $\mathfrak{R}$; that shows the multiple causes (or diseases) that can undergo a firm, a reduction mechanism of the map of causes through aggregation operators is introduced.

## 4.1 The Aggregation Operators

The aggregation operators on fuzzy subsets produce a simple and representative fuzzy subset (Klir and Yuan 1995). Formally any aggregation operator over n fuzzy subsets (n > 2), is defined by the function h: $[0,1]^n \rightarrow [0,1]$. When h is applied to fuzzy sets $A_1$, $A_2$, $A_3$, ..., $A_n$, defined in X, the function h operates on the degrees of membership of these sets, producing an aggregate fuzzy subset A. Then, $A(x) = h(A_1(x), A_2(x), \ldots\ldots\ldots, A_n(x)) \forall x \varepsilon X$ and h must satisfy the following axiomatic requirements to be considered an aggregation operator:

- **Axiom 1.** Boundary condition: h $(0, 0, 0, \ldots, 0) = 0$ y h$(1, 1, 1, \ldots, 1) = 1$
- **Axiom 2.** h is monotonically increasing in all its arguments. That is, given $(a_1, a_2, a_3, .., a_n)$ and $(b_1, b_2, b_3, \ldots, b_n)$, such that, $a_i$, $b_i$ $\varepsilon$ [0,1] if $a_i \leq b_i$, for all i = 1, 2,..., n, then, h $(a_1, a_2, a_3, .., a_n) \leq h(b_1, b_2, b_3, \ldots, b_n)$
- **Axiom 3.** h is a continuous function
  Also, usually, h must satisfy two additional axiomatic requirements.
- **Axiom 4.** h is a symmetric function in its arguments; that is h $(a_1, a_2, a_3, .., a_n) = h(a_{p(1)}, a_{p(2)}, a_{p(3)}, .., a_{p(n)})$. This means that the result of the function h in any permutation of the $a_i$ is unchanged.
- **Axiom 5.** h is an idempotent function, that is, h $(a, a, a, .., a) = a \forall a \varepsilon [0, 1]$. So aggregation of equal fuzzy subsets must provide the same fuzzy subset.

Given any aggregation operator h which satisfies the axioms 2–5, satisfies the inequality: $Min(a_1, a_2, \ldots, a_n) \leq h(a_1, a_2, \ldots, a_n) \leq Max(a_1, a_2, \ldots, a_n)$; $\forall n -$ uplas $\varepsilon [0, 1]^n$.

All operations of aggregation between the standard fuzzy intersection and the standard fuzzy union are idempotent. Furthermore, the h functions that satisfy the inequality condition are the only aggregate operations that are idempotent and are represented by the averaging aggregation operators.

**Averaging operator**: covers the full range between minimum and maximum operations and consists in the general mean of all the arguments. That is $h_\alpha(a_1, a_2, \ldots, a_n) = ((a_1^\alpha + a_2^\alpha + \ldots + a_n^\alpha)/n)^{1/\alpha}$, where $\alpha \varepsilon P'(\alpha \neq 0)$ y $a_i \neq 0, \forall i$. When $\alpha$ approaches to 0, $h_\alpha(a_1, a_2, \ldots, a_n)$ tends to the geometric mean; when $\alpha$ approaches to $(-1)$, $h_\alpha(a_1, a_2, \ldots, a_n)$ tends to the harmonic mean; when $\alpha$ approaches to 1, $h_\alpha(a_1, a_2, \ldots, a_n)$ tends to the arithmetic mean. The $h_\alpha$ function represents a parameterized class of continuous, symmetric and idempotent aggregation operations.

**OWA Operator** (Ordered Weighted Averaging). It is another class of aggregation operators that covers the full range of operations between minimum and maximum, where $w = (w_1, w_2, \ldots, w_n)$, $w_j \varepsilon [0, 1]$ for all $j = 1, 2, \ldots, n$.; and also, $\sum_{j=1}^{n} w_j = 1$; then, an OWA operation associated with w, can be associated to the function: $h_w(a_1, a_2, \ldots, a_n) = w_1 b_1 + w_2 b_2 + \ldots + w_n b_n$, where vector $(b_1, b_2, \ldots, b_n)$ is vector $(a_1, a_2, \ldots, a_n)$ in which the elements are sorted from highest to lowest, $b_i \geq b_j$ if i < j. According to Klir and Yuan (1995); Klir and Folger

(1992); Dubois and Prade (1985); Yager (1983), etc., the OWA operators ($h_w$) satisfied the axioms 1–5 and consequently the inequality functions. Yager (2002) discusses how to characterize the OWA operators and incorporates a new attribute called divergence and introduces the HOWA (Heavy OWA). These HOWA operators differ from the ordinary OWA operators by relaxing the constraints on the associated weighting vector. Also, Yager (2004) presented a generalization of OWA operators called GOWA (Generalized OWA) that add an additional parameter controlling the power to which the argument values are raised. Then Yager (2009) and Yager and Ajajlan (2014), among others, criteria are proposed to prioritizing and characterize OWA operators.

## 4.2 Motion for Monitoring and Grouping of Causes

In this section, is proposed the use of the OWA operators, that satisfy the inequality $\text{Min}(a_1, a_2, \ldots, a_n) \leq h(a_1, a_2, \ldots, a_n) \leq \text{Max}(a_1, a_2, \ldots, a_n); \forall$ n-tuplas $\varepsilon[0, 1]^n$; considering $n = 1$.

Given the estimated matrix $\Re(r_{ij})$; considering key areas $A = \{1, 2,.., a\}$ defined according to the Balanced Scorecard and the estimated causes, are built three $\Re$ matrices for minimum, maximum and mean incidence values of causes ($\Re_{ik}^{Min}; \Re_{ik}^{Max}; \Re_{ik}^{Mean}$). The causes of the aggregate matrix $\Re$ are grouped for these three possible levels of incidence to synthesize the factors that generate diseases.

This grouping methodology of causes from values of maximum, minimum and average $r_{ij}$ is useful to detect the factors of greatest incidence and complement the task of the expert in the monitoring. Given $\Re = \{\Re_{ij}\}$, is obtained the $\Re^{Mean}$ matrix applying the arithmetic mean to the causes within the monitoring area ($\Re_{ia}^{Mean} = (1/h) \sum r_{ij}$). While the matrices $\Re$ with maximum and minimum membership values ($\Re_{ia}^{Max}$ y $\Re_{ia}^{Min}$) are calculated through the maximum ($\Re_{ia}^{Max} = \text{Max}(r_{ij})$) and the minimum ($\Re_{ia}^{Min} = \text{Min}(r_{ij})$) selection of $r_{ij}$ incidence levels within each group of causes. That is,

- $r_{ia}^{Mean} = (1/h) \sum (r_{i1} + r_{i2} + \ldots + r_{ij})$;
- $r_{ia}^{Max} = \text{Max}(r_{i1}; r_{i2}; \ldots; r_{ij})$;
- $r_{ia}^{Min} = \text{Min}(r_{i1}; r_{i2}; \ldots; r_{ij})$.

This grouping allows prediction $P' = Q \propto \Re$; that is, $p'_{hj} = \max(\min(q_{hi}, r_{ij}))$ for the three possible levels of incidence of causes within each key area. And thus, the membership matrices of causes (or diseases) are estimated in the three levels of incidence (minimum ($P'^{Min}$), maximum ($P'^{Max}$) and mean ($P'^{Mean}$)).

- $p_{ia}'^{Min} = \Lambda[(q_{ih}\alpha\, r^{min}_{h1}), (q_{ih}\,\alpha\, r^{min}_{h2}), \ldots\ldots, (q_{ih}\,\alpha\, r^{min}_{ha})$,
- $p_{ia}'^{Max} = \Lambda[(q_{ih}\alpha\, r^{max}_{h1}), (q_{ih}\,\alpha\, r^{max}_{h2}), \ldots\ldots, (q_{ih}\,\alpha\, r^{max}_{ha})$,
- $p_{ia}'^{Mean} = \Lambda[(q_{ih}\alpha\, r^{Mean}_{h1}), (q_{ih}\,\alpha\, r^{Mean}_{h2}), \ldots\ldots, (q_{ih}\,\alpha\, r^{Mean}_{ha})$,

Thus, the multiple causes detected in diagnosis or in prediction of firms' situation are grouped into key areas that facilitate the analyst's task by reducing the necessary information for the analysis. Once is detected a warning indicator in some area of the firm, is possible to disaggregate this key area into each of the causes or factors that generate problems to evaluate and correct the situation. This option of monitoring through key areas (consistent with the disaggregated estimation of the model) allows a continuous and comprehensive tracking of the business areas.

## 5  Goodness Index

As discussed in Sect. 1, Vigier and Terceño model's (2008), like others fuzzy logic, doesn't have a mechanism to verify its prediction capacity. In this sense, is introduced an index of approximate solutions to check if the estimated causes by the model represent the true situation of the firm. For this, the equality index of Brignole et al. (2001) that proposes the comparison of two fuzzy sets is adapted. That is, the comparison between the original set of causes (P*) grouped into minimum $\left(P_{ha}^{*Min} = Min(p_{ha})\right)$, maximum $\left(P_{ha}^{*Max} = Max(p_{ha})\right)$ and mean $\left(P_{ha}^{*Mean} = (1/h) \sum (p_{ha})\right)$ incidence's levels and the set of estimated causes (P').

$$[P^* - P'] = 1 - 1/n \sum_{x \in X} |\mu_p(x) - \mu_p(x)| \tag{1}$$

Therefore,

$$\left[(P_{ha}^*) = (P_{h'a}')\right]^{Min} = [1 - 1/n \sum (|p_{h1}^* - p_{h1}'| + |p_{h2}^* - p_{h2}'| + \cdots\cdots + |p_{ha}^* - p_{ha}'|)]$$

$$\left[(P_{ha}^*) = (P_{h'a}')\right]^{Max} = [1 - 1/n \sum (|p_{h1}^* - p_{h1}'| + |p_{h2}^* - p_{h2}'| + \cdots\cdots + |p_{ha}^* - p_{ha}'|)]$$

$$\left[(P_{ha}^*) = (P_{h'a}')\right]^{Mean} = [1 - 1/n \sum (|p_{h1}^* - p_{h1}'| + |p_{h2}^* - p_{h2}'| + \cdots\cdots + |p_{ha}^* - p_{ha}'|)]$$

where, $P_{ha}^{*Min}$ selects the minimum degree of incidence within the group of causes for each company, $P_{ha}^{*Max}$ chooses the minimum degree of incidence within the group of causes for each company, and $P_{ha}^{*Mean}$ shows the average of the causes within the group or key area of monitoring.

This test is useful to identify what is the best mechanism for grouping causes (by maximum, minimum or average incidence values) and to estimate the degree of fit of the predictions to the diseases present in the companies. So, is helpful to prove the model's capacity for predict insolvency situations.

# 6   Estimation of the Model

This section shows the application of the business model of Vigier and Terceño (2008) to a sample of small and medium-size companies (SMEs). It is selected the construction sector made up of two subsectors: construction and sale of building materials. On a base of 98 SMEs registered in this sector in the municipalities of Bahía Blanca and Punta Alta (Argentina), are selected 15 firms -representing approximately 15 % of the activity in these two cities.[4] This is a very dynamic activity and has a great variety of (healthy and unhealthy) companies that meet the availability requirements of data necessary to estimate the model. According to the latest estimates, this sector accounts for about 10 % of formal employment, and 1 % of Gross Domestic Product; besides generates multiplier effects that impact on local economy (EAR, 112 and 137).

The next steps are followed to apply the model:

1. Selection of the sector and companies to analyze (E = {$E_h$}, where h = 1, 2, ...,15)
2. Data collection for the three periods used in the estimation (T = {$T_k$}, where k = 2008; 2009; 2010) through the design of a standardized questionnaire to detect potential causes of diseases in companies, by using linguistic labels. Data collection was done through interviews with 15 experts of the SMEs, who maintained a long relationship with the analyzed company. In Argentine SMEs, this role is fulfilled accounting advisors, managers and business owners.
3. Systematization and information analysis of the interviews with experts and the financial statements.
4. Selection and estimation of economic- financial ratios (S = {$S_i$}, i = 1, 2, ..., 41). The ratios are selected to reflect aspects of profitability, productivity, liquidity, leverage, solvency, financial structure, debt coverage, economic structure, activity, turnover, efficiency and self-financing according to the classification proposed by many authors and the frequency of use in prediction models (see Terceño et al. 2012).
5. Construction of the cardinal matrix of symptoms for the 15 companies and 41 selected ratios ($S_{ih}$ = 41 × 15).
6. Estimation of the membership matrix of symptoms ($Q_{ih}$ = 41 × 15) according to the methodology presented in Sect. 2.
7. Detection of disease-generating causes following the proposed fuzzy and BSC methodology presented in Terceño et al. (2009, 2012) (C = {$C_j$}, where j = 1,2, ...,72)
8. Construction of the membership matrix of causes objectively and subjectively measurable ($P_{hj}$ = 15 × 72), following the methodology presented in Sect. 2.

---

[4]If only the firms constituted in formal societies are considered, as they are the only ones required to submit economic- financial statements, the percentage of representation is around 30 %.

9. Estimation of matrix R of economic-financial knowledge for the 3 periods $\left(R_{ij}^k = 41 \times 72; k = 1, 2, 3\right)$. This is done through the operation between the transposed membership matrix of symptoms and causes membership matrix that meets the lower ratio $(R_{ij} = (Q_{ih}^{-1} \alpha P_{hj}))$ and applying the $\alpha$ defined operator.

10. According to the recommendations presented by Vigier and Terceño (2012), once the incidence coefficients $r_{ij}$ are obtained, the filtering method for the treatment of inconsistencies is applied. Two levels of filter $(\phi^* = 0.75 \text{ y } \phi^* = 0.50)$ are applied to remove firms. In the first case firms with lower incidence level are eliminated until $\phi < 0.75$; then when observing very little variability of each cause in the firms, it is applied a higher filtering factor $(\phi^* = 0.50)$, which reduces the error and discards a higher percentage of inconsistent responses. Applying filtering method does not show elimination firm patterns according to the activity developed within the sector or the size of the firm but simply removes cases considered as "anomalous" for the model.

11. Also to evaluate the responses, the coefficient of variation $(C_V)$, not included in the model, is applied to determine the variability of the causes. The higher the value, the greater the heterogeneity of responses, relatively high index values being observed. $Cv = \sqrt{1/N - 1 \sum_{i=1}^{n} (x_i - \bar{x})^2 / 1/n \sum_{i=1}^{n} x_i}$.

12. Estimation of the aggregate matrix $\mathfrak{R}(\mathfrak{R}_{ij} = 41 \times 72)$ used to forecasting. This supposes repeating steps (3) to (11) for the three periods ($T_1$, $T_2$ y $T_3$). The calculation of $\mathfrak{R}$ involves aggregating the matrices $R_k = [r_{ij}]_k = \left(R_{ij}^1, R_{ij}^2, R_{ij}^3\right)$ and the correction for temporal trends that distort the validity of the results. As mentioned in Vigier and Terceño (2008), if an upward trend is verified in time of $r_{ij}$, the use of an "average" aggregation procedure would underestimate the true relationship. However, if the trend is decreasing, the average would overestimate the relationship. For this situation, the behavior of each $r_{ij}$ is evaluated in order to determine the aggregation process depending on the trend experienced by each component.

- If $\sum \left| r_{ij}^t - r_{ij}^{t-1} \right| = 0$; $r_{ij}^t$ = aggregate $r_{ij}$

- If $\sum \left| r_{ij}^t - r_{ij}^{t-1} \right| \neq 0$; $r_{ij}$ is determined by the indicator $\xi$, which varies between $-1$ and $1$.

$$\xi = \sum_{k=2}^{t} \left( [r_{ij}]_k - [r_{ij}]_{k-1} \right) / \sum_{k=2}^{t} \left| [r_{ij}]_k - [r_{ij}]_{k-1} \right| \quad (2)$$

If $\xi = 1$, $r_{ij} = \text{Max} \left( \text{Min } [r_{ij}]_k \right)$; there is an increasing trend.
If $\xi = -1$, $r_{ij} = \text{Min} \left( \text{Max } [r_{ij}]_k \right)$; there is a decreasing trend.
If $-1 < \xi < 1$, $r_{ij} = 1/t \sum [r_{ij}]_k$; there is no trend.

The aggregation and the correction of temporal trend is carried out for obtain $\mathfrak{R}$ aggregate matrix. An increasing trend is observed in 349 $r_{ij}$, in 337 $r_{ij}$ the trend is decreasing and in the remaining 2,266 $r_{ij}$ no such trend is observed, performing the aggregation through the arithmetic mean. Thus, the estimated $r_{ij}$ coefficients are consistent and significant, and explain the true relationship between causes and symptoms according to the model.

# 7  Empirical Verification of the Grouping

It is done by applying the following steps:

1. Proposed minimum, maximum and mean incidence aggregation operators are applied to P for obtaining the grouped matrices that reflect this three categories $\left(P^{*Min}; P^{*Max}; P^{*Mean}\right)$. Appendix A shows the three matrices (P*).

   For example, taken as reference the causes' membership matrix of firm 1, the causes are grouped ($p_1$, $p_2$,...., $p_{10}$,..., $p_{15}$) in a single cause ($p^*_1$) that reflects the problems related to entrepreneurial learning. That is,

   • $P_{11}^{*Mean} = (1/15) \sum (0.25 + 1.00 + 0.50 + 0.20 + 0.20 + 0.50 + 0.86 + 0.33 + 0.29 + 0.43 + 0.29 + 0.36 + 1.00 + 0.20 + 0.43) = 0.46$

   • $P_{11}^{*Max} = Max(0.25; 1.00; 0.50; 0.20; 0.20; 0.50; 0.86; 0.33; 0.29; 0.43; 0.29; 0.36; 1.00; 0.20; 0.43) = 1.00$

   • $P_{11}^{*Min} = Min(0.25; 1.00; 0.50; 0.20; 0.20; 0.50; 0.86; 0.33; 0.29; 0.43; 0.29; 0.36; 1.00; 0.20; 0.43) = 0.20$

2. Aggregation operators are applied to $\mathfrak{R}$ to estimate companies diseases grouped into key areas $\left(\mathfrak{R}_{ij}^{Min} = Min\left(r_{ij}\right)\right); \mathfrak{R}_{ij}^{Max} = Max\left(r_{ij}\right); \mathfrak{R}_{ij}^{Mean} = (1/n) \sum r_{ij}$. Through $P' = Q \alpha \mathfrak{R}$; being; $p'_{hj} = max\left(min\left(q_{hi}, r_{ij}\right)\right)$ are estimated three new matrices of causes in minimum $\left(P_{ha}^{'Min}\right)$, maximum $\left(P_{ha}^{'Max}\right)$ and mean $\left(P_{ha}^{'Mean}\right)$ membership values (see Appendix B).

   For example, the membership level of the cause $P_{11}^{'Max}$ for the aggregate is calculated through the operation of the first row of Q matrix with the first column of $\mathfrak{R}^{Max}$ matrix, that is $p'_{11} = Q(1 \times 41) \alpha \mathfrak{R}^{Max}(41 \times 1) = P'_{11}(1 \times 1)$.

   • $P_{11}^{'Max} = \Lambda[(0.60 \, \alpha \, 0.50), (0.67 \alpha 0.50), \ldots\ldots, (0.80 \, \alpha \, 0.60), \ldots\ldots, (0.80 \, \alpha \, 0.57)]$
   • $P_{11}^{'Max} = Max[(0.50), (0.50), \ldots, (0.60), \ldots\ldots\ldots, (0.57)] = 0.60$

In the same way are estimated $p_{11}^{'Mean}$ and $p_{11}^{'Min}$ operating of the first row of Q matrix with the first column of $\mathfrak{R}^{Max}$.

- $p_{11}^{'Mean} = \Lambda[(0.60\,\alpha\,0.29), (0.67\,\alpha\,0.28), \ldots\ldots, (0.93\,\alpha\,0.28), \ldots., (0.80\,\alpha\,0.29)]$
- $p_{11}^{'Mean} = Max[(0.29), (0.28), \ldots., (0.28), \ldots\ldots, (0.29)] = 0.34$

And,

- $p_{11}^{'Min} = \Lambda[(0.60\,\alpha\,0.14), (0.67\,\alpha\,0.14), \ldots\ldots, (0.13\,\alpha\,0.14), \ldots., (0.80\,\alpha\,0.14)]$
- $p_{11}^{'Mean} = Max[(0.14), (0.14), \ldots., (0.14), \ldots\ldots, (0.14)] = 0.14$

3. The goodness index of grouping causes is estimated through the comparison of the comparison of the two fuzzy subsets P* y P'. Thus, is verified the reduction of causes in monitoring key areas and it is evaluated its degree of fit to the answers given by the experts. For example, for firm 1:

- $[(P*) = (P')]^{Min} = 1 - 1/14\ \Sigma\ (|0.20 - 0.14| + |0.12 - 0.13| + |0.57 - 0.56| + |0.20 - 0.20| + |0.13 - 0.19| + |0.20 - 0.40| + |0.20 - 0.60| + |0.13 - 0.13| + |0.20 - 0.26| + |0.43 - 0.43| + |0.20 - 0.20| + |0.20 - 0.20| + |0.20 - 0.14| + |0.27 - 0.24|) = 0.94$
- $[(P*) = (P')]^{Max} = 1 - 1/14\ \Sigma\ (|1.00 - 0.60| + |0.57 - 0.57| + |0.83 - 0.61| + |0.67 - 0.67| + |1.00 - 0.93| + |1.00 - 0.62| + |1.00 - 0.80| + |1.00 - 1.00| + |0.37 - 0.31| + |0.83 - 0.73| + |0.50 - 0.51| + |1.00 - 0.93| + |1.00 - 0.67| + |0.86 - 0.80|) = 0.86$
- $[(P*) = (P')]^{Mean} = 1 - 1/14\ \Sigma\ (|0.46 - 0.34| + |0.27 - 0.35| + |0.71 - 0.57| + |0.46 - 0.47| + |0.43 - 0.47| + |0.70 - 0.51| + |0.60 - 0.70| + |0.47 - 0.47| + |0.29 - 0.26| + |0.61 - 0.58| + |0.29 - 0.37| + |0.72 - 0.61| + |0.64 - 0.35| + |0.55 - 0.49|) = 0.91$

This analysis concludes that the best fit is obtained with minimum rates, which reflect a higher goodness index, with a degree of fit of 93 % (see Table 1). The best fit through minimal incidence rates is consistent with less risk adverse theories in terms of warnings and analysing of results. Also, are validated the properties of the minimum t- norm over others decision rules.

The goodness index is applied to the results of the last period and the aggregate finding a similar degree of fit (93 %), therefore the incidence relationships have almost no changes in not long periods (3 years). Hence, the R matrix (compare with $\mathfrak{R}$) is valid for diagnostic and forecast in the medium term.

# 8 Conclusions

All prediction models that used fuzzy tools could be benefited by the contributions presented in this paper Those are a grouping method of causes through aggregate operators and a testing goodness index for the evaluation of prediction results. In this paper is proposed the enriched of the Vigier and Terceño model (2008) by introducing a mechanism of grouping causes through aggregation operators. This approach allows synthesize firms' diseases and facilitate the expert's task. Also, this analysis supposes an advantage in predicting disease since it involves the reduction of the general map of firms' diseases into key areas. These are detected through the Balanced Scorecard and are grouped according to the aggregation operators (minimum, maximum and mean). Thus, once a warning indicator is detected in some area of the firm, is possible to disaggregate this key area in the multiple factors that generate problems to evaluate and correct the situation. This methodology of grouping causes from values of maximum, minimum and average is useful to detect the factors of greatest incidence for a comprehensive and continuous monitoring of the company.

Furthermore of the grouping proposal is developed a goodness measure, through approximate solutions, to verify the grouping. This index tests the model's capacity to predict diseases and validate the properties of the minimum t- norm over others decision rules.

**Table 1** Degree of fit considering several periods $[(P^*) = (P')]$

| Firm | Min. ($P'^{Min}$) | Max. ($P'^{Max}$) | Mean ($P'^{Mean}$) |
|------|------|------|------|
| 1 | 0.94 | 0.86 | 0.91 |
| 2 | 0.98 | 0.82 | 0.93 |
| 3 | 0.97 | 0.86 | 0.94 |
| 4 | 0.93 | 0.89 | 0.93 |
| 5 | 0.95 | 0.89 | 0.92 |
| 6 | 0.92 | 0.84 | 0.88 |
| 7 | 0.91 | 0.80 | 0.88 |
| 8 | 0.92 | 0.82 | 0.87 |
| 9 | 0.96 | 0.85 | 0.88 |
| 10 | 0.91 | 0.82 | 0.87 |
| 11 | 0.88 | 0.80 | 0.90 |
| 12 | 0.91 | 0.85 | 0.87 |
| 13 | 0.90 | 0.86 | 0.91 |
| 14 | 0.93 | 0.89 | 0.89 |
| 15 | 0.95 | 0.90 | 0.92 |
| Mean | 0.93 | 0.85 | 0.90 |

Also an empirical estimation of Vigier and Terceño (2008) diagnostic model is presented. So its performance and capacity to diagnose and predict future situations is verified. Thus, the estimated $r_{ij}$ coefficients are consistent and significant, and explain the true relationship between symptoms and diseases (effects and causes) because the relationships between causes and effects are significant with a high degree of approximation. And therefore the $\Re$ matrix is useful to simulate or model the expert knowledge when do forecasting.

In summary, this paper proposes advance in the analysis of causes of business failure through synthesizing the causes in monitoring key areas, introducing a methodology to evaluate the goodness of the grouping of causes and finally applying and testing the improvement in the empirical estimation.

# Appendix A

## Grouped membership matrix of causes ($P*^{Max}$)

| Areas | Learning and growth | | | | Business process | | | Customers perspective | | | | Finance | | |
|---|---|---|---|---|---|---|---|---|---|---|---|---|---|---|
| | 1 | 2 | 3 | 4 | 5 | 6 | 7 | 8 | 9 | 10 | 11 | 12 | 13 | 14 |
| E1 | 1.00 | 0.57 | 0.83 | 0.67 | 1.00 | 1.00 | 1.00 | 1.00 | 0.37 | 0.83 | 0.50 | 1.00 | 1.00 | 0.86 |
| E2 | 0.83 | 0.57 | 0.75 | 0.33 | 1.00 | 0.71 | 0.60 | 1.00 | 0.31 | 0.57 | 0.33 | 1.00 | 0.36 | 0.67 |
| E3 | 0.73 | 0.53 | 0.79 | 0.83 | 0.73 | 0.71 | 0.80 | 1.00 | 0.40 | 0.71 | 0.75 | 1.00 | 0.83 | 0.71 |
| E4 | 0.83 | 0.29 | 0.67 | 0.83 | 0.87 | 0.71 | 0.80 | 0.80 | 0.42 | 0.57 | 0.70 | 1.00 | 0.83 | 1.00 |
| E5 | 0.71 | 0.57 | 0.83 | 0.50 | 0.73 | 0.64 | 0.80 | 0.79 | 0.37 | 0.73 | 0.63 | 1.00 | 0.67 | 0.72 |
| E6 | 0.80 | 0.57 | 0.60 | 0.20 | 0.67 | 1.00 | 0.80 | 0.71 | 0.48 | 0.90 | 0.81 | 0.80 | 0.86 | 0.71 |
| E7 | 0.80 | 0.57 | 0.60 | 0.20 | 0.63 | 1.00 | 0.80 | 0.71 | 0.48 | 0.90 | 0.81 | 0.80 | 0.86 | 0.80 |
| E8 | 0.80 | 0.57 | 0.60 | 0.20 | 0.63 | 1.00 | 1.00 | 0.71 | 0.48 | 0.90 | 0.81 | 0.80 | 0.86 | 0.80 |
| E9 | 1.00 | 0.63 | 0.80 | 0.67 | 1.00 | 0.71 | 1.00 | 1.00 | 0.40 | 1.00 | 0.75 | 0.80 | 1.00 | 1.00 |
| E10 | 1.00 | 0.61 | 0.72 | 0.20 | 0.80 | 0.71 | 0.80 | 0.63 | 1.00 | 0.73 | 0.75 | 1.00 | 0.67 | 1.00 |
| E11 | 1.00 | 0.57 | 1.50 | 0.83 | 0.93 | 0.79 | 0.80 | 0.80 | 0.47 | 0.86 | 0.58 | 0.81 | 0.83 | 1.00 |
| E12 | 1.00 | 0.63 | 1.00 | 0.83 | 0.93 | 1.00 | 0.80 | 0.80 | 0.60 | 0.59 | 0.75 | 0.88 | 0.65 | 1.00 |
| E13 | 1.00 | 0.63 | 0.80 | 1.00 | 0.80 | 0.71 | 0.80 | 1.00 | 0.43 | 0.83 | 0.75 | 1.00 | 0.83 | 0.69 |
| E14 | 1.00 | 0.63 | 0.75 | 1.00 | 0.87 | 0.60 | 0.80 | 0.80 | 0.40 | 0.83 | 0.75 | 0.88 | 0.83 | 1.00 |
| E15 | 0.8 | 0.57 | 0.80 | 0.67 | 0.80 | 0.80 | 0.80 | 0.80 | 0.48 | 0.83 | 0.61 | 1.00 | 0.65 | 1.00 |

## Grouped membership matrix of causes ($P*^{Min}$)

| Areas | Learning and growth | | | Business process | | | | Customers perspective | | | Finance | | | |
|---|---|---|---|---|---|---|---|---|---|---|---|---|---|---|
| | 1 | 2 | 3 | 4 | 5 | 6 | 7 | 8 | 9 | 10 | 11 | 12 | 13 | 14 |
| E1 | 0.20 | 0.12 | 0.57 | 0.20 | 0.13 | 0.20 | 0.20 | 0.13 | 0.20 | 0.43 | 0.20 | 0.20 | 0.20 | 0.27 |
| E2 | 0.14 | 0.13 | 0.57 | 0.17 | 0.07 | 0.40 | 0.60 | 0.13 | 0.20 | 0.39 | 0.20 | 0.20 | 0.14 | 0.27 |
| E3 | 0.14 | 0.13 | 0.60 | 0.17 | 0.07 | 0.53 | 0.60 | 0.13 | 0.26 | 0.29 | 0.20 | 0.22 | 0.14 | 0.17 |
| E4 | 0.14 | 0.13 | 0.14 | 0.20 | 0.13 | 0.40 | 0.40 | 0.25 | 0.40 | 0.43 | 0.20 | 0.20 | 0.14 | 0.27 |
| E5 | 0.14 | 0.13 | 0.71 | 0.33 | 0.25 | 0.53 | 0.60 | 0.13 | 0.20 | 0.57 | 0.20 | 0.20 | 0.14 | 0.20 |
| E6 | 0.14 | 0.13 | 0.56 | 0.17 | 0.25 | 0.47 | 0.80 | 0.20 | 0.40 | 0.43 | 0.13 | 0.11 | 0.55 | 0.27 |
| E7 | 0.14 | 0.13 | 0.56 | 0.17 | 0.25 | 0.47 | 0.80 | 0.20 | 0.40 | 0.43 | 0.13 | 0.11 | 0.55 | 0.27 |
| E8 | 0.14 | 0.13 | 0.56 | 0.17 | 0.25 | 0.47 | 0.80 | 0.20 | 0.40 | 0.43 | 0.13 | 0.11 | 0.55 | 0.27 |
| E9 | 0.14 | 0.14 | 0.43 | 0.17 | 0.13 | 0.40 | 0.80 | 0.13 | 0.37 | 0.43 | 0.20 | 0.20 | 0.14 | 0.27 |
| E10 | 0.14 | 0.38 | 0.57 | 0.17 | 0.25 | 0.20 | 0.60 | 0.25 | 0.42 | 0.57 | 0.38 | 0.33 | 0.14 | 0.27 |
| E11 | 0.14 | 0.24 | 0.14 | 0.20 | 0.14 | 0.20 | 0.20 | 0.38 | 0.40 | 0.29 | 0.20 | 0.33 | 0.14 | 0.24 |
| E12 | 0.20 | 0.57 | 0.71 | 0.20 | 0.14 | 0.40 | 0.60 | 0.13 | 0.52 | 0.57 | 0.20 | 0.20 | 0.14 | 0.27 |
| E13 | 0.17 | 0.53 | 0.29 | 0.20 | 0.14 | 0.40 | 0.60 | 0.29 | 0.40 | 0.57 | 0.20 | 0.40 | 0.14 | 0.27 |
| E14 | 0.20 | 0.53 | 0.57 | 0.40 | 0.25 | 0.40 | 0.60 | 0.13 | 0.32 | 0.43 | 0.20 | 0.20 | 0.14 | 0.27 |
| E15 | 0.20 | 0.25 | 0.57 | 0.20 | 0.25 | 0.40 | 0.60 | 0.13 | 0.40 | 0.57 | 0.20 | 0.33 | 0.14 | 0.27 |

## Grouped membership matrix of causes ($P*^{Mean}$)

| Areas | Learning and growth | | | | Business process | | | Customers perspective | | | | Finance | | |
|---|---|---|---|---|---|---|---|---|---|---|---|---|---|---|
| | 1 | 2 | 3 | 4 | 5 | 6 | 7 | 8 | 9 | 10 | 11 | 12 | 13 | 14 |
| E1 | 0.46 | 0.27 | 0.71 | 0.46 | 0.43 | 0.70 | 0.60 | 0.47 | 0.29 | 0.61 | 0.29 | 0.72 | 0.64 | 0.55 |
| E2 | 0.29 | 0.40 | 0.67 | 0.23 | 0.46 | 0.50 | 0.60 | 0.50 | 0.25 | 0.46 | 0.26 | 0.49 | 0.26 | 0.47 |
| E3 | 0.35 | 0.36 | 0.65 | 0.40 | 0.46 | 0.61 | 0.75 | 0.47 | 0.33 | 0.48 | 0.40 | 0.78 | 0.41 | 0.41 |
| E4 | 0.45 | 0.19 | 0.52 | 0.51 | 0.46 | 0.54 | 0.65 | 0.54 | 0.41 | 0.48 | 0.38 | 0.59 | 0.48 | 0.58 |
| E5 | 0.46 | 0.36 | 0.80 | 0.41 | 0.50 | 0.59 | 0.65 | 0.53 | 0.29 | 0.62 | 0.39 | 0.43 | 0.51 | 0.59 |
| E6 | 0.39 | 0.39 | 0.57 | 0.18 | 0.45 | 0.66 | 0.80 | 0.45 | 0.44 | 0.63 | 0.35 | 0.41 | 0.71 | 0.52 |
| E7 | 0.39 | 0.39 | 0.57 | 0.18 | 0.47 | 0.66 | 0.80 | 0.45 | 0.44 | 0.63 | 0.35 | 0.41 | 0.71 | 0.55 |
| E8 | 0.39 | 0.39 | 0.57 | 0.18 | 0.47 | 0.66 | 0.85 | 0.45 | 0.44 | 0.63 | 0.35 | 0.41 | 0.71 | 0.55 |
| E9 | 0.51 | 0.41 | 0.64 | 0.34 | 0.53 | 0.55 | 0.90 | 0.56 | 0.38 | 0.67 | 0.43 | 0.46 | 0.50 | 0.57 |
| E10 | 0.42 | 0.52 | 0.65 | 0.18 | 0.51 | 0.53 | 0.75 | 0.51 | 0.71 | 0.62 | 0.61 | 0.55 | 0.41 | 0.59 |
| E11 | 0.45 | 0.35 | 0.94 | 0.62 | 0.40 | 0.48 | 0.60 | 0.62 | 0.43 | 0.51 | 0.35 | 0.63 | 0.58 | 0.56 |
| E12 | 0.59 | 0.59 | 0.85 | 0.62 | 0.43 | 0.70 | 0.70 | 0.57 | 0.56 | 0.58 | 0.43 | 0.69 | 0.42 | 0.56 |
| E13 | 0.41 | 0.57 | 0.65 | 0.62 | 0.42 | 0.57 | 0.70 | 0.59 | 0.41 | 0.66 | 0.52 | 0.77 | 0.38 | 0.45 |
| E14 | 0.48 | 0.57 | 0.67 | 0.80 | 0.52 | 0.53 | 0.75 | 0.47 | 0.36 | 0.61 | 0.43 | 0.61 | 0.49 | 0.57 |
| E15 | 0.44 | 0.39 | 0.63 | 0.46 | 0.48 | 0.61 | 0.65 | 0.52 | 0.44 | 0.66 | 0.45 | 0.59 | 0.42 | 0.60 |

# Appendix B

## Estimated membership matrix of diseases ($P'^{Max}$)

| | 1 | 2 | 3 | 4 | 5 | 6 | 7 | 8 | 9 | 10 | 11 | 12 | 13 | 14 |
|---|---|---|---|---|---|---|---|---|---|---|---|---|---|---|
| E1 | 0.57 | 0.57 | 0.61 | 0.67 | 0.87 | 0.80 | 0.80 | 0.93 | 0.31 | 0.83 | 0.51 | 1.00 | 0.83 | 0.67 |
| E2 | 0.57 | 0.57 | 0.60 | 0.60 | 1.00 | 0.60 | 0.60 | 0.60 | 0.40 | 0.73 | 0.60 | 0.93 | 0.67 | 1.00 |
| E3 | 0.67 | 0.43 | 0.61 | 0.80 | 1.00 | 0.60 | 0.80 | 0.80 | 0.27 | 0.59 | 0.63 | 0.80 | 0.55 | 1.00 |
| E4 | 0.67 | 0.57 | 0.61 | 0.73 | 0.78 | 0.60 | 0.80 | 0.73 | 0.40 | 0.73 | 0.51 | 0.93 | 0.73 | 0.80 |
| E5 | 0.83 | 0.57 | 0.60 | 0.67 | 0.63 | 0.60 | 0.80 | 0.80 | 0.40 | 0.73 | 0.63 | 1.00 | 0.67 | 0.92 |
| E6 | 0.73 | 0.57 | 0.60 | 0.83 | 0.63 | 0.80 | 0.80 | 0.80 | 0.40 | 0.80 | 0.63 | 0.87 | 0.67 | 0.67 |
| E7 | 0.57 | 0.57 | 0.60 | 0.67 | 0.63 | 0.80 | 0.80 | 1.00 | 0.40 | 0.83 | 0.63 | 1.00 | 0.83 | 0.73 |
| E8 | 0.83 | 0.57 | 0.60 | 0.83 | 0.63 | 0.80 | 0.80 | 0.80 | 0.40 | 0.80 | 0.63 | 1.00 | 0.67 | 0.80 |
| E9 | 0.67 | 0.57 | 0.61 | 1.00 | 1.00 | 0.60 | 0.80 | 0.87 | 0.33 | 0.83 | 0.63 | 1.00 | 0.83 | 1.00 |
| E10 | 0.57 | 0.57 | 0.60 | 0.80 | 1.00 | 0.60 | 0.80 | 0.60 | 0.40 | 0.73 | 0.63 | 1.00 | 0.67 | 1.00 |
| E11 | 0.67 | 0.57 | 0.61 | 0.67 | 0.89 | 0.67 | 0.67 | 0.67 | 0.40 | 0.67 | 0.51 | 0.80 | 0.67 | 0.89 |
| E12 | 0.80 | 0.57 | 0.61 | 0.83 | 1.00 | 0.60 | 0.80 | 0.80 | 0.40 | 0.59 | 0.63 | 0.93 | 0.53 | 1.00 |
| E13 | 0.67 | 0.57 | 0.61 | 0.93 | 1.00 | 0.60 | 0.80 | 0.80 | 0.40 | 0.83 | 0.63 | 0.93 | 0.83 | 1.00 |
| E14 | 0.67 | 0.57 | 0.61 | 0.87 | 1.00 | 0.60 | 0.80 | 0.80 | 0.31 | 0.60 | 0.63 | 0.93 | 0.60 | 1.00 |
| E15 | 0.83 | 0.57 | 0.60 | 0.67 | 0.80 | 0.80 | 0.80 | 0.80 | 0.31 | 0.83 | 0.51 | 1.00 | 0.50 | 1.00 |

## Estimated membership matrix of diseases ($P'^{Min}$)

|     | 1    | 2    | 3    | 4    | 5    | 6    | 7    | 8    | 9    | 10   | 11   | 12   | 13   | 14   |
|-----|------|------|------|------|------|------|------|------|------|------|------|------|------|------|
| E1  | 0.14 | 0.13 | 0.56 | 0.20 | 0.14 | 0.40 | 0.60 | 0.13 | 0.20 | 0.43 | 0.20 | 0.20 | 0.14 | 0.24 |
| E2  | 0.14 | 0.13 | 0.56 | 0.20 | 0.14 | 0.40 | 0.60 | 0.13 | 0.26 | 0.43 | 0.20 | 0.20 | 0.14 | 0.24 |
| E3  | 0.14 | 0.13 | 0.57 | 0.20 | 0.14 | 0.40 | 0.60 | 0.13 | 0.26 | 0.29 | 0.20 | 0.20 | 0.14 | 0.17 |
| E4  | 0.14 | 0.13 | 0.57 | 0.20 | 0.14 | 0.40 | 0.60 | 0.13 | 0.26 | 0.43 | 0.20 | 0.20 | 0.14 | 0.24 |
| E5  | 0.14 | 0.13 | 0.57 | 0.20 | 0.14 | 0.40 | 0.60 | 0.13 | 0.26 | 0.43 | 0.20 | 0.20 | 0.14 | 0.24 |
| E6  | 0.14 | 0.13 | 0.56 | 0.20 | 0.14 | 0.40 | 0.60 | 0.13 | 0.26 | 0.43 | 0.20 | 0.20 | 0.14 | 0.24 |
| E7  | 0.14 | 0.13 | 0.56 | 0.20 | 0.14 | 0.40 | 0.60 | 0.13 | 0.26 | 0.43 | 0.20 | 0.20 | 0.14 | 0.24 |
| E8  | 0.14 | 0.13 | 0.56 | 0.20 | 0.14 | 0.40 | 0.60 | 0.13 | 0.26 | 0.43 | 0.20 | 0.20 | 0.14 | 0.24 |
| E9  | 0.14 | 0.13 | 0.57 | 0.20 | 0.14 | 0.40 | 0.60 | 0.13 | 0.26 | 0.43 | 0.20 | 0.20 | 0.14 | 0.24 |
| E10 | 0.14 | 0.13 | 0.56 | 0.20 | 0.14 | 0.40 | 0.60 | 0.13 | 0.26 | 0.43 | 0.20 | 0.20 | 0.14 | 0.24 |
| E11 | 0.14 | 0.13 | 0.57 | 0.20 | 0.14 | 0.33 | 0.60 | 0.13 | 0.26 | 0.29 | 0.20 | 0.20 | 0.14 | 0.24 |
| E12 | 0.14 | 0.13 | 0.56 | 0.20 | 0.14 | 0.40 | 0.60 | 0.13 | 0.26 | 0.29 | 0.20 | 0.20 | 0.14 | 0.24 |
| E13 | 0.14 | 0.13 | 0.57 | 0.20 | 0.14 | 0.40 | 0.60 | 0.13 | 0.26 | 0.43 | 0.20 | 0.20 | 0.14 | 0.24 |
| E14 | 0.14 | 0.13 | 0.57 | 0.20 | 0.14 | 0.40 | 0.60 | 0.13 | 0.20 | 0.33 | 0.20 | 0.20 | 0.14 | 0.24 |
| E15 | 0.14 | 0.13 | 0.56 | 0.20 | 0.14 | 0.40 | 0.60 | 0.13 | 0.20 | 0.43 | 0.20 | 0.20 | 0.14 | 0.24 |

## Estimated membership matrix of diseases ($P'^{Mean}$)

|     | 1    | 2    | 3    | 4    | 5    | 6    | 7    | 8    | 9    | 10   | 11   | 12   | 13   | 14   |
|-----|------|------|------|------|------|------|------|------|------|------|------|------|------|------|
| E1  | 0.33 | 0.33 | 0.57 | 0.47 | 0.51 | 0.53 | 0.70 | 0.42 | 0.25 | 0.58 | 0.37 | 0.64 | 0.38 | 0.49 |
| E2  | 0.32 | 0.33 | 0.57 | 0.51 | 0.47 | 0.51 | 0.60 | 0.42 | 0.33 | 0.58 | 0.35 | 0.65 | 0.38 | 0.49 |
| E3  | 0.33 | 0.33 | 0.60 | 0.57 | 0.51 | 0.51 | 0.70 | 0.42 | 0.27 | 0.53 | 0.37 | 0.64 | 0.35 | 0.49 |
| E4  | 0.32 | 0.33 | 0.60 | 0.57 | 0.47 | 0.53 | 0.65 | 0.42 | 0.33 | 0.58 | 0.33 | 0.61 | 0.38 | 0.49 |
| E5  | 0.33 | 0.33 | 0.60 | 0.57 | 0.51 | 0.51 | 0.65 | 0.42 | 0.33 | 0.58 | 0.35 | 0.65 | 0.38 | 0.49 |
| E6  | 0.33 | 0.33 | 0.57 | 0.51 | 0.41 | 0.53 | 0.70 | 0.42 | 0.33 | 0.58 | 0.35 | 0.65 | 0.38 | 0.49 |
| E7  | 0.32 | 0.33 | 0.57 | 0.57 | 0.41 | 0.53 | 0.70 | 0.42 | 0.33 | 0.58 | 0.37 | 0.65 | 0.38 | 0.49 |
| E8  | 0.33 | 0.33 | 0.57 | 0.57 | 0.40 | 0.53 | 0.70 | 0.42 | 0.33 | 0.58 | 0.37 | 0.65 | 0.35 | 0.49 |
| E9  | 0.33 | 0.33 | 0.60 | 0.57 | 0.51 | 0.53 | 0.70 | 0.42 | 0.33 | 0.58 | 0.37 | 0.64 | 0.38 | 0.49 |
| E10 | 0.32 | 0.33 | 0.57 | 0.53 | 0.47 | 0.51 | 0.70 | 0.42 | 0.33 | 0.58 | 0.37 | 0.61 | 0.38 | 0.49 |
| E11 | 0.33 | 0.33 | 0.60 | 0.57 | 0.51 | 0.53 | 0.65 | 0.42 | 0.33 | 0.56 | 0.35 | 0.54 | 0.38 | 0.49 |
| E12 | 0.33 | 0.33 | 0.57 | 0.57 | 0.51 | 0.53 | 0.70 | 0.42 | 0.33 | 0.53 | 0.37 | 0.64 | 0.36 | 0.49 |
| E13 | 0.33 | 0.33 | 0.60 | 0.57 | 0.51 | 0.51 | 0.70 | 0.42 | 0.33 | 0.58 | 0.37 | 0.64 | 0.38 | 0.49 |
| E14 | 0.33 | 0.33 | 0.60 | 0.57 | 0.51 | 0.51 | 0.70 | 0.42 | 0.25 | 0.56 | 0.37 | 0.64 | 0.36 | 0.49 |
| E15 | 0.33 | 0.33 | 0.57 | 0.46 | 0.47 | 0.53 | 0.65 | 0.42 | 0.25 | 0.58 | 0.37 | 0.65 | 0.35 | 0.49 |

# References

Altman, E.: Financial ratios, discriminant analysis and the prediction of corporate bankruptcy. J. Fin. **23**(4), 589–609 (1968)

Arias- Aranda, D., Castro, J.L., Navarro, M., Sánchez, J.M., Zurita, J.M.: A fuzzy expert system for business management. Expert Syst. Appl. **37**(12), 7570–7580 (2010)

Balcaen, S., Ooghe, H.: 35 years of studies on business failure: an overview of the classic statistical methodologies and their related problems. Br. Account. Rev. **38**(1), 63–93 (2006)

Beaver, W.: Financial ratios as predictors of failure. J. Account. Res. (Sel. Stud.) **4**, 71–111 (1966)

Behbood, V., Lu, J., Zhang, G.: Adaptive inference-based learning and rule generation algorithms in fuzzy neural network for failure prediction. In: IEEE International Conference of Intelligent Systems and Knowledge Engineering, pp. 33–38 (2010)

Behbood, V., Lu, J.: Intelligent financial warning model using fuzzy neural network and case-based reasoning. In: IEEE Symposium on Computational Intelligence for Financial Engineering and Economics, pp. 1–6 (2011)

Brignole, D., Entizne, R., Vigier, H.: Análisis de Soluciones Aproximadas de Relaciones Binarias Fuzzy, pp. 20–21. VIII SIGEF Congress, Naples (Italy) (2001)

Delcea, C., Scarlat, E.: The diagnosis of firm's disease using the grey systems theory methods. In: IEEE International Conference of GSIS. Nanjing (China), pp. 755–762 (2009)

Delcea, C., Scarlat, E., Maracine, V.: Grey relational analysis between firm's current situation and its possible causes: A bankruptcy syndrome approach. Grey Syst. Theory Appl. **2**(2), 229–239 (2012)

Dubois, D., Prade, H.: A review of fuzzy set aggregation connectives. Inf. Sci. **36**(1/2), 85–121 (1985)

Gil Aluja, J.: Ensayo sobre un Modelo de Diagnóstico Económico-Financiero. V Jornadas Hispano- Lusas de Gestión Científica. Vigo (Spain), pp. 26–29 (1990)

Gil Lafuente, J.: El Control de las Actividades de Marketing. III SIGEF Congress. Buenos Aires (Argentina), vol. 244, pp. 1–21 (1996)

Jones, F.L.: Current techniques in bankruptcy prediction. J. Account. Lit. **6**(1), 131–164 (1987)

Keasey, K., Watson, R.: Non financial symptoms and the prediction of small company failure: a test of Argenti's hypothesis. J. Bus. Fin. Account. **14**(3), 335–354 (1987)

Klir, G., Yuan, B.: A Review of Fuzzy Sets and Fuzzy Logic: Theory and Applications. Prentice Hall PTR, Upper Saddle River (1995)

Klir, G., Folger, A.: Fuzzy Sets, Uncertainty, and Information. Prentice-Hall International, New York (1988)

Maracine, V., Delcea, C.: How we can diagnose the firm's diseases using grey systems theory. Econ. Comput. Econ. Cybern. Stud. Res. **3**, 39–55 (2009)

Regional Center for Economic Studies Bahía Blanca- Argentina (CREEBBA), Economic Activity Report, 112/ 137, last accessed 12 Jan 2015. http://www.creebba.org.ar/main/index.php?op= archivo_iae

Sánchez, E.: Solution of Fuzzy Equations with Extended Operations, Electronics Research Laboratory, pp. 8–10. University of California, Berkeley (1982)

Scarlat, E., Delcea, C., Maracine, V.: Genetic-fuzzy-grey algorithms: a hybrid model for establishing companies' failure reasons. In: IEEE International Conference (SMC), Istanbul (Turkey), pp. 955–962 (2010)

Scherger, V., Vigier, H., Barberá- Mariné, G.: Finding business failure reasons through a fuzzy model of diagnosis. Fuzzy Econ. Rev. **19**(1), 45–62 (2014)

Tascón, M., Castaño, F.: Variables y modelos para la identificación y predicción del fracaso empresarial: Revisión de la investigación empírica reciente. RC- SAR **15**(1), 7–58 (2012)

Thapar, A., Pandey, D., Gaur, S.K.: Optimization of linear objective function with max-t fuzzy relation equations. Appl. Soft Comput. **9**(3), 1097–1101 (2009)

Terceño, A., Vigier, H., Barberá-Mariné G., Scherger, V.: Hacia una integración de la Teoría del Diagnóstico Fuzzy y del Balanced Scorecard. In: XV SIGEF Conference on Economic and Financial Crisis. New Challenges and Perspectives, Lugo (Spain), pp. 364–379 (2009)

Terceño, A., Vigier, H.: Economic—financial forecasting model of businesses using fuzzy relations. Econ. Comput. Econ. Cybern. Stud. Res. **45**(1), 215–232 (2011)

Terceño, A., Vigier, H., Scherger, V.: Application of a fuzzy model of economic—financial diagnosis to SMEs. In: Gil Aluja., J., Terceño, A. (eds.) Methods for Decision Making in an Uncertain Environment. Series on Computer Engineering and Information Science, vol. 6, pp. 146–162. World Scientific Publishing, Singapore (2012)

Terceño, A., Vigier, H., Scherger, V.: Identificación de las causas en el diagnóstico empresarial mediante relaciones Fuzzy y el BSC. Actualidad Contable Fases **17**(28), 101–118 (2014)

Vigier, H., Terceño, A.: A model for the prediction of diseases of firms by means of fuzzy relations. Fuzzy Sets Syst. **159**(17), 2299–2316 (2008)

Vigier, H., Terceño, A.: Analysis of the inconsistency problem in the model for predicting "diseases" of firms. Fuzzy Econ. Rev. **17**(1), 73–88 (2012)

Yager, R.R.: Families of OWA operators. Fuzzy Sets Syst. **59**, 125–148 (1993)

Yager, R.R.: Heavy OWA operators. Fuzzy Optim. Decis. Making **1**(4), 379–397 (2002)

Yager, R.R.: Generalized OWA aggregation operators. Fuzzy Optim. Decis. Making **3**(1), 93–107 (2004)

Yager, R.R.: Prioritized OWA aggregation. Fuzzy Optim. Decis. Making **8**(3), 245–262 (2009)

Yager, R.R., Alajlan, N.: On characterizing features of OWA aggregation operators. Fuzzy Optim. Decis. Making **13**(1), 1–32 (2014)

Zimmermann, H.J.: Fuzzy Sets, Decision Making and Expert Systems, pp. 41–44. Kluwer Academic Publishers, Boston, Norwell (1987)

Zmijewski, M.: Methodological issues related to the estimation of financial distress prediction models. J. Account. Res. 59–86 (1984)

# Forecasting Global Growth in an Uncertain Environment

Nigar A. Abdullayeva

**Abstract** Forecasting is an integral part of economic decision making. However, the current forecasting algorithms offer poor precision in solution of uncertainty problems. In this paper, a novel forecasting model for world GDP growth rate using fuzzy regression was proposed.

**Keywords** Forecasting · Fuzzy regression · GDP growth rate · Uncertainty

## 1 Introduction

Expectations about future GDP can be the primary determinant of investments, employment, wages, profits and even stock market activities. First of all, GDP represents economic production and growth. So it gives a signal about the future employment and wages. GDP also determines stock market return rates. If the GDP growth rate is positive, then investors may expect to gain revenue. The GDP statistics can also help economists in solving the problem of inflation. If the GDP is really expected to increase, then inflation may pick up and the Banks may need to raise interest rates. If the GDP is likely to continue to shrink, the Banks may need to pursue further quantitative easing. The interest rate changes can take up to 18 months to have an effect. Therefore, monetary authorities are trying to set the optimal rates for the future economic situation.

Considering its large impact on social and economic processes, forecasting GDP has a great importance both theoretically and practically. Traditionally, GDP forecasts have been produced and utilized by several international organizations (United Nations – UN, International Monetary Fund – IMF, World Bank – WB and etc.) and agencies. They estimate projections about the expected GDP and make

N.A. Abdullayeva (✉)
Department of Innovative Projects Coordination, Azerbaijan National Academy of Science,
H.Cavid Av., 119, Baku, Azerbaijan
e-mail: nigyar.a@gmail.com

© Springer International Publishing Switzerland 2015       159
J. Gil-Aluja et al. (eds.), *Scientific Methods for the Treatment of Uncertainty in Social Sciences*, Advances in Intelligent Systems and Computing 377,
DOI 10.1007/978-3-319-19704-3_13

appropriate changes pertaining to spending, capital utilization and leverage. Spending on GDP analytics forms is an important part of research spending in these organizations. Such forecasts have been made from time to time to reflect dynamic changes in the economy.

Uncertainty factors and risks (oil price shock, interstate and civil wars, political and social instability, climate change, extreme weather events, natural disasters like earthquakes and etc., corruption, pandemics) have a significant impact on GDP forecasting (Kaufmann and Gil Aluja 1986; Gil Lafuente et al. 1992; Gil Aluja 2004).

For example, in projections presented by IMF experts, data for Syria and Ukraine are excluded because of the uncertain political situation and ongoing crisis. As mentioned in Report "Legacies, Clouds, Uncertainties" (World Economic Outlook 2014), "...uncertain investment prospects in Russia had already lowered growth before the Ukraine crisis, and the crisis has made growth worse. Uncertain prospects and low investment are also weighing on growth in Brazil." With weaker than expected global growth for 2014 and increased downside and medium-term risks, the projected pickup in growth may again fail to materialize or fall short of expectations. In regard to the components underlying uncertainty around the forecasts, risks to global growth due to oil prices have increased compared with the April 2014, and notably so for 2015. Brazil, China, India and Russia (the BRICs), whose share in global GDP at PPP (purchasing-power-parity) weights is about 28 %, account for about half of the overall forecast error.

The UN WESP report (World Economic Situation and Prospects 2015) likewise recorded, that many risks and uncertainties could dash efforts to get the global economy on track and moving forward. This report identifies a number of risks and uncertainties for the world economy, including international spill-overs from the future unwinding of the monetary easing by major developed economies; vulnerabilities of emerging economies on both external and domestic fronts; remaining fragilities in the euro area; unsustainable public finance in the longer run for many developed countries; and risks associated with geopolitical tensions.

The World Bank's most recent report (Global Economic Prospects 2015) stated, that challenges are increasingly of a medium term nature including those related to fiscal sustainability uncertainties and an orderly exit from unconventional monetary police (Europe, US and Japan), deflation risks (in Europe) and the need for structural reforms to boost productivity growth. Among developing countries, heightened risks in one or more of these economies (Brazil, India, Indonesia, Turkey and South Africa) could spark contagion to other countries.

In the existing literature, forecasting GDP is widely studied with different methods. In this paper forecasting model for world GDP growth rate using fuzzy regression was discussed and the relationship between GDP and other variables was studied over the period 2003–2013.

## 2 Literature Review

There are several models for forecasting GDP. Tkacz and Hu (1999) have determined whether more accurate indicator models of output growth, based on monetary and financial variables, can be developed using neural networks. The authors have used artificial neural network (ANN) model to forecast GDP growth for Canada. The main findings of this study are that, at the 1-quarter forecasting horizon, neural networks yield no significant forecast improvements. At the 4-quarter horizon, however, the important forecast accuracy is statistically significant. The root mean squared forecast errors of the best neural network models are about 15–19 % lower than their linear model counterparts.

Marcellino (2007) has calculated whether complicated time series models can outperform standard linear models for forecasting GDP growth and inflation for the United States. In the study, it is considered as a large variety of models and evaluation criteria, using a bootstrap algorithm to evaluate the statistical significance of the results. The main conclusion is that in general linear time series, models can be hardly beaten if they are carefully specified.

Schumacher and Breitung (2008) have employed factor models to forecast German GDP using mixed-frequency real-time data, where the time series are subject to different statistical publication lags. In the empirical application, the authors have used a novel real-time dataset for the German economy. Employing a recursive forecast experiment, they have evaluated the forecast accuracy of the factor model with respect to German GDP.

Guegan and Rakotomarolahy (2010) have conducted an empirical forecast accuracy comparison of the non-parametric method, known as multivariate Nearest Neighbor method, with parametric VAR modeling on the euro area of GDP. By using both methods for now casting and forecasting the GDP, through the estimation of economic indicators plugged in the bridge equations, the authors have got more accurate forecasts when using nearest neighbor method. It is also proven the asymptotic normality of the multivariate k-nearest neighbor regression estimator for dependent time series, providing confidence intervals for point forecast in time series.

Mirbagheri (2010) has investigated the supply side economic growth of Iran by estimating GDP growth. In this study, the predictive results of fuzzy logic and neural-fuzzy methods are also compared. According to the findings of the study, forecasting by neural-fuzzy method is recommended.

Liliana and Napitupulu (2010) have also used ANN method in forecasting GDP. In this study, authors have forecasted GDP for Indonesia and they put forward many advantages and disadvantages of the method. According to the results, the authors have concluded that the ANN model has better ability in forecasting the macroeconomic indicators.

Ge and Cui (2011) have used process neural network (PNN) into the GDP forecast and established the forecast model based on PNN by choosing the main factors influencing GDP and using the dual extraction capacity on time and space cumulative effect on PNN. By means of comparing and analyzing with traditional neural network forecast model, the result shows that GDP forecast model, which bases on PNN, has a better performance.

## 3 Fuzzy Regression Model

Since Tanaka first introduced the fuzzy regression models in 1982 (Tanaka and Watada 1989), they have been applied in many fields for explaining the relation between fuzzy independent and dependent variables. The one of purposes of the fuzzy regression analysis is to estimate the fuzzy output, which has lower error between the predicted and observed output than any other estimated fuzzy number. We can do this by forecasting $\alpha$-level set of the observed fuzzy output.

Fuzzy regression is also a powerful method to derive the fuzzy function for variables with uncertain nature. For selecting the best fuzzy function between input and output variables, a model fitting measurement has to be defined. This approach is an essential tool to estimate the goodness of a defined model where the similarity between each of predicted and calculated outputs is computed.

Fuzzy regression analysis can be seen as an optimization problem where the aim is to derive a model which fits the given dataset.

Consider the set of observed data $\{(Y_i, X_{ij}), i = 1,\ldots, n\}$, where $X_{ij}$ is the explanatory variable, which may be with crisp/fuzzy numbers and $Y_i$ is the response variable, which is either with symmetric/non-symmetric triangular or trapezoidal fuzzy numbers. Our aim is to fit a fuzzy regression model with fuzzy coefficients to the data sets as follows:

$$Y_i = A_0 + A_1 \cdot X_{i1} + \cdots + A_k \cdot X_{ik}, \ i = 1, \ldots, n,$$

where $A_j = (a_j, c_j)$, $j = 0, 1, \ldots, k$ are fuzzy numbers, that are characterized by the membership functions. The membership functions are defined as:

$$\mu_a(x) \begin{cases} 0 & x < a^l \\ \frac{x - a^l}{a^m - a^l} & a^l \leq x \leq a^m \\ \frac{a^r - x}{a^r - a^m} & a^m \leq x \leq a^r \\ 0 & x > a^r \end{cases}$$

The fuzzy regression model with fuzzy coefficients can be expressed with lower boundary, center and upper boundary. The fuzzy regression model is formulated to minimize its width including relations between data and the model. This problem is solved with the following linear programming method:

$$\min \sum_{i=1}^{n} \sum_{j=0}^{k} c_j x_{ij}$$

$$\sum_{j=0}^{k} \left(a_j + (1-\alpha)c_j\right)x_{ij} \geq y_i + (1-\alpha)e_i$$

$$\sum_{j=0}^{k} \left(a_j - (1-\alpha)c_j\right)x_{ij} \leq y_i - (1-\alpha)e_i$$

$$a_j = \text{ free, } c_j \geq 0, \ j = 0, 1, \ldots, k; \ i = 1, 2, \ldots, n.$$

where $a_j$ is the center and $c_j$ is the spread of fuzzy coefficient $A_j$, $j = 0, 1, \ldots, k$; $x_{ij}$ is the crisp independent variable ($x_{i0} = 1$), $y_i$ is the center and $e_i$ is the spread of fuzzy dependent variable $Y_i(i = 1, 2, \ldots, n)$.

# 4 Forecasting Model of GDP Growth Rate

GDP growth rate is times series data. Parameters influencing GDP growth rate are many, but it is necessary to be selective to ensure that the developed model is effective. The indicators, which influence the GDP growth rate in this study, are Brent crude oil price (OP), volume of exports (EXP), volume of imports (IMP) and consumer price index (CPI). Note these are not the only criteria influencing GDP growth rate.

Correlations drawn between export and import volumes, inflation, oil prices and GDP have been widely researched in Economics and Econometrics literature. There also exist few quantitative models that seek to make the observed correlation between these measures. Meanwhile, the nature of this dependency is not very clear, and no one mathematical model exists, that explicitly relate all of this quantities. Hence, the problem is solved with using fuzzy regression method.

Above mentioned determinants were used as independent variables in fuzzy regression model. GDP growth rate is a dependent variable in this model. The data used in this study are collected from IMF World Economic Outlook 2014. In order to calculate future GDP growth rate, the related past trend data for a period from 2003 to 2013 has extracted.

So, in this task $n = 11$, $k = 4$. The calculated parameters in fuzzy regression model are triangular fuzzy numbers, which include the central value, upper and lower limit of value, the degree of ambiguity and their possible uncertainty (Table 1).

**Table 1** Statistical data

| Years | Gross domestic product world, percent change (GDP) | Brent crude oil price world, dollars per barrel (OP) | World average consumer prices, percent change (CPI) | World volume of exports of goods and services, percent change (EXP) | World volume of imports of goods and services, percent change (IMP) |
|-------|------|---------|-------|---------|---------|
| 2003  | 4.036 | 28.853  | 3.982 | 5.847   | 5.960   |
| 2004  | 5.437 | 38.298  | 3.844 | 10.673  | 11.302  |
| 2005  | 4.883 | 54.434  | 4.122 | 7.580   | 7.674   |
| 2006  | 5.557 | 65.390  | 4.152 | 9.514   | 8.970   |
| 2007  | 5.673 | 72.713  | 4.373 | 7.927   | 8.231   |
| 2008  | 3.038 | 97.660  | 6.428 | 2.966   | 3.060   |
| 2009  | 0.013 | 61.861  | 2.803 | −10.267 | −10.855 |
| 2010  | 5.432 | 79.631  | 3.852 | 12.561  | 12.543  |
| 2011  | 4.143 | 110.952 | 5.204 | 6.556   | 6.790   |
| 2012  | 3.365 | 111.960 | 4.238 | 2.950   | 2.866   |
| 2013  | 3.279 | 108.844 | 3.883 | 3.153   | 2.849   |

Forecasting of GDP growth rate is carried out in three possible scenarios – pessimistic (lower-case), baseline (most likely) and optimistic (upper-case) - for the next three years (2014, 2015 and 2016). Predicted values of GDP global growth rate based on the information available for the 2003–2013 periods are presented in Table 2.

For fuzzy forecasts, the baseline value, optimistic value and pessimistic value are respectively defined as follows:

$$B(A) = (a + 2b + c)/4$$

$$O_\lambda(x) = \begin{cases} 2\lambda b + (1 - 2\lambda)c, & if\ \lambda \le 0.5 \\ (2\lambda - 1)a + (2 - 2\lambda)b, & else \end{cases}$$

$$P_\lambda(x) = \begin{cases} (1 - 2\lambda)a + 2\lambda b, & if\ \lambda \le 0.5 \\ (2 - 2\lambda)b + (2\lambda - 1)c, & else \end{cases}$$

where $A = (a, b, c)$ - triangular fuzzy number, $\lambda \in [0, 1]$ - credibility level (Liu and Liu 2002).

This paper has found GDP growth rate baseline forecasts lower than the calculated estimates of International Monetary Fund and approximately near to projections of United Nations and World Bank. The global economy is expected to pick up speed as the year progresses and is projected to expand by 3.0 % in 2014, strengthening to 3.1 and 3.2 % in 2015 and 2016, respectively.

**Table 2** Forecasted values of GDP growth rate

| GDP world growth rate, % change | Fuzzy forecasting through three scenarios | | | IMF forecast | UN forecast | WB forecast |
|---|---|---|---|---|---|---|
| Years | Pessimistic (lower-case) | Baseline | Optimistic (upper-case) | | | |
| 2014 | 1.4488 | **3.0008** | 4.5291 | **3.313** | 2.6 | 2.8 |
| 2015 | 1.6527 | **3.0895** | 6.4948 | **3.847** | 3.1 | 3.4 |
| 2016 | 2.1434 | **3.1978** | 7.7548 | **4.040** | 3.3 | 3.5 |

# 5 Conclusion

The socio-economic variables are extremely difficult to estimate accurately. Steps should be taken to improve measures of these indicators. In particular, substantial effort should be devoted to developing and implementing robust, reliable measures of socio-economic variables that can be shown to predict GDP.

Unforeseen events – wars, natural disasters, advances in technology, the macroeconomic cycle – can have drastic impacts on GDP and defy even the most thorough modeling effort. Fuzzy regression model appears a promising approach for capturing uncertainty factors. Nowadays, fuzzy regression models have shown great potential for modeling complex, ill-defined and not well understood systems with approximate, vague, imprecise values and lack of exact knowledge.

In order to develop a fuzzy regression model, GDP growth rate is used as the dependent variable. In the study of the relationship between economic variables (OP, EXP, IMP, CPI) and GDP, this structure can be considered as a fuzzy function, in which parameters are determined by fuzzy set theory using alpha values based operations. The results of fitting the fuzzy regression model suggest that there is a relationship between these economic variables and GDP.

The regression model that has explanatory variables as crisp numbers can be a good model only if the shape of the membership function for the fuzzy regression coefficient is identical with that of the response variable, i.e. the predicted values should coincide almost with the mode of the observed response. The predicted upper and lower limit values are with very small width from the center denotes that predicted model has effective performance in forecasting.

# References

Abbasov, A.M., Mamedova, M.H.: Application of fuzzy time series to population forecasting. In: Proc. of the 8th Symposium on Information Technology in Urban and Spatial Planning, Vienna University of Technology, pp. 545–552, 25 Feb – 1 Mar 2003

Abdullayeva, N.A.: Sustainable social environment: fuzzy assessment and forecasting. Fuzzy Econ. Rev. **XVI**(2), 21–32 (2011)

Aliev, R.A., Fazlollahi, B., Aliev, R.R.: Soft Computing and its Application in Business and Economics. Springer, Berlin (2004)

De Andres, J., Terceño, A.: Estimating a term structure of interest rates for fuzzy financial pricing by using fuzzy regression methods. Fuzzy Sets Syst. **139**, 339–361 (2003)

Ge, L., Cui, B.: Research on forecasting GDP based on process neural network. IEEE **2011**(7), 821–824 (2011)

Gil Aluja, J.: Towards a new concept of economic research. Fuzzy Econ. Rev. **0**, 5–23 (1995)

Gil Aluja, J.: Investment in Uncertainty. Kluwer, Dordrecht (1999)

Gil Aluja, J.: Fuzzy Sets in the Management of Uncertainty. Springer, Berlin (2004)

Gil Lafuente, A.M., Gil Aluja, J., Teodorescu, H.N.: Periodicity and chaos in economic forecasting. In: Fuzzy systems Proc. ISKIT'92, Japan, pp. 85–93 (1992)

Gil Lafuente, A.M.: Fuzzy Logic in Financial Analysis. Springer, Berlin (2005)

Gil Lafuente, J. (ed.): Algorithms for Excellence Keys for Being Successful in Sport Management (in Spanish). Milladoiro, Vigo (2002)

Guegan, D., Rakotomarolahy, P.: Alternative methods for forecasting GDP. Working Paper, University of Paris, France (2010)

Imanov, Q.C.: The fuzzy approach to estimation of the index of population life quality. Fuzzy Econ. Rev. **XII**(2), 85–93 (2007)

Imanov, Q.C.; Yusifzade, R.A.: Fuzzy approach to assessment of financial sustainability of corporation. J. Financ. Decis. Mak. (Special Issue—Fuzzy Logic in the Financial Uncertainty) **5**(2), 61–66 (2009)

Kaufmann, A., Gil Aluja, J. (eds.): Introduction to the Theory of Fuzzy Subsets in Business Management (in Spanish). Milladoiro, Santiago de Compostela (1986)

Liu, B., Liu, Y.-K.: Expected value of fuzzy variable and fuzzy expected value models. IEEE Trans. Fuzzy Syst. **10**(4), 445–450 (2002)

Marcellino, M.: A comparison of time series models for forecasting GDP growth and inflation. http://www.eui.eu/Personal/Marcellino/1.pdf (2007)

Savic, D.A., Pedrycz, W.: Evaluation of fuzzy linear regression models. Fuzzy Sets Syst. **39**, 51–63 (1991)

Schumacher, C., Breitung, J.: Real-time forecasting of German GDP based on large factor model with monthly and quarterly data. Int. J. Forecast. **24**, 386–398 (2008)

Tanaka, H., Watada, J.: Possibilistic linear systems and their application to the linear regression models. Fuzzy Sets Syst. **27**, 275–289 (1989)

Zadeh, L.A.: Fuzzy sets. Inf. Control **8**, 338–353 (1965)

Zadeh, L.A.: A note on Z-numbers. Inf. Sci. **181**, 2923–2932 (2011)

Zadeh, L.A., Abbasov, A.M., Yager, R.R., Shahbazova, S.H.N., Reformat, M.Z.: Recent developments and new directions in soft computing. Springer, Berlin (2014)

# Fuzzy NN Time Series Forecasting

Juan J. Flores, Federico González-Santoyo,
Beatriz Flores and Rubén Molina

**Abstract** The kNN time series forecasting method is based on a very simple idea. kNN forecasting is base on the idea that similar training samples most likely will have similar output values. One has to look for a certain number of nearest neighbors, according to some distance. The first idea that comes to mind when we see the nearest neighbor time series forecasting technique is to weigh the contribution of the different neighbors according to distance to the present observation. The fuzzy version of the nearest neighbor time series forecasting technique implicitly weighs the contribution of the different neighbors to the prediction, using the fuzzy membership of the linguistic terms as a kind of distance to the current observation. The training phase compiles all different scenarios of what has been observed in the time series' past as a set of fuzzy rules. When we encounter a new situation and need to predict the future outcome, just like in normal fuzzy inference systems, the current observation is fuzzyfied, the set of rules is traversed to see which ones of them are activated (i.e., their antecedents are satisfied) and the outcome of the forecast is defuzzyfied by the common center of gravity rule.

**Keywords** Time series forecasting · Fuzzy reasoning · Automatic machine learning

J.J. Flores (✉)
Division de Estudios de Posgrado, Facultad de Ingenieria Electrica, Universidad Michoacana,
Morelia, Mexico
e-mail: juanf@umich.mx

F. González-Santoyo · B. Flores
Facultad de Ciencias Contables y Administrativas, Universidad Michoacana, Morelia,
Mexico
e-mail: fsantoyo@umich.mx

B. Flores
e-mail: betyf@umich.mx

R. Molina
Instituto de Investigaciones Económicas y Empresariales, Universidad Michoacana,
Morelia, Mexico
e-mail: ruben.molinam@gmail.com

© Springer International Publishing Switzerland 2015                                    167
J. Gil-Aluja et al. (eds.), *Scientific Methods for the Treatment of Uncertainty
in Social Sciences*, Advances in Intelligent Systems and Computing 377,
DOI 10.1007/978-3-319-19704-3_14

# 1   Introduction

A time series is defined as a set of quantitative observations arranged in chronological order (Gebhard Kirchgssner 2013), where we generally assume that time is a discrete variable. Examples of time series are commonly found in the fields of engineering, science, sociology, and economics, among others (Brookwell and Davis 2002); time series are analyzed to understand the past and to predict the future (Cowpertwait and Metcalfe 2009). This forecast must be as accurate as possible, since it can be linked to activities involving marketing decisions, product sales, stock market indices, and electricity load demand, among others.

In many occasions, it is fairly simple and straightforward to forecast the next value of the time series. But the further we delve into the future, the more uncertain we are and the bigger forecast errors we get.

Some statistical models such as autoregressive models (Shumway and Stoffer 2011) can be used for time series modeling and forecasting. These traditional forecasting techniques are based on linear models, however many of the time series encountered in practice exhibit characteristics not shown by linear processes. If the time series is confirmed to be nonlinear, very rich dynamic possibilities can emerge, including sensibility to initial conditions, known as chaotic behavior (Leven and Koch 1981).

In the past, several methods have been proposed to predict chaotic time series. One of the first methods proposed was the modeling of the system using Nonlinear Models (NLM) (Farmer and Sidorowich 1987), but the main problems of this approach are proper model selection and data-dependency.

Another approach to solve the time series forecasting problem is Artificial Neural Networks (ANNs) (Maguire et al. 1998). The ANNs are trained to learn the relationships between the input variables and historical patterns, however the main disadvantage of ANNs is the required learning procedure.

More recently, classification techniques based on the nearest neighbors have been successfully applied in different areas from the traditional pattern recognition. The k-Nearest-Neighbors (kNN) (Casdagli 1989) is an algorithm that can be employed in time series forecasting due to its simplicity and intuitiveness. kNN searches for similar instances recovering them from large dimensional feature spaces and incomplete data (Farmer and Sidorowich 1987). The kNN algorithm assumes that subsequences of the time series that emerged in the past are likely to have a resemblance to the future subsequences and can be used for generating kNN-based forecasts.

One significant drawback of the kNN forecasting method is the sensitivity to changes in the input parameters, (i.e. the number of nearest neighbors and the embedding dimension). If the input parameters are not selected appropriately, it could decrease the accuracy of the forecasts.

One way to select the best possible input parameters for the kNN method is using an algorithm of the Evolutionary Computation (EC) family (Palit and Popovic 2005). EC is a computational technology made up of a collection of randomized global search paradigms for finding the optimal solutions to a given problem. In particular, Differential Evolution (DE) is a population-based search strategy that has recently proven to be a valuable method for optimizing real valued multi-modal objective functions (Storn and Price 1995; Storn and Lampinen 2005). It is a parallel direct search method having good convergence properties and simplicity in implementation.

The design of models for time series prediction has traditionally been done using statistical methods. In modeling time series, we find the ARIMA (Auto-Regressive Integrated Moving Average), ARMA, and AR, among others (Wheelwright and Makridakis 1985). These models are defined in terms of past observations and prediction errors. Statistical techniques like auto-correlation, and partial auto-correlation, help scientists identify which of the past observations and/or errors are significant in the construction of the forecasting models.

In the last decade, artificial neural networks have been used successfully to model and forecast time series. Designing an artificial neural network (ANN) that provides a good approximation is an optimization problem. Given the many parameters to choose from in the design of an ANN, the search space in this design task is enormous. On the other hand, the learning algorithms used to train ANNS are only capable of determining the weights of the synaptic connections, and do not include architectural design issues. So, a scientist in need of an ANN model has to design the network on a trial and error basis. When designing an ANN by hand, scientists can try only a few of them, selecting the best one from the set they tested.

The first idea that comes to mind when we see the nearest neighbor time series forecasting technique is to weigh the contribution of the different neighbors according to distance to the present observation. In this paper we propose the fuzzy version of the nearest neighbor time series forecasting technique. This model, implicitly weighs the contribution of the different neighbors to the prediction, using the fuzzy membership of the linguistic terms as a kind of distance to the current observation. The training phase compiles all different scenarios of what has been observed in the time series' past as a set of fuzzy rules. When we encounter a new situation and need to predict the future outcome, just like in normal fuzzy inference systems, the current observation is fuzzyfied, the set of rules is traversed to see which ones of them are activated (i.e., their antecedents are satisfied) and the outcome of the forecast is defuzzyfied by the common center of gravity rule.

The rest of the paper is organized as follows. Section 2 describes the nearest neighbor time series forecasting method. Section 3 exposes the basic concepts needed to perform fuzzy reasoning. Section 4 proposes the fuzzy nearest neighbor method. Section 4 proposes the use of the fuzzy nearest neighbor method to reduce noise – an inherent component of time series. Section 5 presents a study case where the use of the proposed method is illustrated. Finally, Sect. 6 concludes the work.

## 2 Nearest Neighbour Forecasting Method

The kNN approximation method is a very simple, but powerful one. It has been used in many different applications and particularly in classification tasks (Sorjamaa and Lendsasse 2006). For this purpose, kNN uses the average of the forecast of the k objects, with- out taking into account assumptions on the distribution of predicting variables during the learning process.

The key idea behind the kNN is that similar training samples most likely will have similar output values. One has to look for a certain number of nearest neighbors, according to some distance. The distance usually used to determine the similarity metric between the objects is the Euclidean distance. However, the method also permits the use of other distance measures like Chebyshev, Manhattan, and Mahalanobis (Baldi and Brunak 2001). Once we find the neighbors, we compute an estimation of the output simply by using the average of the outputs of the neighbors in the neighborhood.

In contrast to statistical methods that try to identify a model from the available data, the kNN method uses the training set as the model (Lora et al. 2003). The main advantage of kNN is the effectiveness in situations where the training dataset is large and contains deterministic structures.

### 2.1 One-Step-Ahead Forecasting

Let $\mathbb{S} = \{s_1, s_2, \ldots, s_t \ldots, s_N\}$ be a time series, where $s_t$ is the recorded value of variable s at time t. The forecasting problem targets the estimation of $\Delta n$ consecutive future values, i.e. $\{s_{N+1}, s_{N+2}, \ldots s_{N+\Delta n}\}$, using any of the currently available observations from $\mathbb{S}$ (Yankov and Keogh 2006).

If we choose a delay time $\tau$ and an embedding dimension m, it is possible to construct delay vectors of the form $S_t = [s_{t-(m-1)\tau}, s_{t-(m-2)\tau}, \ldots, s_{t-\tau}, s_t]$, where $m > 0$ and $t > 0$. The time series is then organized in a training set by running a sliding window $S_N$ of size m along each delay vector. To retrieve the k nearest neighbors of $S_N$ we choose the parameter $\varepsilon$ and calculate the distance to every delay vector $S_t$, where $t = (m-1)\tau + 1, (m-1)\tau + 2, \ldots, N-1$.

For all the k vectors $S_t$ that satisfy Eq. (1) we look up the individual values $s_{t+\Delta n}$.

$$|s_N - s_t| \le \varepsilon \tag{1}$$

The forecast is then the average of all these individual values, expressed by Eq. (2),

$$\hat{s}_{N+\Delta n} = \frac{1}{k} \sum_{j=1}^{k} s_{t+\Delta n} \tag{2}$$

where $\hat{s}_{N+\Delta n}$ is the forecasted value at time N. In the case where we do not find any neighbors, we just pick the closest vector $S_N$ to. In order to forecast a sequence of future values we use real data as input, that is, $FS(\hat{s}_t) = \hat{s}_{t+\Delta n}$, where $FS$ is the one-step-ahead forecasting function.

Note that when $\Delta n = 1$, Eq. (2) produces only one forecast, however in many occasions, it is necessary to forecast more than one value at the time. For this purpose, two strategies have been used in the past, the iterative or recursive scheme and the simultaneous or direct scheme (Sorjamaa and Lendsasse 2006).

# 3 Fuzzy Logic

## 3.1 Classical and Fuzzy Sets

A classical set A is defined as a collection of elements or objects. Any element or object x either belongs or does not belong to A. The membership $\mu_A(x)$ of x in A is a mapping $\mu_A : X \to \{0, 1\}$. From the membership mapping follows that $A \cap \bar{A} = \emptyset$.

Fuzzy logic is a logic based on fuzzy sets, i.e. sets of elements or objects characterized by truth values in the [0, 1] interval rather than crisp 0 and 1, as in the conventional set theory. A set's membership function assigns a number in [0, 1] to each element of the universe of discourse of a fuzzy set. Fuzzy set A is completely characterized by its membership function $\mu_A : X \to [0, 1]$, and is defined as a set of pairs: $A = \{(x, \mu_A(x))\}$. Although fuzzy set's membership functions can take several forms, in this work we use the triangular form, which is the most common. This form is given by Eq. (3) – see Fig. 1 (Pappis and Siettos 2014; Siler and Buckley 2005).

$$\text{Tri}[x, \alpha, \beta, \gamma] = \begin{bmatrix} 0 & x < \alpha \\ \frac{x-\alpha}{\beta-\alpha} & \alpha \le x \le \beta \\ -\frac{x-\gamma}{\gamma-\beta} & \beta \le x \le \gamma \\ 0 & \gamma \le x \end{bmatrix} \qquad (3)$$

## 3.2 Fuzzy Operations

The fuzzy set operations used in this work are fuzzification, and, implication, and defuzzification. Since there are many definitions for these fuzzy operations, we will define them precisely in this section, so there is a common ground with the readers from where we can establish the Fuzzy Nearest Neighbor forecasting method.

**Fig. 1** Triangular fuzzy set

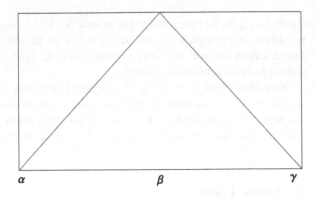

To fuzzify a real value x, given a set of linguistic terms or sets, we just compute the membership function of x in each of those fuzzy sets.

If A and B are fuzzy sets with membership functions $\mu_A(x)$ and $\mu_B(x)$, A and B is also a fuzzy set whose membership function can be computed as $\forall x \in X$ $\mu_A$ and $_B(x) = \min(\mu_A(x), \mu_B(x))$.

The fuzzy model computed by FNN contains a set of fuzzy inference rules. Rule j has the form (see Sect. 4) given by Eq. 4.

If $X_n$ is $A_{j,0}$ and $X_{n-1}$ is $A_{j,1}$ and ... and $X_{n-m}$ is $A_{j,m}$ then $X_{n+1}$ is $A_{j,n+1}$

$$(4)$$

where $X_t$ is the value of the time series at time t and $A_{j,k}$ are the linguistic term to which they must belong to satisfy the antecedents of the rule. $A_{j,n+1}$ is the consequent or result of the inference. Meaning that if previous observed values belong to the sets required by the rule (the rule is fired), we can assert that the predicted value of the time series (i.e., $X_{n+1}$) will belong to the set $A_{j,n+1}$ with a membership value equal to the extent the antecedents are satisfied (fire strength). Let us call this value $\alpha_j$.

In any given situation, more than one rule may fire. In those situations, we need to defuzzify those fuzzy results as to provide a single real value for the forecast. The center of gravity is the most common and perhaps the most intuitive defuzzification method. If n rules are fired with strengths $\alpha_j$, and the center of the triangular membership function $A_{j,n+1}$ is $\beta_j$ of the forecasted value is given by (see Eq. 5).

$$X_{n+1} = \frac{\sum_j^n \alpha_j \beta_j}{\sum_j^n \alpha_j} \tag{5}$$

Note that by assuming triangular fuzzy sets, we imply that their centers of gravity lie at the center of the triangle, i.e., $\beta$ in Fig. 1.

# 4 Fuzzy Nearest Neighbours

This section presents the time series forecasting method proposed in this paper. FNN is the fuzzy version of the nearest neighbour forecasting method (NN). NN takes no training phase (see Sect. 2); when the latest observations of the time series are available, the method just compares it with the system's history, searching for similar patterns, which occurred in the past. Those similar patterns are used to predict the magnitude of the following observation(s). Unlike NN, FNN does require a training phase; in that phase, a set of fuzzy rules is learned, i.e., a model is produced. Once the set of fuzzy rules has been produced, it is used every time a prediction is required.

## 4.1 Fuzzy Linguistic Terms

The main idea behind FNN is to determine a set of fuzzy rules that allow us to predict a time series behavior. A fuzzy rule has the form given by Eq. 4. Additionally, we can assume that the sampling step is not necessarily 1, but in general, let us call it $\tau$. Equation 4 then becomes Eq. 6.

$$\text{If } X_n \text{ is } A_o \text{ and } X_n - \tau \text{ is } A_1 \text{ and } X n - 2\tau \text{ is } A_1 \text{ and } \dots \text{ and } X_n - m\tau \text{ is } A_m$$
$$\text{then } X_n + 1 \text{ is } A_n + 1 \tag{6}$$

where the $A_i$ are the fuzzy linguistic terms (FLT). For the time being, we assume we know m and $\tau$.

To obtain the FLT, just divide the time series range into 3, 5, or 11 values, to form the fuzzy scales, corresponding to linguistic terms or fuzzy sets. The more linguistic terms we use, the more precise the prediction will be. At the same time, the process will be more time consuming as the number of scales increases. Figure 2 shows a time series and its corresponding linguistic terms. Note that the FLT overlap in a way that every real value belongs to exactly two terms and its membership function is never zero.

## 4.2 Fuzzy Rules

A fuzzy rule, like the one in Eq. 6, captures the behavior of the time series in a given window of time. Figure 3 illustrates a portion of the time series that would contribute to the formation of the rule shown in Eq. 7.

If $X_n$ is $LT_0$ and $X_{n-1}$ is $LT_2$ and $X_{n-2}$ is $LT_4$ and $X_{n-3}$ is $LT_1$ and $X_{n-4}$ is $LT_1$

$$\text{then } X_{n+1} \text{ is } LT_3 \tag{7}$$

**Fig. 2** Time series and its corresponding linguistic terms

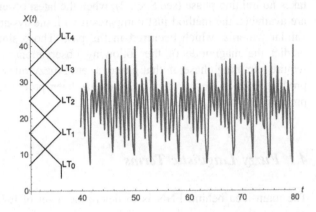

**Fig. 3** Part of a time series and their corresponding fuzzy values

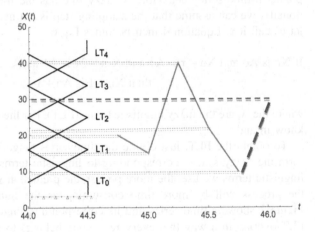

The form of this rule is based on two parameters window size, m, and sampling interval, $\tau$. The window size, indicates how many elements from the time series are considered to predict the following value (m = 7 for the rule in the example of Eq. 7). $\tau$ indicates how far apart from each other those elements of the time series are ($\tau = 1$ for Eq. 7).

## 4.3 Learning Phase

This section proposes an algorithm to learn the set of fuzzy rules from the time series. Algorithm 1 shows pseudocode of the procedure to learn the fuzzy rules.

Learn Rules takes as arguments X – the time series, m – the window size, τ – the sampling size, and returns the set of fuzzy rules derived from the time series.

To acquire the set of fuzzy rules (FR) we traverse the TS, recognizing patterns described by the time series, and verifying those places where similar patterns are found. Those recognized patterns are collected to form the rules. Line 1 initializes the set of rules and computes the size of the time series. Lines 2 and 3 mark every location of the time series as white, meaning the pattern starting at that location has not been discovered. The loop starting at Line 4 traverses the time series. The conditional statement at Line 5 makes sure all windows are processed only once, marking the window that ends at time n as processed. Lines 7 and 8 compile the fuzzy values of the time window into their corresponding linguistic terms (LT).[1] The pattern described by those linguistic terms is searched forward in the time series (loop in Line 10). All found matches are recorded in MFP – Matching Fuzzy Patterns (Line 11). The following value and the firing strength of the window are recorded (Line 12, initialized previously in Line 9). The set MFP is averaged using the Center of Gravity (COG) method. COG computes a weighted average and fuzzyfies the resulting value. The set of learned rules are returned in Line 14.

---

LearnRules(X, m, τ)
1    Rules = φ; N = size(TS)
2    For i = 1 to size(X)
3        Colori = white
4    For n = m to N
5        If Colorn = white
6            Colorn = black
7            For i=0 to m-1
8                LTi = Fuzzyfy(Xn-i)
9            MFP = φ
10           For j = n+1 to N
11               If Xj+mτ is LT0 & … & Xj is LTm-1 with membership μj>0
12                   MFP += (Xn+1, μj)
13           Rules += (Xj+mτ is LT0 & … & Xj is LTm-1 → X n+1 is COG(MFP)
14   Return Rules

---

**Algorithm 1.** Learning Fuzzy Rules from a Time Series

The set of FR produced by the procedure outlined above characterizes the behavior of the time series. Obtaining the set of rules is equivalent to producing a fuzzy model of the time series, or in terms of machine learning, to training the model.

---

[1] In the fuzzification process, we assume that the FLT selected is the one for which the maximum membership value is produced. Ties are resolved randomly.

## 4.4 Fuzzy Forecasting

Given a set of FR, performing a 1-step ahead forecasting reduces to deploying those rules to produce the predicted value. For the given set of FR and the observed $m\tau$ values in the TS's history, more than one rule may fire. As in any Fuzzy Inference System (FIS), the result (i.e., the forecast) is defuzzyfied using the center of gravity method. Algorithm 2 shows pseudocode of the procedure to forecast one value from last observations and the set of fuzzy rules.

```
FuzzyForecasting(FR, X, m, τ)
Fired = φ; N = size(TS)
2    For i = 1 to size(FR)
3        If (μi = Satisfies(TS, FRi)) > 0
4            Fired += (Consequent(FRi), μi)
5    Forecast = COG(Fired)
6    Return Forecast
```

**Algorithm 2.** Forecasting using the set of Fuzzy Rules

Procedure Fuzzy Forecasting takes as arguments the set of fuzzy rules – FR, the time series – X, up to the point where the forecast needs to be produced, m, and $\tau$ – the window size and sampling step, respectively. Line 1 initializes the set of fired rules and determines the size of the time series. Loop in Line 2 traverses the set of fuzzy rules. Line 3 verifies if the last observations of the time series satisfy the ith fuzzy rule. If so, the rule's consequent and the firing strength are recorded (Line 4). When all rules have been traversed, those that were fired are defuzzyfied using the Center of Gravity method (Line 5). The forecasted value is returned (Line 6).

## 4.5 Multi-step Simultaneous Forecasting

Given $m\tau$ values in the TS's history, the procedure described before produces a forecasted value for the magnitude of variable X at time n + 1, i.e., $X_{n+1}$. We can use the method described in the previous sections to produce r independent models to forecast $X_{n+i}$, $i \in [1, r]$. See Eq. (8).

$$X_{n+i} = f_i(X_n, X_{n-\tau}, X_{n-2\tau}, ..., X_{n-(m-1)\tau})$$  (8)

where $f_i$ is the i-th fuzzy model.

**Fig. 4** Scenario Similar to the one present in Fig. 3

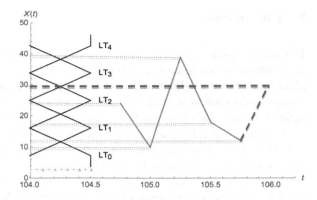

## 5 Study Case

As an illustrative example, we assume in a given time series, the situation of Fig. 3 is being analyzed. We try to produce a rule that reflects that scenario. The first scenario produces the rule given in Eq. 7, the membership functions of the different readings are: $\mu_{LT_0}(X_{45.75}) = 0.68$, $\mu_{LT_2}(X_{45.5}) = 0.6$, $\mu_{LT_4}(X_{45.25}) = 0.71$, $\mu_{LT_1}(X_{45.0}) = 0.90$, $\mu_{LT_1}(X_{44.75}) = 0.53$, the membership of the rule is min(0.68, 0.6, 0.71, 0.9, 0.53). The algorithm adds the consequent ($LT_2$, 0.53).

Traversing the time series, we find the scenario given in Fig. 4 at a subsequent time. The membership functions of the different readings in that patter are: $\mu_{LT_0}(X_{105.75}) = 0.45$, $\mu_{LT_2}(X_{105.5}) = 0.15$, $\mu_{LT_4}(X_{105.25}) = 0.55$, $\mu_{LT_1}(X_{105.0}) = 0.25$, $\mu_{LT_1}(X_{104.75}) = 0.09$, the membership of the rule is min(0.68, 0.6, 0.71, 0.9, 0.53). The algorithm adds the consequent ($LT_3$, 0.51).

The middle points and the firing strengths of the original and matching patterns are used to compute a weighted average (Eq. 9), and the result fuzzyfied again.

$$X_{n+1} = (24.94 * 0.53 + 33.81 * 0.51)/(0.53 + 0.51) = 29.28$$
$$X_{n+1} \text{ is } LT_3 \tag{9}$$

The fuzzy model corresponding to this time series is the set of all fuzzy rules collected by the algorithm. Assuming the following reading is present: (22, 12.5, 41, 20, 11). Their membership functions are: $\mu_{LT_0}(X_n) = 0.49$, $\mu_{LT_2}(X_{n-1}) = 0.49$, $\mu_{LT_4}(X_{n-2}) = 1.0$, $\mu_{LT_1}(X_{n-3}) = 0.25$, $\mu_{LT_1}(X_{n-4}) = 0.31$. These values fire the rule (perhaps among other rules) with a firing strength of min(0.49, 0.49, 1.0, 0.25, 0.31) = 0.25. This fact indicates that the final result for the forecasting is 33.81 * 0.25/0.25. = 33.81. If more rules had fired, the center of gravity method would have had more terms.

# 6 Conclusions

This paper presents a fuzzy method to forecast a time series. This fuzzy method is a fuzzy version of the nearest neighbor method for time series forecasting. The main idea of the nearest neighbor method is to verify what scenarios in the past are close enough to the current scenario and average their outcomes, assuming similar scenarios in history must have produce similar outcomes. This fuzzy version compiles a set of fuzzy rules from the whole time series. Those rules capture the different scenarios observed in the time series and turn them into fuzzy rules. When a forecast is needed, we present the latest observations to the set of rules and take into account the result of those rules that fire. The final outcome is a weighted average that takes into account the firing strength of each rule, which is somehow a measure of similarity of the current situation with the rule, which in turn was extracted from the time series according to the same similarity criteria. Similarity here is provided by the membership function of the linguistic terms.

One difference between the nearest neighbor method and fuzzy nearest neighbor is that nearest neighbor does not require training, but when a forecast is requested, it has to search for all similar past behaviors in the time series. FNN, on the other hand, requires training to learn the set of fuzzy rules. Nevertheless, when a forecast is required, its production just traverses the set of rules, which are much less in number than the observations in the time series. It can be easily shown that the time complexity of both, the search procedure in NN and the training phase in FNN are quadratic.

Similar to the NN method, FNN takes advantage of long time series, where a wide variety of scenarios have been exhibited by the system. This situation is in some cases prohibitive for traditional approaches (e.g., ARIMA), where generally, the computer resources are not enough to derive an appropriate model. Nowadays, where data gathering devices are ubiquitous, and storage is not an issue anymore, we need to be prepared to mine information from large repositories. This idea is know as big data.

This method has been tested manually and is currently under implementation. Once implemented, it has to be tested with different time series, real and synthetic. Its results have to be measured statistically and compared against NN, ARIMA, ANN, etc.

# References

Baldi, P., Brunak, S.: Bioinformatics: The Machine Learning Approach. MIT press (2001)
Brookwell, P.J., Davis, R.A.: Introduction to Time Series and Forecasting. Springer (2002)
Casdagli, M.: Nonlinear prediction of chaotic time series. Physica D **35**(3), 335–356 (1989)
Cowpertwait, P.S., Metcalfe, A.V.: Introductory Time Series with R. Springer (2009)
Farmer, J.D., Sidorowich, J.J.: Predicting chaotic time series. Phys. Rev. Lett. **59**(8), 845 (1987)

Gebhard Kirchgssner, J.W., Hassler, U.: Introduction To Modern Time Series Analysis. Springer (2013)

Leven, R., Koch, B.: Chaotic behaviour of a parametrically excited damped pendulum. Phys. Lett. A **86**(2), 71–74 (1981)

Lora, A.T., Riquelme, J.C., Ramos, J.L.M., Santos, J.M.R., Exposito, A.G.: Influence of knn-based load forecasting errors on optimal energy production. In: Progress in Artificial Intelligence, pp. 189–203. Springer (2003)

Maguire, L.P., Roche, B., McGinnity, T.M., McDaid, L.: Predicting a chaotic time series using a fuzzy neural network. Inf. Sci. **112**(1), 125–136 (1998)

Palit, A.K., Popovic, D.: Computational Intelligence in Time Series Forecasting. Springer (2005)

Pappis, C.P., Siettos, C.I.: Fuzzy reasoning. In: Burke, E.K., Kendall, G. (eds.) Introductory Tutorials in Optimization and Decision Support Techniques, 2nd edn. (2014)

Shumway, R.H., Stoffer, D.S.: Time Series Analysis and its Applications. Springer (2011)

Siler, W., Buckley, J.J.: Fuzzy Expert Systems and Fuzzy Reasoning. Wiley, New Jersey, USA (2005)

Sorjamaa, A., Lendsasse, A.: Time series prediction using dir-rec strategy. In: ESANN proceedings-European Symposium on ANN's (2006)

Storn, R., Price, K.: Differential evolution—a simple and efficient heuristic for global optimization over continuous spaces. Global Optim. **11**, 341–359 (1995)

Rainer Storn, K.P., Lampinen, J.: Differential Evolution. A Practical Approach to Global Optimization. Springer, Berlin (2005)

Wheelwright, S., Makridakis, S.: Forecasting Methods for Management (1985)

Dragomir Yankov, D.D., Keogh, E.: Ensembles of Nearest Neighbor Forecasts. ECMIL 1, pp. 545–556 (2006)

# Part IV
# Fuzzy Logic and Fuzzy Sets

# A Methodology for the Valuation of Quality Management System in a Fuzzy Environment

José M. Brotons-Martínez and Manuel E. Sansalvador-Selles

**Abstract** The needs and demands of the present markets have favoured the development of Quality Management. Among the different strategies related to quality management that an organisation can develop, the establishment of quality systems in accordance with the requirements established in ISO 9000 standards has acquired a special importance. The worldwide implementation and certification of ISO 9001 quality management systems have increased significantly during the lasts years. However, the contribution of ISO 9000 certification is a controversial issue. While firms continue to seek certification and some researchers support its value, other studies suggest that it has little value to companies. This work provides a tool that facilitates the valuation of the ISO 9001 quality system in large and small companies. To make it, given the uncertainty this process involves, the use of fuzzy math is very useful.

**Keywords** Quality system · Company valuation · Case study

## 1 Introduction

Even though some authors considered the "quality management movement" a fad, after more than two decades, it is still an important area of research in management as demonstrated by many papers still published on this topic. One of the most important areas in this field is the study of issues relating to ISO 9000 standards.

J.M. Brotons-Martínez (✉) · M.E. Sansalvador-Selles
Department of Economic and Financial Studies, Miguel Hernández University Edificio La Galia, Avda. de la Universidad s/n, 03202 Elche, Alicante, Spain
e-mail: jm.brotons@umh.es

M.E. Sansalvador-Selles
e-mail: manri@umh.es

© Springer International Publishing Switzerland 2015
J. Gil-Aluja et al. (eds.), *Scientific Methods for the Treatment of Uncertainty in Social Sciences*, Advances in Intelligent Systems and Computing 377,
DOI 10.1007/978-3-319-19704-3_15

The phenomenal growth in the number of companies attaining ISO 9001 certification worldwide suggests certification will yield benefits to the firm. According to the latest report published by the International Organization for Standardization (ISO), the 46,571 certifications granted throughout the world at the end of 1993 grew by 2013 to 1,129,446 ISO 9001 certifications in 187 different countries (ISO 2014). There are several reasons for this, but maybe the most important are two: the external pressures to be certified – mainly from industrial customers – and the belief in the benefits derived from the application of the norm (Rodríguez et al. 2006).

Notwithstanding, Juran (1999), one of the pioneers in quality management, has always been critical of the ISO 9000 standards. Juran warns about the lack of research actually capable of yielding evidence on the practical effects from the implementation and certification of ISO 9001 quality systems. Although several very interesting studies have been published in recent years analyzing the relationship between business performance and ISO 9001 certification, they hardly advance in the same direction, obtaining differing and even contradictory results. The contribution from ISO 9000 standards currently remains a controversial issue.

Therefore, the objective of this paper is to propose a valuation of the ISO 9001 quality management system certifications in large and small companies by developing a tool that makes this possible.

## 2 Proposed Methodology

Two valuations will be prepared, the first one assuming that the company has implemented the ISO 9001 quality system and the other assuming that the company has not implement it.

### 2.1 Valuation of the Company with the ISO 9001 Quality System

The valuation of the company will be performed by the discounted free cash flow model. To apply the discounted free cash flow model, a 5-year time period is considered, upon which distinct growth assumptions are applied according to the cases along with a residual period, to which a constant cash flow growth rate is applied for all assumptions.

$$V_0 = \sum_{t=1}^{5} \frac{CF_t}{(1+i)^t} + \frac{CF_T(1+g)}{(1+i)^5(i-g)} \tag{1}$$

where $V_0$ is the value of the company, $CF_t$ is the cash flow of the period t, i is the interest rate, and g is the cash flow growth rate in the residual period.

**Estimation of net sales.** The first step in the valuation of the company using discounted cash flows is the estimation of net sales, from which the evolution of the other items can be projected. Starting with the average turnover of the three years prior to 2012, it's possible estimating an equivalent growth rate for the expected time horizon and another that is different for the residual period.

**Growth in the time horizon.** For Spanish companies belonging to the same sector with ISO 9001 certification was calculated the growth rates. Taking the sector's full variation range offers a very wide interval. Because of this, the extreme values were eliminated, and taken were the average rate of change for the company occupying the middle value, the median of the values preceding the first quartile (lower extreme), and the median of the values beyond the third quartile (upper extreme).

We are aware that the uncertainty and subjectivity inherent in the valuation of the company advise treating them properly. In this respect, fuzzy logic is a particularly suitable tool, since it allows treating the present information, and not in specific terms, but rather incorporating the existing ambiguity and uncertainty into the model.

Because of this, the growth rate could be considered a triangular fuzzy number (TFN) that collects information from both the company as well as the sector, and considers three scenarios:

- Good case. This consists in considering the performance of a company similar to the best companies within the sector.
- Base case. This considers that the company's growth rate remains at the same levels as in recent years.
- Bad case. In the worst cases, this assumes behaviour that is similar to the worst companies within the sector with ISO 9001 certification, having negative growth rates.

From these three scenarios we can define the TFN growing rate with its membership function. Considering this, we are not in front of three punctual values, but in front of all of the intermediate values, being the information much more complete. The TFN is defined by three values $\tilde{A} = (a, \alpha, \beta)$, where a is the central point, $\alpha$ and $\beta$ are the left and the right radius (Zeng and Li 2007). The membership function is defined as,

$$
\mu = (x) = \begin{cases} \frac{x - a + \alpha}{\alpha} & a - \alpha < x < a \\ 1 & x = a \\ \frac{a + \beta - x}{\beta} & a < x < a + \beta \\ 0 & \text{otherwise} \end{cases}
\tag{2}
$$

**Growth during the residual period.** There is a pseudo axiom that must be considered in the long-term evolution of a company: the company's sales growth cannot be superior to the nominal gross domestic product (GDP) (Casanovas 2009). Initially considering a rate similar to the GDP growth rate could be acceptable, but as a precaution, a growth rate that is 50 % of this will be considered.

**Projected income statements.** Once the rate of change for net sales in upcoming years is estimated, we proceeded to estimate the income statements. The various components were grouped into material, personnel expenses, depreciation and amortization of fixed assets, other expenses, and financial performance.

**Company value.** The valuation of the company is made by discounting the expected cash flow in the time horizon and residual period in each of the three scenarios. The applicable discount rate is the risk-free rate plus a premium discount. For each company three values are obtained. Because of this, the value of the company is obtained by a TFN.

**Defuzzyfication.** Starting from a TFN $\tilde{A} = (a, \alpha, \beta)$ the crisp number containing all the information of that TFN can be obtained according to the expression

$$A = a + \frac{\beta - \alpha}{6} \tag{3}$$

Once the value of the company is obtained, we will set out again to determine its value, but this time under the assumption that it had not implemented and certified the ISO 9001 quality management system. In this manner, by the difference between both valuations, it will be possible to approximate the value that the commitment to quality benefits the firm.

## 2.2 Valuation of the Company Without the ISO 9001 Quality System

The starting point will be the average turnover of the three years prior to 2012, but with different growth rates applied in the time horizon. For the TFNs corresponding to the growth rates, again, we are going to consider three possible scenarios (good case, base case, and bad case). In the good case and bad case the analysis was focused exclusively on those companies within the sector without ISO 9001 certified quality systems. In the base case, we have used the company' growth in the last year without ISO 9001 certificate corrected by the change in the growth of the sector.

The same assumption will be considered for the residual period as for the company valuation with the quality system. While for the income statement projection identical assumptions will be applied to avoid distortions due to methodological differences, it must be remembered that the company will avoid some costs associated with the implementation and maintenance of the certified quality system, so they should be removed from the projections made. Finally, the discount rate applied will be the same.

In the determination of the costs deriving from the ISO 9001 quality system certification, we will distinguish between two components:

- Certification cost itself. This cost is easily quantifiable by corresponding to the amounts billed annually by the firm responsible for auditing the implemented system and certifying its adaptation to the contents of the ISO 9001 standard.
- Hidden quality costs. Costs for whose estimation, subjective and occasionally unconventional criteria must be resorted to. These consist in the costs for the internal staff involved in the implementation and maintenance of the quality system. Often, company employees, above all at certain times and especially in small businesses, will have to dedicate a significant part of their working hours to collaborate on these issues. Also noteworthy is the increase in bureaucracy that, in some cases, the acquisition and maintenance of ISO 9001 certification may involve (Vázquez et al. 2001).

From data supplied by the company, the quality costs were estimated. While the cost of certification corresponds exactly to that budgeted by the certifying company, in order to determine the hidden quality costs, we had to resort to estimations based on past experience. Thus, the personnel cost was obtained from the gross hourly cost of the employee (or employees) who assumed responsibility for the quality function, multiplied by the number of estimated hours spent managing it.

## 2.3 Valuation of the Quality System

The quality system can be obtained as the difference between the valuation of the company with the ISO 9001 quality system minus the valuation of the company without the ISO 9001 quality system.

## 3 Case Study

In order to prove the validity of the instrument developed we decided to look at a real case and apply the proposed methodology. To do so, we looked for the collaboration of a company; to be exact, a small engineering and environmental company (NACE code 3320) situated in Madrid (Spain). This is a sector where the characteristics of implementation the ISO 9001 quality system can have an especial relevance and where precisely small and medium-sized enterprises (SMEs) abound. The analysed company has obtained ISO 9001 certificate in 2004.

The key figures in the company's income statement, available on the date the valuation was made, are shown in Table 1 (SABI 2012).

Table 1 Company data for the 2004–2011 period (SABI 2012)

| Year | 2011 | 2010 | 2009 | 2008 | 2007 | 2006 | 2005 | 2004 |
|---|---|---|---|---|---|---|---|---|
| Net sales | 1.400.391 | 1.187.523 | 1.113.289 | 1.327.798 | 1.585.310 | 1.396.758 | 1.768.097 | 1.721.332 |
| Material | 6.843 | 6.489 | 30.336 | 10.928 | 8.528 | -2.179 | -166 | 1.218 |
| Personnel costs | 578.790 | 493.845 | 466.724 | 506.888 | 661.412 | 584.767 | 795.117 | 812.862 |
| Depreciation fixed assets | 505.041 | 492.864 | 494.040 | 527.920 | 500.010 | 472.369 | 509.681 | 511.080 |
| Other expenses | 52.137 | 51.344 | 58.479 | 51.107 | 52.108 | 31.395 | 29.814 | 31.384 |
| Financial performance | 183.972 | 199.988 | 225.343 | 244.439 | 209.483 | 209.487 | 204.879 | 162.519 |
| Extraordinary results | | | | 2.093 | | 2.441 | -6.017 | 2.594 |
| Corporate tax | | | | | 45.237 | 29.701 | 74.190 | 68.794 |
| Net profit | 87.294 | -44.029 | -100.961 | 6.279 | 125.588 | 69.301 | 148.233 | 138.505 |
| Cash flow | 4.698 | 67.841 | 67.554 | 39.952 | 60.465 | 47.007 | 56.570 | 61.464 |

## 3.1 Valuation Assuming that It Has the Quality System Implementation

The company implements a quality management system at the beginning of 2004, and may see an annual sales growth rate change from 7.62 % for the 2000-2003 period to −4.71 % for the 2004–2007 period. The growth rates for Spanish companies belonging to the same sector (NACE Code 3320) with ISO 9001 certification was calculated for the 2004–2007 period (the information contained within the SABI database was used for this). Taking the sector's full variation range offers a very wide interval (−22.4, +71.3 %), being the median of the sector, 4.3 %. So the number (4.3, 18.0, 75.68) can be considered as a triangular fuzzy number, where 4.3 is the centre, 18.0 is the left radius and 75.68 the right radius. The support of this fuzzy number is excessively large, so we have taken as extreme values the median of the first and third quartile, and as the centre point the median of the sector (−4.3, 8.1, 9.3).

The growth considered during the residual period of 1.55 % which is the 50 % of the growth rate of the GPD for the period 1980–2007 (INE 2012)

Once the rate of change for net sales in upcoming years is estimated, we proceeded to estimate the income statements (base case, Table 2). The different items were projected based on regressions with the net sales according to the accounting information available in the SABI database for 1996–2007. The various components were grouped into material, personnel expenses, depreciation and amortization of fixed assets, other expenses, and financial performance.

The applicable discount rate is the risk-free rate plus a premium discount. For the risk-free rate, the 10-year state bond is considered. On December 29, 2007, and in accordance with the rates offered by the Bank of Spain for that date, this was 4.43 %, while for the risk premium, given the dispersion existing in the literature, a range between 4.2 and 8.5 % will be considered (the range obtained in the aforementioned literature) with a most possible value of 5.25 % (average for managers,

**Table 2** Income statement estimation in the base case for the 2008–2013 periods

|  | 2008 | 2009 | 2010 | 2011 | 2012 | 2013 |
|---|---|---|---|---|---|---|
| **Net sales** | **1.334.467** | **1.271.647** | **1.211.784** | **1.154.739** | **1.100.380** | **1.114.791** |
| Material | 568.954 | 542.702 | 517.686 | 493.847 | 471.130 | 477.153 |
| Personnel costs | 432.200 | 412.921 | 394.550 | 377.043 | 360.361 | 364.783 |
| Dep. fixed assets | 36.522 | 35.063 | 33.672 | 32.347 | 31.084 | 31.419 |
| Other expenses | 156.719 | 148.370 | 140.415 | 132.835 | 125.611 | 127.526 |
| Financial performance | −4.815 | −4.751 | −4.690 | −4.632 | −4.577 | −4.592 |
| **Profit before taxes** | **135.258** | **127.840** | **120.771** | **114.035** | **107.616** | **109.318** |
| Corporate tax | 33.815 | 31.960 | 30.193 | 28.509 | 26.904 | 27.330 |
| **Net profit** | **101.444** | **95.880** | **90.579** | **85.526** | **80.712** | **81.989** |
| **Cash flow** | **137.966** | **130.943** | **124.250** | **117.873** | **111.796** | **113.407** |

analysts, and university professors). Thus, the TFN risk premium $\tilde{i} = (5.25, 1.05, 3.25)$, by which the discount rate is $\tilde{i} = (9.68, 1.05, 3.25)$.

The valuation of the company is made by discounting the expected cash flow in the time horizon and residual period of the aforementioned discount rate. With all this information, the value of the company is not a TFN, but it has been approximated by Zeng and Li (2007) getting the TFN (€1,241,478, €590,972, €932,186) which point out the most possible value and the left and right deviations, estimation acceptable according to Jimenez and Rivas (1996). Applying (3), we get the value €1,298,347.

Once the value of the company in early 2012 is obtained, we will set out again to determine its value at that time, but this time under the assumption that it had not implemented and certified the ISO 9001 quality management system. In this manner, by the difference between both valuations, it will be possible to approximate the value that the commitment to quality benefits the firm.

## 3.2 Valuation Assuming that It Has not the Quality System Implemented

Here the company will be valued at the same date, early 2012, but under the assumption that in 2004 it had not implemented the ISO 9001 quality system. In order to make the valuation, the starting point will be the sales for 2011, but with different growth rates applied in the time horizon. The same assumption will be considered for the residual period as for the company valuation with the quality system. While for the income statement projection identical assumptions will be applied to avoid distortions due to methodological differences, it must be remembered that the company will avoid some costs associated with the implementation and maintenance of the certified quality system, so they should be removed from the projections made, a reason for which a section is devoted to the estimation of those costs. Finally, the discount rate applied will be the same.

By not considering changes in the residual period with respect to the previous case, we are going to focus exclusively on the time horizon. Therefore, our objective is to estimate a growth rate that contains both the information available for the company (4.76 % in the last period that the company did not have the certified quality system, 2001–04) as well as for those companies within the sector without ISO 9001 certified quality systems (SABI 2012).

The median of the values preceding the first quartile (lower extreme), the median of the values beyond the third quartile (upper extreme) and the median of the of the part of the sector that do not have implemented the quality systems for the period 2004–07 give us the TFN (−9.3 %, 7.7 %, 7.7 %).

The TFN of the growing rates of the company we have considered the extreme values of the TFN of the sector, and as a central point we have considered the sum of the growing rate of the company for the last period for which it has not implemented the quality system plus the increase of the median sector growing (without system quality), $-10.13$ %. So The TFN used for estimating the growth of the company is $(-10.13, 6.90, 8.45)$.

The valuation of the company, just as in the previous case, is made by discounting the expected cash flow from the time horizon and residual period in each one of the three scenarios considered according to the discount rate $\tilde{i} = (9.68, 1.05, 3.25)$. The growth rate values for the cash flow in the residual period (g) are the same as in the previous case. The results show the FN value of the company. It can be approximated by the TFN (€955,823, €436,029, €656,716), where €955,823 is the most possible value of the company, but in can increase till €1,612,539 or decrease until 519,794.

The defuzzification of the preceding fuzzy number according to the expression (3), provides the company's value in early 2012 assuming that it did not have the ISO 9001 certified quality system in place. This value is €992,604.

The previous estimation has been made considering the quality systems costs. So, we have to estimate these costs and to take away from the previous value.

Estimated costs for the implementation and certification of the quality system are supplied by the company. While the cost of certification corresponds exactly to that budgeted by the certifying company (1.274 €/year in 2012), in order to determine the hidden quality costs, we had to resort to estimations based on past experience. Thus, the personnel cost was obtained from the gross hourly cost of the employee who assumed responsibility for the quality function, multiplied by the number of estimated hours spent managing it. Moreover, it was decided to approximate the cost of those materials, such as office supplies, consumed by the ISO 9001 quality system. For 2012, the quality systems costs were estimated in €1,971 for the optimistic case, €1,664 for the base case and €1,433 for the pessimistic case. According to the rates budgeted by the certifying company and their anticipated evolution in upcoming years, triennial growth of 2 % was calculated, which for simplicity was also extrapolated to the remaining estimated costs. Appling (3) maintaining the ISO 9001 certified represents a cost of €62,353.

Finally, the value of the company, assuming that it has not implemented the ISO 9001 quality system is the value of the company obtained previously (€992,604) less the cost of the quality system (€62,353), that is, €927,132.

## 3.3 Valuation of the ISO 9001 Quality System

As a result, the value of the quality system for 2012 is the difference between the value of the company assuming that it has the quality system ISO 9001 implemented (€1,298,347) and the value of the company assuming that it has not the quality system ISO 9001 implemented (€927,132). The result shows that the value of the quality systems has a positive value of €371,216.

## 4  Conclusions

This paper introduces a methodological proposal for the valuation of the ISO 9001 quality system in a fuzzy environment. In order to test the proposal's validity, it was decided to create a real case application of the methodology, for which collaboration with a company in possession of the ISO 9001 quality system certification was requested.

One method used widely in business valuation is discounted cash flow. However, its use presents high levels of uncertainty and subjectivity in both future cash flow estimation and the type of interest used for discounting, so the use of fuzzy math is very useful.

We have valuated the company twice, once assuming the company has implemented the quality system and the other considering that it has not implemented it.

The first valuation has been considering the company has implemented the ISO 9001 quality system. To determine the growth rate in the time horizon, TFN were used that are capable on the one hand of collecting the company's evolution in recent years, and on the other, the evolution of companies within the sector that have implemented ISO 9001 quality systems, getting the TFN (€1,241,478, €590,972, €932,186), where €1,241,478 is the centre, and €590,972 and €932,186 are the left and right radius respectively. The defuzzified value is €1,298,347.

The valuation process was repeated, but now by assuming that the company did not possess ISO 9001 certification. The same discount rate was used for this, but with different growth rates: on the one hand, the company's growth rate before implementing the ISO 9001 quality system corrected by the difference between the median of the sector (that has implemented the quality system) between 2004 (the year that the company has implemented the quality system) and 2012, and on the other, the growth rate for companies within the sector but without ISO 9001 certification. In this section, the estimation made for the costs deriving from the certified quality system is remarkable. These costs must be removed from the projections made in the company's income statement under the hypothesis of an absence of ISO 9001 certification. By comparing the two estimated values, with and without the quality, it is possible to make an approximation of the contribution that ISO 9001 certification benefits the company. As a result, the value of the company is the value of the company obtained previously (€992,604) less the cost of the quality system (€ 62,353), that is, €927,132.

Consequently, the final value of the quality management system is the difference between these two valuations: €1,298,347 and €927,132, that is, €371.216.

## References

Casanovas, M.: Valoración de empresas: bases conceptuales y aplicaciones prácticas. Profit Editorial, Barcelona (2009)
INE (Spanish Statistical Office): Contabilidad nacional trimestral de España. http://www.ine.es/jaxiBD/tabla.do?per=03&type=db&divi=CNTR&idtab=9 (2012)

ISO: The ISO Survey of Management System Standard Certifications – 2013. http://www.iso.org (2014)

Jiménez, M., Rivas, J.A.: Aproximación de números borrosos. Paper presented at the III Congreso de la Sociedad Internacional de Gestión de empresas y Economía Fuzzy (SIGEF), vol. I, pp. 2–12. Buenos Aires, Argentina (1996)

Juran, J.M.: Juran urges research. What they're saying about standards. Qual. Prog. **32**(7), 30–31 (1999)

Rodríguez-Escobar, J.A., Gonzalez-Benito, J., Martínez-Lorente, A.R.: An analysis of the degree of small companies' dissatisfaction with ISO 9000 certification. Total Qual. Manag. Bus. Excellence **17**(04), 507–521 (2006)

SABI-Iberian Balance sheet Analysis System: https://sabi.bvdinfo.com/version-2013514/home.serv?product=sabineo (2012) (referred on December 2013)

Vázquez, C., Fernández, E., Escanciano, C.: La relación entre el coste y los beneficios de la certificación ISO 9000: Resultados de un estudio empírico. Inv. Europeas Dirección Economía Empresa **7**(1), 135–146 (2001)

Zeng, W., Li, H.: Weighted triangular approximation of fuzzy numbers. Int. J. Approximate Reasoning **46**, 137–150 (2007)

# A Qualitative Study to Strong Allee Effect with Fuzzy Parameters

Xavier Bertran, Dolors Corominas and Narcis Clara

**Abstract** The Allee effect is related to those aspects of dynamical of populations Connected with a decreasing in individual fitness when the population size diminishes to very low levels. In this work we propose a fuzzy approach to Allee effect that permits to deal with uncertainty. This fuzzy proposal is considering the Allee effect with fuzzy parameters from two points of view.

**Keywords** Allee effect · Fuzzy number · Level sets · Hukuhara difference · Fuzzy differential equations

## 1 Introduction

In single population Dynamics the rate of change of the number of individuals $x(t)$ is often expressed as a function of the per capita rate of change $\varphi(x)$ such as

$$\frac{dx}{dy} = x\varphi(x) \tag{1}$$

The Malthus equation considers a constant per capita growth and is the simplest way to model the growth of a specie. This equation has only one equilibrium point at $x = 0$ and the population tends to infinity if $k > 0$ (0 is unstable) or vanish if $k < 0$ (0 is stable). This point of view implies an exponential growth of the population.

X. Bertran (✉) · D. Corominas · N. Clara
Departament D'Empresa, Departament D'Informàtica I Matemàtica Aplicada,
Universitat de Girona, Campus de Montilivi S/N, 17071 Girona, Spain
e-mail: Xavier.bertran@udg.edu

D. Corominas
e-mail: dolors.corominas@udg.edu

N. Clara
e-mail: narcis.clara@udg.edu

© Springer International Publishing Switzerland 2015
J. Gil-Aluja et al. (eds.), *Scientific Methods for the Treatment of Uncertainty in Social Sciences*, Advances in Intelligent Systems and Computing 377,
DOI 10.1007/978-3-319-19704-3_16

The logistic equation

$$\frac{dx}{dy} = g(x) = rx\left(1 - \frac{x}{k}\right) \tag{2}$$

Solves the question of the existence of a positive equilibrium point because $x = k$ is an asymptotically stable point and the size of the population tends to k. Figure 1 describes this very well known equation.

One feature of Eq. (2) often criticized in ecological Dynamics is that for small values the rate growth is always positive. Several equations have been proposed to deal with this problem. We will focus our attention to the strong Allee effect what implies that bellow a positive number the population goes to extinction. A differential equation that governs the population growth taking into account these aspects is

$$\frac{dx}{dy} = g(x) = rx\left(1 - \frac{x}{k}\right)(x - a) \tag{3}$$

where, r is the intrinsic rate of increase, k is the carrying capacity and a is the extinction threshold. Equation (3) has three equilibrium points, two are stable (0 and k) and one is unstable (a). Figure 1 shows the behaviour of function f and its differences with the logistic equation.

**Fig. 1** Logistic equation vs Allee effect

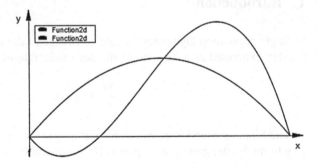

The per capita growth of the logistics equation is a linear decreasing function instead of the Allee effect is increasing for small values although negative.

The Allee effect has strong evidence in ecological Systems (Courchamp et al. 1999) and is associated to a wide range of Systems focusing in conspecific interactions, rarity and animal sociality (Allee 1938; Amarasekare 1998; Stephens and Sutherland 1999; Sutherland 1998). Conspecific interaction can help to understand aggregative behaviour (Reed and Dobson 1993), Allee effect can explain some properties about the Distribution of individuals between patches (Fretwell 1972), it is obviously related to extinction (Poggiale 1998) and influence any system of maximum sustainable yields as some fisheries (Lande et al. 1994). Bidimensional models also benefit with the introduction of the Allee effect (Aguirre et al. 2009).

The analysis of environmental conditions which influence the growth of population is subject to imprecise factors. In practice, the coefficients in the Allee effect equation are calculated taking into account many uncertain factors such as natural and social resources, age structure, assistance. This uncertainty makes it appropriate to express the equation in a fuzzy context and leads to the introduction of fuzzy parameters in the Allee effect equation. From this point of view, the population at instant t, $x(t)$, is considered to be a fuzzy number for every $t \geq 0$, that is, the number of individuals is not known exactly, but subject to a degree of accuracy $\alpha \in [0, 1]$.

In this paper, we consider the Allee effect equation $x' = rx\left(1 - \frac{x}{k}\right)(x - a)$, where $r, k, a > 0$ and rewrite this real Allee effect equation into its equivalent form $x' + rax + \frac{r}{k}x^3 = \left(r + \frac{ra}{k}\right)x^2$, and consider $x, a \in \mathcal{R}$, in such a way that operations are fuzzy numbers operations.

We point out that there are several approaches to evaluate fuzzy expressions and one obtains, in general, different results. Nieto and Rodriguez-López (2008), analyse the logistic equation from two different points of view: Extending the real logistic function by using Zadeh's Extension Principle and writing the real logistic equation into its equivalent form $x' + \frac{r}{k}x^2 = rx$ and we consider $x, r, k \in \mathcal{R}$. Bertran, Clara i Ferrer (2010) analyse the Allee effect equation extending the real Allee effect function by using Zadeh's Extension Principle.

## 2 Allee Effect Equation with Fuzzy Parameters

From the analysis in the Bertran, Clara and Ferrer, we deduce that Zadehs Extension Principle leads to a situation which extends the classical behaviour of the Allee effect equation, in the sense that crisp solutions are the same as in the ordinary case. However, if the initial condition is not crisp, solutions have a complex expression. In this section we present an alternative approach to the analysis of fuzzy Allee effect equation. We analyse the Allee effect equation from two different points of view: using the Minkowski difference and using the Hukuhara difference.

In our context we must consider that the growth of a population follows the allee effect law and it is given by a fuzzy real number $\tilde{x}(t)$ for every $t$ and the extinction threshold is given by a fuzzy real number $\tilde{a}$. Solving the levelwise equations, we obtain that the population can be calculated subject to a degree of accuracy $\alpha \in [0, 1]$.

In order to simplify the notation we note the level sets of fuzzy real numbers $\tilde{x}(t)$ and $\tilde{a}$: $\underline{x}(\alpha, t) = x_1$, $\bar{x}(\alpha, t) = x_2$, $\underline{a}(\alpha) = a_1$ and $\bar{a}(\alpha) - u_2$.

First of all, we must find an appropriate way to express the equation in order to deal with fuzzy parameters.

If we consider fuzzy Allee effect equation taking into account the Minkowski difference,

$$\tilde{x}'(t) = \frac{r}{k}\tilde{x}(t)(k + (-\tilde{x}(t)))(\tilde{x}(t) + (-\tilde{a}))$$

this equation can be written levelwise as

$$\begin{cases} x_1' = \frac{r}{k}x_1(k - x_2)(x_1 - a_2) \\ x_2' = \frac{r}{k}x_2(k - x_1)(x_2 - a_1) \end{cases}$$

which is a coupled system.

Notice that, in this case we have three equilibrium points: $P_1 = (0,0)$, $P_2 = (0,a_1)$, $P_3 = (k,k)$.

Analysing qualitatively the equilibrium points, we obtain,

- If $0 < x_1 \le x_2 \le a_1 \le a_2 \le k$, then $x_1' < 0$ and $x_2' < 0$, and the trajectories tend to the point $P_1 = (0,0)$.
- If $x_1 = 0$ and $0 < x_2 \le a_1 \le a_2 \le k$, then $x_1' = 0$ and $x_2' < 0$, and the trajectories tend to the point $P_1 = (0,0)$.
- If $x_1 = 0$ and $0 \le a_1 \le x_2 \le a_2 \le k$, then $x_1' = 0$ and $x_2' > 0$.
- If $0 < x_1 \le a_1 \le x_2 \le a_2 \le k$ then $x_1' < 0$ and $x_2' > 0$.
- If $x_2 = a_1$ and $0 < x_1 \le a_1$, then $x_1' < 0$ and $x_2' = 0$.
- If $0 \le a_1 \le a_2 \le x_1 \le x_2 \le k$, then $x_1' > 0$ and $x_2' > 0$.
- If $x_2 = k$ and $0 \le a_1 \le a_2 \le x_1 \le k$, then $x_1' > 0$ and $x_2' > 0$.

Hence, $P_1 = (0,0)$ is an attractor and $P_2 = (0,a_1)$ and $P_3 = (k,k)$ are not attractors

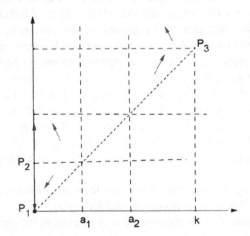

In the other hand, if we consider the equation in terms of the Hukuhara difference we get,

$$x' + rax + \frac{r}{k}x^3 = \left(r + \frac{ra}{k}\right)x^2$$

And this equation can be written levelwise as

$$\begin{cases} x_1' = \frac{r}{k}x_1(k - x_1)(x_1 - a_1) \\ x_2' = \frac{r}{k}x_2(k - x_2)(x_2 - a_2) \end{cases}$$

Notice that, in this case we have six equilibrium points: $P_1 = (0,0)$, $P_2 = (0, a_2)$, $P_3 = (0, k)$, $P_4 = (a_1, a_2)$, $P_5 = (a_1, k)$, $P_6 = (k, k)$. Analysing qualitatively the equilibrium points, we obtain,

- If $0 < x_1 \le x_2 \le a_1 \le a_2 \le k$, then $x_1' < 0$ and $x_2' < 0$, and the trajectories tend to the point $P_1 = (0,0)$.
- If $x_1 = 0$ and $0 < x_2 \le a_1 \le a_2 \le k$, then $x_1' = 0$ and $x_2' < 0$, and the trajectories tend to the point $P_1 = (0,0)$.
- If $x_1 = 0$ and $0 \le a_1 \le x_2 \le a_2 \le k$, then $x_1' = 0$ and $x_2' < 0$.
- If $x_2 = a_1$ and $0 < x_1 \le a_1$, then $x_1' < 0$ and $x_2' < 0$.
- If $0 \le x_1 \le a_1 \le x_2 \le a_2 \le k$, then $x_1' < 0$ and $x_2' < 0$.
- If $x_2 = a_2$ and $0 < x_1 \le a_1 \le a_2 \le k$, then $x_1' < 0$ and $x_2' = 0$.
- If $x_1 = 0$ and $0 \le a_1 \le a_2 \le x_2 \le k$, then $x_1' = 0$ and $x_2' > 0$.
- If $x_1 = 0$ and $0 \le a_1 \le a_2 \le k \le x_2$, then $x_1' = 0$ and $x_2' < 0$.
- If $0 \le x_1 \le a_1 \le a_2 \le x_2 \le k$, then $x_1' < 0$ and $x_2' > 0$.
- If $x_2 = k$ and $0 < x_1 \le a_1 \le a_2 \le k$, then $x_1' < 0$ and $x_2' = 0$.
- If $0 \le x_1 \le a_1 \le a_2 \le k \le x_2$, then $x_1' < 0$ and $x_2' < 0$.
- If $x_1 = a_1$ and $0 \le a_1 \le x_2 \le a_2 \le k$, then $x_1' = 0$ and $x_2' < 0$.
- If $x_1 = a_1$ and $0 \le a_1 \le a_2 \le x_2 \le k$, then $x_1' = 0$ and $x_2' > 0$.
- If $0 < a_1 \le x_1 \le x_2 \le a_2 \le k$, then $x_1' > 0$ and $x_2' < 0$.
- If $0 < a_1 \le x_1 \le a_2 \le x_2 \le k$, then $x_1' > 0$ and $x_2' > 0$.
- If $x_2 = a_2$ and $0 < a_1 \le x_1 \le a_2 \le k$, then $x_1' > 0$ and $x_2' = 0$.
- If $x_1 = a_1$ and $0 \le a_1 \le a_2 \le k \le x_2$, then $x_1' = 0$ and $x_2' < 0$.
- If $x_2 = k$ and $0 < a_1 \le x_1 \le a_2 \le k$, then $x_1' > 0$ and $x_2' = 0$.
- If $0 < a_1 \le x_1 \le a_2 \le k \le x_2$, then $x_1' > 0$ and $x_2' < 0$.
- If $0 < a_1 \le a_2 \le x_1 \le x_2 \le k$, then $x_1' > 0$ and $x_2' > 0$.
- If $0 < a_1 \le a_2 \le x_1 \le k \le x_2$, then $x_1' > 0$ and $x_2' < 0$.
- If $x_2 = k$ and $0 < a_1 \le a_2 \le x_1 \le k$, then $x_1' > 0$ and $x_2' = 0$.

Hence, $P_1 = (0,0)$ and $P_6 = (k, k)$ are attractors and $P_2 = (0, a_2)$, $P_3 = (0, k)$, $P_4 = (a_1, a_2)$, and $P_5 = (a_1, k)$ are not attractors.

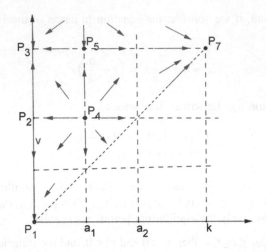

## 3 Examples

In this section we realise a quality study and compare the behaviour of the population using Minkowski difference and Hukuhara difference. To this purpose we take $r = 0.8$, $k = 10$ and the triangular fuzzy number $\tilde{a} = (5, 7, 9)$. Solving the levelwise equations, we obtain that the population can be calculated subject to a degree of accuracy $\alpha \in [0, 1]$.

Using Minkowski difference we get,

$$\begin{cases} x_1' = \frac{0.8}{10} x_1 (10 - x_2)(x_1 - 9 + 2\alpha) \\ x_2' = \frac{0.8}{10} x_2 (10 - x_1)(x_2 - 5 - 2\alpha) \end{cases}$$

In this case, the equilibrium points are $P_1 = (0, 0)$, $P_2 = (0, 5 + 2\alpha)$, $P_3 = (10, 10)$. In the next graphs we can see the vector field in the following zones:

$zone\ 1 = \{(x_1, x_2)\ tq\ \forall \alpha \in [0, 1], 0 \leq x_1 \leq x_2 \leq 5 + 2\alpha\}$

$zone\ 2 = \{(x_1, x_2)\ tq\ \forall \alpha \in [0, 1], 0 \leq x_1 \leq 5 + 2\alpha \leq x_2 \leq 9 - 2\alpha\}$

$zone\ 3 = \{(x_1, x_2)\ tq\ \forall \alpha \in [0, 1], 0 \leq x_1 \leq 5 + 2\alpha \leq 9 - 2\alpha \leq x_2 \leq 10\}$

$zone\ 4 = \{(x_1, x_2)\ tq\ \forall \alpha \in [0, 1], 0 \leq 5 + 2\alpha \leq x_1 \leq x_2 \leq 9 - 2\alpha\}$

$zone\ 5 = \{(x_1, x_2)\ tq\ \forall \alpha \in [0, 1], 0 \leq 5 + 2\alpha \leq x_1 \leq 9 - 2\alpha \leq x_2 \leq 10\}$

$zone\ 6 = \{(x_1, x_2)\ tq\ \forall \alpha \in [0, 1], 0 \leq 5 + 2\alpha \leq 9 - 2\alpha \leq x_1 \leq x_2 \leq 10\}$

$zone\ 7 = \{(x_1, x_2)\ tq\ \forall \alpha \in [0, 1], 0 \leq 5 + 2\alpha \leq 9 - 2\alpha \leq x_1 \leq 10 \leq x_2\}$

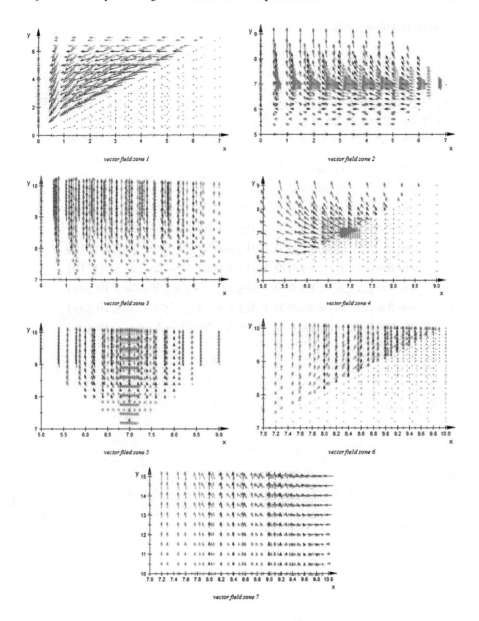

vector field zone 1

vector field zone 2

vector field zone 3

vector field zone 4

vector filed zone 5

vector field zone 6

vector field zone 7

Using Hukuhara difference we get,

$$\begin{cases} x_1' = \frac{0.8}{10} x_1 (10 - x_1)(x_1 - 5 - 2\alpha) \\ x_2' = \frac{0.8}{10} x_2 (10 - x_2)(x_2 - 9 + 2\alpha) \end{cases}$$

In this case, the equilibrium points are $P_1 = (0,0)$, $P_2 = (0, 9 - 2\alpha)$, $P_3 = (0, 10)$, $P_4 = (5 + 2\alpha, 9 - 2\alpha)$, $P_5 = (5 + 2\alpha, 10)$, $P_6 = (10, 10)$.

In the next graphs we can see the vector field in the following zones:

$zone\ 1 = \{(x_1, x_2)\ tq\ \forall \alpha \in [0, 1], 0 \le x_1 \le x_2 \le 5 + 2\alpha\}$

$zone\ 2 = \{(x_1, x_2)\ tq\ \forall \alpha \in [0, 1], 0 \le x_1 \le 5 + 2\alpha \le x_2 \le 9 - 2\alpha\}$

$zone\ 3 = \{(x_1, x_2)\ tq\ \forall \alpha \in [0, 1], 0 \le x_1 \le 5 + 2\alpha \le 9 - 2\alpha \le x_2 \le 10\}$

$zone\ 4 = \{(x_1, x_2)\ tq\ \forall \alpha \in [0, 1], 0 \le 5 + 2\alpha \le x_1 \le x_2 \le 9 - 2\alpha\}$

$zone\ 5 = \{(x_1, x_2)\ tq\ \forall \alpha \in [0, 1], 0 \le 5 + 2\alpha \le x_1 \le 9 - 2\alpha \le x_2 \le 10\}$

$zone\ 6 = \{(x_1, x_2)\ tq\ \forall \alpha \in [0, 1], 0 \le 5 + 2\alpha \le 9 - 2\alpha \le x_1 \le x_2 \le 10\}$

$zone\ 7 = \{(x_1, x_2)\ tq\ \forall \alpha \in [0, 1], 0 \le x_1 \le 5 + 2\alpha \le 9 - 2\alpha \le 10 \le x_2\}$

$zone\ 8 = \{(x_1, x_2)\ tq\ \forall \alpha \in [0, 1], 0 \le 5 + 2\alpha \le x_1 \le 9 - 2\alpha \le 10 \le x_2\}$

$zone\ 9 = \{(x_1, x_2)\ tq\ \forall \alpha \in [0, 1], 0 \le 5 + 2\alpha \le 9 - 2\alpha \le x_1 \le 10 \le x_2\}$

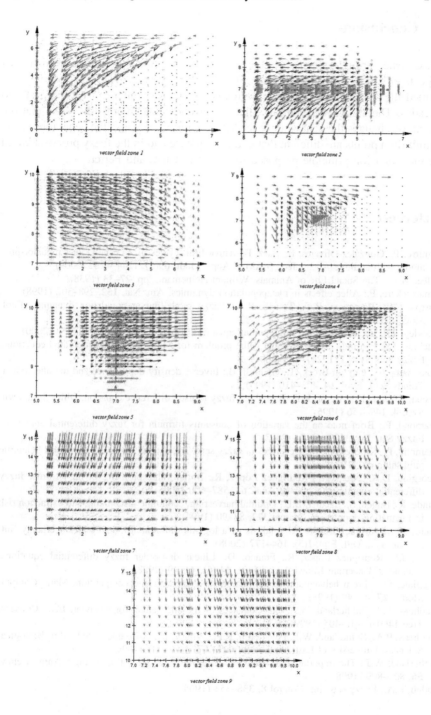

vector field zone 1

vector field zone 2

vector field zone 3

vector field zone 4

vector field zone 5

vector field zone 6

vector field zone 7

vector field zone 8

vector field zone 9

# 4 Conclusions

W. C. Allee brought attention to the possibility of a positive relationship between aspects of fitness and population size over fifty years ago. This phenomenon, termed the Allee effect, has been the focus of increased interest over the past two decades. The study of the strong Allee effect under uncertainty shows us that the fuzziness changes the behaviour of the set of solutions because the behaviour of the equilibrium points are different that in the crisp case and in the fuzzy proposal based on the extension principle proposed for Bertran, Clara and Ferrer.

# References

Aguirre, P., González-Olivares, E., Saez, E.: Three límit cycles in a Leslie-Gower predator-prey model with additive Allee effect. SIAM J. Appl. Math. **69**(5), 1244–1262 (2009)

Allee, W.C.: The Social Life of Animals. William Heinemann, pp. 27–33 (1938)

Amarasekare, P.: Allee effects in metapopulation dynamics. Am. Nat. **152**, 298–302 (1998)

Barros, L.C., Bassanezi, R.C., Tonelli, P.A.: Fuzzy modelling in population dynamics. Ecol. Model. **128**, 27–33 (2000)

Buckley, J.J., Feuring, T.: Fuzzy differential equations. Fuzzy Sets Syst. **110**, 43–54 (2000)

Buckley, J.J., Feuring, T.: Fuzzy initial value problem for Nth-order linear differential equations. Fuzzy Sets Syst. **121**, 247–255 (2001)

Courchamp, F., Clutton-Brock, T., Grenfell, B.: Inverse density dependence and the allee effect. Trends Ecol. Evol. **14**, 405–410 (1999)

Deeba, E.Y., de Korvin, A., Koh, E.L.: On a fuzzy logistic difference equation. Differ. Eqn. Dyn. Syst. **4**, 149–156 (1996)

Diamond, P.: Brief note on the variation of constants formula for fuzzy differential equations. Fuzzy Sets Syst. **129**, 65–71 (2002)

Diamond, P., Kloeden, P.: Metric spaces of fuzzy sets. Theory and Applications. World Scientific publishing co. (1994)

Georgiou, D.N., Nieto, J.J., Rodríguez López, R.: Initial value problem for higher-order fuzzy differential equations. Nonlinear Anal. **63**, 587–600 (2005)

Lande, R., Engen, S., Saether, B.E.: Optimal harvesting, economic discounting and extinction risk in fluctuating populations. Nature **372**, 88–90 (1994)

Nieto, J.J., Rodríguez López, R.: Analysis of a logistic differential model with uncertainty. Int. J. Dyn. Syst. Diff. Eqn. **1**(3), 164–177 (2008)

Nieto, J.J., Rodríguez López, R., Franco, D.: Linear first-order fuzzy differential equations, Internat. J. Uncertain Knowl.-Based Syst. **14** (6), 41–49 (1998)

Poggiale, J.C.: From behavioural to population level: growth and competition. Math. Comput. Model. **27**, 41–49 (1998)

Stephens, P.A., Sutherland, W.J.: Consequences of the allee effect for behavior. Ecol. Conserv. Tree **14**(10), 401–405 (1999)

Stephens, P.A., Sutherland, W.J., Freckleton, R.P.: What is the Allee effect? School of Biological Sciences, University of East Anglia, Norwich, UK NR4 7TJ (1999)

Sutherland, W.J.: The importance of behavioural studies in conservation biology. Anim. Behav. **56**, 801–809 (1998)

Zadeh, L.A.: Fuzzy sets. Inf. Control **8**, 338–353 (1965)

# Distribution of Financial Resources Using a Fuzzy Transportation Model

Luisa L. Lazzari and Patricia I. Moulia

**Abstract** The classical transportation model refers to the shipment of a product of $m$ sources of supply or origins to $n$ points of demand or destinations. The aim is to assign the offer available at each source, so that demand of destinations be satisfied, both to minimize the total costs of transport or some measure of distance, or else to maximize the profit total. The use of fuzzy numbers makes it possible to consider the aspects of an imprecise environment. The application areas of the transportation problem can be extended when some parameters are fuzzy. In this paper we present an application of a fuzzy transportation model to obtain the best distribution of the means of financing available by a company, to meet its needs, with the goal of minimizing the costs when they are expressed by triangular fuzzy numbers.

**Keywords** Uncertainty · Fuzzy transportation model · Financial resources

## 1 Introduction

The classical transportation model refers to the shipment of a product of $m$ sources of supply or origins to $n$ points of demand or destinations. The aim is to assign the offer available at each source, so that demand of destinations be satisfied, both to

This work is supported by the Secretaría de Ciencia y Técnica of the Universidad de Buenos Aires, Project UBACYT 20020130100083BA.

L.L. Lazzari (✉) · P.I. Moulia
CIMBAGE, IADCOM, Facultad de CienciasEconómicas, Universidad de Buenos Aires, Av. Córdoba 2122, CABA, C1120AAQ Buenos Aires, Argentina
e-mail: luisalazzari@cimbage.com.ar

P.I. Moulia
e-mail: patriciamoulia@cimbage.com.ar

© Springer International Publishing Switzerland 2015
J. Gil-Aluja et al. (eds.), *Scientific Methods for the Treatment of Uncertainty in Social Sciences*, Advances in Intelligent Systems and Computing 377,
DOI 10.1007/978-3-319-19704-3_17

minimize the total costs of transport or some measure of distance, or else to maximize the profit total.

The use of fuzzy numbers makes it possible to consider the aspects of an imprecise environment. The application areas of the transportation problem can be extended when some parameters are fuzzy.

All lineal programming problems can be solved through the simplex method, but transportation problems, due to its special structure, can be solved through more efficient methods. There are several transportation model applications.

World reality nowadays shows problems, in which supplies, demands or cost cannot be precisely specified. In this case, we are in front of a fuzzy transportation problem (Lazzari 2001).

A firm is a productive combination of factors among which the necessary financing availabilities to develop its activities are found (Fernández Pirla 1964). The company's financing lies in the obtainment of resources or means of payment devoted to the purchase of fixed assets to fulfill its aims.

In this paper we present an application of a fuzzy transportation model to obtain the best distribution of the means of financing available by a company, to meet its needs, with the goal of minimizing the costs when they are expressed by triangular fuzzy numbers.

It is structured as follows: In section two we define the theoretical framework, in section three a fuzzy transportation model is presented, in section four the latter is applied to the case of distribution of financial resources available for a construction firm and in the last section some concluding remarks are made.

## 2 Preliminaries

### 2.1 Classic Transportation Problem

The transportation problem can be formulated as follows. A homogeneous product is to be shipped from $m$ origins where quantities $a_i(i = 1, .., m)$ are available to $n$ destinations where amounts $b_i(j = 1, \ldots, n)$ are required. $x_{ij}$ indicates the quantity of units that will embark from origin $i$ to destination $j$ and $c_{ij}$ the corresponding unit cost. It is assumed that $\sum_{i=1}^{m} a_i = \sum_{j=1}^{n} b_j$. The problem is to find the amounts $x_{ij}$ so that the total transportation cost is minimized subject to origin availabilities, destination requirements, and no negativity constraints; that is:

$$\text{Min} \quad z = \sum_{i=1}^{m} \sum_{j=1}^{n} c_{ij} x_{ij}$$

$$\sum_{j=1}^{n} x_{ij} = a_i, \quad i = 1, 2, \ldots, m$$

$$\text{s.t.} \quad \sum_{i=1}^{m} x_{ij} = b_j, \quad j = 1, 2, \ldots, n$$

$$\sum_{i=1}^{n} a_i = \sum_{j=1}^{m} b_j$$

$$x_{ij} \geq 0, \ \forall i, \ \forall j$$

The specific procedure of solution, called transportation algorithm, consists of two phases. In the first one an initial feasible solution is obtained. Different methods can be used to obtain this solution, among others: the method of Russell (1969), the method of Northwest corner, the method of Vogel or the method of minimal cost (Gould et al. 1992).

To enter the second phase, a non-degenerate initial solution must be obtained (calculated in the previous stage), i.e. the amount of basic positions be equal to $m+n-1$.

In the second phase it is checked that the solution obtained in the first phase is the optimal one and if it is not, a way to improve it is looked for. The methods which are mostly used are stepping-stone and modified distribution (MODI).

The transportation model can be extended directly to cover practical situations in the area of inventory control, job scheduling, allocation of personnel, planning of production and resources (Taha 2004).

## 2.2 Elements of Fuzzy Sets

Let $X$ be a universe continuous or discrete. A fuzzy subset $\tilde{A}$ is a $\mu_{\tilde{A}} : X \to [0, 1]$ function which assigns to each element of the $X$ a value $\mu_{\tilde{A}}(x)$ that belongs to interval $[0,1]$. $\mu_{\tilde{A}}(x)$ is the membership degree or level of $x$.

*Let* $\tilde{A}$ be a fuzzy set of referential $X$. An $\alpha$-cut of $\tilde{A}$ is the crisp set $A_\alpha = \{x \in X / \mu_{\tilde{A}}(x) \geq \alpha\}$ for every $\alpha \in (0, 1]$. If $\alpha = 0$, the $\alpha$-cut corresponding is the closure of the union of $A_\alpha$, $0 < \alpha \leq 1$ (Lazzari 2010).

A fuzzy set $\tilde{A} \subset \Re$ is normal if and only if $\forall x \in E$, $\max \mu_{\tilde{A}}(x) = 1$, and is convex if and only if, $\forall x \in [x_1, x_2] \subset \Re$ is verified that $\mu_{\tilde{A}}(x) \geq \min\{\mu_{\tilde{A}}(x_1), \mu_{\tilde{A}}(x_2)\}$.

A fuzzy number (FN) is a fuzzy set of the real line convex and normal. A fuzzy number can be expressed through its $\alpha$-cuts in a unique way $A_\alpha = [a_1(\alpha), a_2(\alpha)] \forall \alpha \in [0, 1]$, or through its membership function $\mu_{\tilde{A}}(x)$, $\forall x \in \Re$ (Kaufmann and Gupta 1985). A fuzzy number is continuous if its membership function is continuous.

Triangular fuzzy number (TFN) is a real and continuous fuzzy number $\tilde{A} = (a_1, a_2, a_3)$, which $\alpha$-cuts are $A_\alpha = [a_1 + (a_2 - a_1)\alpha, a_3 + (a_2 - a_3)\alpha]$ and its membership function is:

$$\forall x \in \mathfrak{R}: \quad \mu(x) = \begin{cases} 0 & si \quad x < a_1 \\ \frac{x - a_1}{a_2 - a_1} & si \quad a_1 \leq x \leq a_2 \\ \frac{a_3 - x}{a_3 - a_2} & si \quad a_2 < x \leq a_3 \\ 0 & si \quad x > a_3 \end{cases}$$

Let $\tilde{A} = (a_1, a_2, a_3)$ and $\tilde{B} = (b_1, b_2, b_3)$ be two arbitrary TFN, we define (Kaufmann and Gil Aluja 1978):

$$\tilde{A}(+)\tilde{B} = (a_1 + b_1, a_2 + b_2, a_3 + b_3) \tag{1}$$

$$\tilde{A}(-)\tilde{B} = (a_1 - b_3, a_2 - b_2, a_3 - b_1) \tag{2}$$

$$k.\tilde{A} = (min(k.a_1, k.a_3), k.a_2, min(k.a_1, k.a_3))\forall k \in \mathfrak{R} - \{0\} \tag{3}$$

## 2.3 Ranking Fuzzy Numbers

There are many methods to order fuzzy quantities, a lot of them have been analyzed and compared by different authors using various criteria (Lazzari 2010). The criterion of order, expressed in (4), proposed by Yager (1981) is used in this paper.

$$Y_2(\tilde{A}_i) = \int_0^{h(\tilde{A}_i)} M(A_{i\alpha}) d\alpha \tag{4}$$

where $h(\tilde{A}_i) = \sup_{x \in X} \mu_{\tilde{A}_i}(x)$ and $M(A_{i\alpha}) = \frac{a_1(\alpha) + a_2(\alpha)}{2}$. Then, we define $\tilde{A}_i \geq \tilde{A}_j$ by $Y_2$ if and only if $Y_2(\tilde{A}_i) \geq Y_2(\tilde{A}_j)$. If $\tilde{A} = (a_1, a_2, a_3)$ is a FTN, then $Y_2 = \frac{1}{4}(a_1 + 2a_2 + a_3)$. This value is called the real number associated to $\tilde{A}$ and denoted $\bar{a}$ (Kaufmann and Gil Aluja 1978).

## 3  Fuzzy Transportation Problem (FTP)

Within the fuzzy lineal programming the transportation problem has received special treatment due to the importance of the subject and because its methodology of resolution can be extended and applied to other type of problems, as it is the case of the resource allocation.

Unlike the classical approach there are different models of fuzzy transportation, according to the fuzziness that show the problem, and several approaches to resolve each of them.

FTP was first studied by Zimmermann (1978), Prade (1980), Oheigeartaigh (1982), Verdegay (1983), Delgado and Verdegay (1984) and Chanas et al. (1984). Other studies of this problem are Lai and Hwang (1992), Chanas and Kuchta (1996, 1998), Liu and Kao (2004), Kumar and Kaur (2011), Kumar and Murugesan (2012) and Khalaf (2014).

The mathematical model of FTP used in this paper considers that all the cost coefficients are triangular fuzzy numbers and all supply and demand are crisp numbers, and is given by:

$$\text{Min} \quad \tilde{z} = \sum_{i=1}^{m} \sum_{j=1}^{n} \tilde{c}_{ij} x_{ij}$$

$$\text{s.t.} \quad \sum_{j=1}^{n} x_{ij} = a_i, \quad i = 1, 2, \ldots, m$$

$$\sum_{i=1}^{m} x_{ij} = b_j, \quad j = 1, 2, \ldots, n$$

$$\sum_{i=1}^{n} a_i = \sum_{j=1}^{m} b_j$$

$$x_{ij} \geq 0, \forall i, \forall j$$

where $a_i$ is the amount of material available at source of supply $S_i (i = 1, 2, \ldots, m)$, $b_j$ is the amount of material required at destination $D_j (j = 1, 2, \ldots, n)$ and $\tilde{c}_{ij}$ is the unitary cost of transportation from the source of supply $S_i$ to the destination $D_i$, expressed by a TFN.

The fuzzy transportation problem can be represented by a matrix of order $m \times n$ (Table 1).

| Table 1 Fuzzy cost matrix (FCM) | $D_i$ | ... | $D_n$ | A |
|---|---|---|---|---|
| $S_1$ | $\tilde{c}_{11}$ | ... | $\tilde{c}_{1n}$ | $a_1$ |
| ... | .... | ... | ... | ... |
| $S_m$ | $\tilde{c}_{m1}$ | ... | $\tilde{c}_{mn}$ | $a_m$ |
| D | $b_1$ | ... | $b_n$ | |

In the first phase to find a fuzzy initial basic feasible solution (FIBFS) we will use the approximation method of Russell (1969). In the second phase the modified distribution method (MDM) is used to test if the obtained FIBFS is optimal (adapted by Khalaf 2014, for fuzzy problems).

The steps of unified model used in this paper, are:

Step 1.  Fuzzy unitary cost matrix is obtained (Table 1)

Step 2.  In each fuzzy cost matrix row, the highest fuzzy unitary cost $\tilde{u}_i$ is selected

Step 3.  In each fuzzy cost matrix column, the highest fuzzy unitary cost $\tilde{v}_j$ is selected

Step 4.  For each variable $x_{ij}$ non selected in Steps 2 and 3 $\tilde{w}_{ij} = \tilde{u}_i(+)\tilde{v}_j(-)\tilde{c}_{ij}$ is calculated, applying operations (1) and (2)

Step 5.  The $\tilde{w}_{ij}$ are ordered according to the criteria of order for fuzzy numbers given in 2.2.1. Then, the highest is chosen. If the maximum is not unique, then select the variable $x_{ij}$ so that the highest allocation can be made

Step 6.  The offered and demanded quantities are adjusted according to the selection made. Those rows and columns in which supply and demand have been exhausted are eliminated

Step 7.  If all supply and demand requirements have not been satisfied, then go to the *Step* 1 and calculate the new $\tilde{w}_{ij}$. If all the values from the rows or columns have been satisfied it means that a FIBFS has been obtained

Step 8.  For a basic variable $x_{ij}$ let $\tilde{u}_i$ and $\tilde{v}_j$ be that satisfies the equations $\tilde{c}_{ij} = \tilde{u}_i(+)\tilde{v}_j$ for all $(i, j)$ which correspond to a basic variable. There are $m + n - 1$ basic variables and therefore $m + n - 1$ equations of this type. As the number of unknown is $m + n$, an arbitrary value can be assigned to any of those variables. The choice of this variable and its value does not affect the result of any $\tilde{c}_{ij} = \tilde{u}_i(+)\tilde{v}_j$, although $x_{ij}$ is not basic

Step 9.  We consider any $\tilde{u}_i$ or $\tilde{v}_j$ equal to (0,0,0), then the remaining values from the real numbers associated with each TFN are obtained

Step 10. For all $(i, j)$ such that $x_{ij}$ is not basic, $\bar{d}_{ij} = \bar{u}_i + \bar{v}_j - \bar{c}_{ij}$ is obtained. If all $\bar{d}_{ij} \le 0$, then the FIBFS is the optimal fuzzy solution. If there is at least one $\bar{d}_{ij} > 0$ then the FIBFS is not optimal and go to *Step* 11

Step 11. The entering basic variable corresponding to the highest positive $\bar{d}_{ij}$ value is selected

Step 12. The salient basic variable is determined as follows: build up a closed loop to start and end at the entering variable cell. The loop consists in connecting horizontal and vertical segments only (diagonals do are not allowed). Except for the entering variable cell, each corner of the closed loop must coincide with a basic variable. Identify the chain reaction required to retain feasible solution when the entering basic variable is increased. From donating cells, select the basic variable having the smallest value

*Step* 13.    Determine the new FIBFS. Add the value of salient basic variable to the allocation for each recipient cell. Subtract this value from the allocation for each donating cell

*Step* 14.    Repeat the steps 8, 9 and 10 to analyze if the new FIBFS is the optimal fuzzy solution, until obtaining $\bar{d}_{ij} \leq 0$, $\forall i \forall j$.

# 4  Determining the Optimal Structure in the Financing of a Firm

The establishment of the financial program comprises two phases (Bonilla Musoles 1982).

In the first one, the financing needs for each period are determined. The value of the necessary means should be calculated according to the programs and budgets of activities and investments. In the second one, it is defined for each period the way needs will be satisfied, analysing available financing means, their capacity and cost.

The general statement of the two phases, comprising the financing program, is developed through the application of the transportation problem.

There are financing means $(A_1, A_2, \ldots A_m)$ available in quantities $a_1, a_2, \ldots a_m$, respectively, necessary to finance the needs in the time $(B(t_1), B(t_2), \ldots, B(t_n))$ required in quantities $b_1, b_2, \ldots, b_n$ to be covered.

Financing of needs through the use of available means has a cost $c_{ij}$ (cost of use of means $A_i$ to finance the need $B$ in instant $t_j$). These are expressed as TFN if they cannot be accurately stated.

The problem consists in determining the quantities $x_{ij}$ available from the means $A_i(i = 1, \ldots, m)$ to finance all needs $B(t_j)$, $(j = 1, \ldots, n.)$ with the aim of minimizing the total cost.

We find an application in the following section.

## 4.1 Application Case

A construction company must increase its economic capacity, due to the existence of a new Urban Plan in the area and to the economic growth and industry development in it. Thus, the firm should invest in current and fixed assets during the next two years. These investments are a closed quarter and a group of houses for a number of working families.

We assume that the budgets of such needs to be covered are considered compatible with the company's activity and its capacity of debt entering and that all financing means are available at starting time $(t_0)$.

Available financing means: $A_1 \rightarrow a_1 = 10000000$, $A_2 \rightarrow a_2 = 34000000$, $A_3 \rightarrow a_3 = 42000000$.

Demanded current $(BC)$ and fixed assets $(BF)$: $BF(t_1) \rightarrow b_1 = 12000000$, $BC(t_1) \rightarrow b_2 = 25000000$, $BF(t_2) \rightarrow b_3 = 15000000$, $BC(t_2) \rightarrow b_4 = 34000000$.

Since unitary cost of each mean usage $A_1$, $A_2$ and $A_3$ in order to finance the stated needs is vague, it is expressed through TFN (Table 2).

**Table 2** Fuzzy unitary cost matrix

|       | $BF(t_1)$ | $BC(t_1)$ | $BF(t_2)$ | $BC(t_2)$ |
|-------|-----------|-----------|-----------|-----------|
| $A_1$ | $\tilde{c}_{11} = (0.4, 0.5, 0.58)$ $\tilde{c}_{11} = 0.495$ | $\tilde{c}_{12} = (0.1, 0.135, 0.15)$ $\tilde{c}_{12} = 0.13$ | $\tilde{c}_{13} = (0.5, 0.55, 0.65)$ $\tilde{c}_{13} = 0.5625$ | $\tilde{c}_{14} = (0.12, 0.14, 0.15)$ $\tilde{c}_{14} = 0.1375$ |
| $A_2$ | $\tilde{c}_{21} = (0.18, 0.19, 0.21)$ $\tilde{c}_{21} = 0.1925$ | $\tilde{c}_{22} = (0.17, 0.19, 0.2)$ $\tilde{c}_{22} = 0.1875$ | $\tilde{c}_{23} = (0.19, 0.2, 0.21)$ $\tilde{c}_{23} = 0.2$ | $\tilde{c}_{24} = (0.18, 0.2, 0.21)$ $\tilde{c}_{24} = 0.1975$ |
| $A_3$ | $\tilde{c}_{31} = (0.1, 0.12, 0.125)$ $\tilde{c}_{31} = 0.11625$ | $\tilde{c}_{32} = (0.11, 0.12, 0.13)$ $\tilde{c}_{32} = 0.12$ | $\tilde{c}_{33} = (0.11, 0.13, 0.14)$ $\tilde{c}_{33} = 0.1275$ | $\tilde{c}_{34} = (0.12, 0.13, 0.15)$ $\tilde{c}_{34} = 0.1325$ |

According to Table 2, the FTP consists in obtaining the quantities $x_{ij}(i = 1, 2, 3)$ and $(j = 1, 2, 3, 4)$ available of means $A_1$, $A_2$ and $A_3$ to finance the needs $BF(t_1)$, $BC(t_1)$, $BF(t_2)$, $BC(t_2)$ with the aim of minimizing the total cost.

The problem is:

$$\text{Min} = \tilde{z} = \sum_{i=1}^{3} \sum_{j=1}^{4} \tilde{c}_{ij} \cdot x_{ij}$$

$$\sum_{j=1}^{4} x_{ij} = a_i \quad (i = 1, 2, 3)$$

$$\text{s.t} \quad \sum_{i=1}^{3} x_{ij} = b_j \quad (j = 1, 2, 3, 4)$$

$$\sum_{i=1}^{3} a_i = \sum_{j=1}^{4} b_j = 86000000$$

$$x_{ij} \geq 0, \quad (i = 1, 2, 3; \ j = 1, 2, 3, 4)$$

The model developed in 3 is applied for its resolution.

*Step* 1.   Cost unitary fuzzy matrix is obtained (Table 2)

*Step* 2.   From the fuzzy unitary cost matrix (Table 2) the $\tilde{u}_i$ corresponding to each row is obtained: $\tilde{u}_1 = (0.5,\ 0.55,\ 0.65)$, $\tilde{u}_2 = (0.18, 0.2, 0.21)$ and $\tilde{u}_3 = (0.12, 0.13, 0.15)$

*Step* 3.   From the fuzzy unitary cost matrix (Table 2) the $\tilde{v}_j$ corresponding to each column is obtained: $\tilde{v}_1 = (0.4, 0.5, 0.58)$, $\tilde{v}_2 = (0.17, 0.19, 0.2)$, $\tilde{v}_3 = (0.5, 0.55, 0.65)$ and $\tilde{v}_4 = (0.18,\ 0.2, 0.21)$

*Step* 4.   For each variable $x_{ij}$ non selected in Steps 2 and 3 $\tilde{w}_{ij} = \tilde{u}_i(+)\tilde{v}_j(-)\tilde{c}_{ij}$ is calculated. We obtain $\tilde{w}_{11} = \tilde{u}_1(+)\tilde{v}_1(-)\tilde{c}_{11}$,

$$\tilde{w}_{11} = (0.5, 0.55, 0.65)(+)(0.4, 0.5, 0.58)(-)(0.4, 0.5, 0.58).$$
$$\tilde{w}_{11} = (0.32, 0.55, 0.83) \text{ and } \bar{w}_{11} = 0.5625.$$

The other values are calculated similarly

$\tilde{w}_{12} = (0.52,\ 0.605,\ 0.75)$, $\bar{w}_{12} = 0.62$, $\tilde{w}_{13} = (0.35,\ 0.55,\ 0.8)$, $\bar{w}_{13} = 0.565$

$\tilde{w}_{14} = (0.53,\ 0.61,\ 0.74)$, $\bar{w}_{14} = 0.6225$, $\tilde{w}_{21} = (0.37,\ 0.51,\ 0.61)$, $\bar{w}_{21} = 0.5$

$\tilde{w}_{22} = (0,\ 13,\ 0.2,\ 0.24)$, $\bar{w}_{22} = 0.1925$, $\tilde{w}_{23} = (0.47,\ 0.55,\ 0.67)$, $\bar{w}_{23} = 0.56$

$\tilde{w}_{24} = (0.15,\ 0.2,\ 0.24)$, $\bar{w}_{24} = 0.1975$, $\tilde{w}_{31} = (0.395,\ 0.51,\ 0.62)$, $\bar{w}_{31} = 0.51125$;

$\tilde{w}_{32} = (0.16,\ 0.5,\ 0.24)$, $\bar{w}_{32} = 0.2\ \tilde{w}_{33} = (0.48,\ 0.55,\ 0.69)$, $\bar{w}_{33} = 0.5675$;

$\tilde{w}_{34} = (0,\ 15,\ 0.2,\ 0.24)$, $\bar{w}_{34} = 0.1975$

Step 5.   All $w_{ij}$ are ordered and $\bar{w}_{14}$, the highest is selected. Therefore, $x_{14} = 10000000$ is chosen as the first basic variable. This allocation uses the whole financial mean $A_1$. Then this row is eliminated

Step 6.   The offered and demanded quantities are adjusted according to the selection made

Step 7.   The FIBFS obtained is: $x_{14} = 10000000$, $x_{33} = 15000000$, $x_{31} = 12000000$, $x_{24} = 24000000$, $x_{22} = 10000000$ and $x_{32} = 15000000$. The remaining decision variables take value zero in this solution. The objective function value for its FIBFS is: $\tilde{z} = (11720000,\ 13290000,\ 14240000)$.

Step 8.   It is fulfilled.

Step 9.   $\tilde{u}_1 = (0, 0, 0)$ is allocated and the real numbers associated to the remaining values of $\tilde{u}_i$ and $\tilde{v}_j$ are obtained, using the equation $\bar{c}_{ij} = \bar{u}_i + \bar{v}_j$ for the allocated cells $\bar{u}_2 = 0.03$, $\bar{u}_3 = -0.0075$, $\bar{v}_1 = 0.12375$, $\bar{v}_2 = 0.1275$, $\bar{v}_3 = 0.135$ and $\bar{v}_4 = 0.1375$.

Step 10.   For all $(i,\ j)$ such that $x_{ij}$ is not basic, $\bar{d}_{ij} = \bar{u}_i + \bar{v}_j - \bar{c}_{ij}$ is obtained. $\bar{d}_{11} = -0.37125$, $\bar{d}_{12} = -0.0025$, $\bar{d}_{13} = 0.425$, $\bar{d}_{21} - = -0.00875$, $\bar{d}_{23} = -0.0025$ and $\bar{d}_{34} = -0.0025$. All $\bar{d}_{ij} \leq 0$, then the FIBFS is the optimal fuzzy solution

**Table 3** FIBFS

| | $BF(t_1)$ | $BC(t_1)$ | $BF(t_2)$ | $BC(t_2)$ | A |
|---|---|---|---|---|---|
| $A_1$ | $\bar{c}_{11} = (0.4, 0.5, 0.58)$ <br> $x_{11} = 0$ | $\bar{c}_{12} = (0.1, 0.135, 0.15)$ <br> $x_{12} = 0$ | $\bar{c}_{13} = (0.5, 0.55, 0.65)$ <br> $x_{13} = 0$ | $\bar{c}_{14} = (0.12, 0.14, 0.15)$ <br> $x_{14} = 1000000$ | 10000000 |
| $A_2$ | $\bar{c}_{21} = (0.18, 0.19, 0.21)$ <br> $x_{21} = 0$ | $\bar{c}_{22} = (0.17, 0.19, 0.2)$ <br> $x_{22} = 1000000$ | $\bar{c}_{23} = (0.19, 0.2, 0.21)$ <br> $x_{23} = 0$ | $\bar{c}_{24} = (0.18, 0.2, 0.21)$ <br> $x_{24} = 2400000$ | 34000000 |
| $A_3$ | $\bar{c}_{31} = (0.1, 0.12, 0.125)$ <br> $x_{31} = 12000000$ | $\bar{c}_{32} = (0.11, 0.12, 0.13)$ <br> $x_{32} = 15000000$ | $\bar{c}_{33} = (0.11, 0.13, 0.14)$ <br> $x_{33} = 15000000$ | $\bar{c}_{34} = (0.12, 0.13, 0.15)$ <br> $x_{34} = 0$ | 42000000 |
| D | 12000000 | 25000000 | 15000000 | 34000000 | 86000000 |

**Table 4** α-cuts of the minimum cost

| α | $C_\alpha$ | α | $C_\alpha$ | α | $C_\alpha$ |
|---|---|---|---|---|---|
| 0 | [11720000, 14240000] | 0.4 | [12348000, 13860000] | 0.8 | [12976000, 13480000] |
| 0.1 | [11877000, 14154000] | 0.5 | [12505000, 13765000] | 0.9 | [13133000, 13385000] |
| 0.2 | [12034000, 14050000] | 0.6 | [12662000, 13670000] | 1 | [13290000, 13290000] |
| 0.3 | [12191000, 13955000] | 0.7 | [12819000, 13575000] | | |

Therefore, the needs for fixed assets of the first and second years are wholly covered with $A_3$, for the needs for current assets on the first year $A_2$ and $A_3$ are used, and for those needed in the second year $A_1$ and $A_2$ are used (Table 3).

Total cost is expresses through TFN $\tilde{C} = (11720000, 13290000, 14240000)$ and the α-cuts are $c_\alpha = [11720000 + 1570000\alpha, 14240000 - 950000\alpha]$.

Table 4 shows the cost for different presumption levels.

## 5 Final Comments

The transportation model is a high operative method, very efficient as a quantitative instrument in a firm's decision. It can be considered as a decision-making tool, with possibility of being accepted by all the directing boards of the firm, because it expresses the best use of financing means to cover needs at minimum cost.

The fuzzy lineal programing has the advantage for the decision maker to not be forced to state his problems in accurate terms, it is only enough describe them in fuzzy terms and he can select the suitable model and approach for his problem.

The fuzzy approximation method of Russellis simple to understand and apply, providing a FIBFS optimal or almost optimal.

The optimal solution obtained in the problem of Sect. 4.1 is crisp, even though the minimum value of de objective function is a TFN. The usage of flexible models is a valuable contribution for the decision maker of a firm.

## References

Bonilla Musoles, M.: Aplicaciones del modelo de transporte a la financiación de la empresa. Anales del Instituto de Actuarios Españoles: Colegio Profesional, vol. 22, pp. 31–50 (1982)

Chanas, S., Kuchta, D.: Fuzzy integer transportation problem. Fuzzy Set Syst. **98**, 291–298 (1998)

Chanas, S., Kolodziejczlyk, W., Machaj, A.: A fuzzy approach to the transportation problem. Fuzzy Sets Syst. **13**, 211–221 (1984)

Delgado, M., Verdegay, J.L.: Resolución del problema de transporte fuzzy con números borrosos triangulares, pp. 748–758. Actas del XIV Congreso Nacional de Estadística, Investigación Operativa e Informática, Granada (España) (1984)

Fernández Pirla, J.: Economía de la Empresa. El autor, Madrid (1964)

Gould, F., Eppen, G., Schmidt, C.: Investigación de operaciones en la ciencia administrativa. Prentice-Hall, México (1992)

Kaufmann, A., Gupta, M.: Introduction to Fuzzy Arithmetic. Van Nostrand Reinhold Company, New York (1985)

Kaufmann, A., Gil Aluja, J.: Técnicas operativas de gestión para el tratamiento de la incertidumbre. Editorial Hispano Europea, Barcelona (1987)

Khalaf, W.: Solving fuzzy transportation problems using a new algorithm. J. Appl. Sci. **14**, 252–258 (2014)

Kumar, A., Kaur, A.: Application of classical transportation methods to find the fuzzy optimal solution of fuzzy transportation problems. Fuzzy Inf. Eng. **3**, 81–99 (2011)

Kumar, B., Murugesan, S.: On fuzzy transportation problem using triangular fuzzy numbers with modified revised simples method. Int. J. Eng. Sci. Technol. **4**, 285–294 (2012)

Lai, Y., Hwang, C.: Fuzzy Mathematical Programming: Methods and Applications. Springer, Berlín (1992)

Lazzari, L.: El comportamiento del consumidor desde una perspectiva fuzzy. Una aplicación al turismo. EDICON, Buenos Aires (2010)

Lazzari, L. (comp.): Los conjuntos borrosos y su aplicación a la programación lineal. Facultad de Ciencias Económicas, Universidad de Buenos Aires, Buenos Aires (2001)

Liu, S., Kao, C.: Solving fuzzy transportation problems based on extension principle. Eur. J. Oper. Res. **153**, 661–674 (2004)

Oheigeartaigh, M.: A fuzzy transportation algorithm. Fuzzy Sets Syst. **8**, 235–243 (1982)

Prade, H.: Operations research with fuzzy data. In: Wang, P.P., Chang, S.K. (eds.) Fuzzy sets. Theory and applications to policy analysis and information systems, pp. 155–170. Plenum Press, New York (1980)

Russell, E.: Letters to the editor-extension of Dantzig´s algorithm to finding an initial near-optimal basis for the transportation problem. Oper. Res. **17**, 187–191 (1969)

Taha, H.: Investigación de Operaciones. Pearson, México (2004)

Verdegay, J.L.: Problema de transporte con parámetros fuzzy. Revista de la Real Academia de Ciencias Matemáticas, Físico Químicas y Naturales de GranadaII, pp. 47–56 (1983)

Yager, R.: A procedure for ordering fuzzy sets of unit interval. Inf. Sci. **24**, 143–161 (1981)

Zimmermann, H.: Fuzzy programming and linear programming with several objective functions. Fuzzy Sets Syst. **1**, 45–55 (1978)

# Clustering Variables Based on Fuzzy Equivalence Relations

Kingsley S. Adjenughwure, George N. Botzoris
and Basil K. Papadopoulos

**Abstract** We develop a method of grouping (clustering) variables based on fuzzy equivalence relations. We first compute the pairwise relationship (correlation) matrix between the variables and transform the matrix into a fuzzy compatibility relation. Then a fuzzy equivalence relation is constructed by computing the transitive closure of the compatibility relation. Finally, by taking all appropriate α-cuts, we obtain a hierarchical type of variable clustering. As examples, we use the proposed method first as a variable clustering tool in a regression model and secondly as a new way of performing factor analysis.

**Keywords** Fuzzy sets · Fuzzy equivalence relations · Clustering variables · Feature selection · Factor analysis · Regression analysis

## 1  Introduction

When modelling a system or a dependent variable, there is the problem of too many independent variables. In regression analysis for example, too many variables might lead to overfitting the data and render the model inefficient for the purpose of prediction. Also too many variables lead to a complex model and hence it is difficult to explain the interactions between all variables used in the model and how each variable affects the system (Sanche and Lonergan 2006). The problem of many variables is also known to affect cluster analysis. This is because for certain

K.S. Adjenughwure (✉) · G.N. Botzoris · B.K. Papadopoulos
Department of Civil Engineering, Democritus University of Thrace, 12, Vas. Sofias St.,
67100 Xanthi, Greece
e-mail: kingadje@civil.duth.gr

G.N. Botzoris
e-mail: gbotzori@civil.duth.gr

B.K. Papadopoulos
e-mail: papadob@civil.duth.gr

© Springer International Publishing Switzerland 2015            219
J. Gil-Aluja et al. (eds.), *Scientific Methods for the Treatment of Uncertainty
in Social Sciences*, Advances in Intelligent Systems and Computing 377,
DOI 10.1007/978-3-319-19704-3_18

distributions, distances between data points become relatively uniform as the dimension increases (Beyer et al. 1999). Variable reduction is therefore useful when there are many observable variables and one need to choose a subset of these variables (feature selection) or find a reduced set of latent variables (dimensionality reduction) which explain a significant amount of the variance in the data.

Feature selection techniques are divided into wrapper, filters and embedded methods. Wrapper methods assign scores based on the predictive performance of each selected feature subset while filter methods use measures like correlation, mutual information, etc., to assign scores to feature subsets. Wrapper methods are known to be more accurate that filter methods but are computationally more expensive (Guyon and Elisseeff 2003). Embedded methods perform feature selection as part of the model calibration process. Their goal is to shrink the regression coefficients of all the variables in the model by constraining their sum to a specific value (Tibshirani 1996). This procedure effectively reduces most of the coefficients to zero leaving only the most important ones. Details about various variable selection techniques can be found in (Guyon and Elisseeff 2003). One obvious disadvantage of feature selection is that we only use a very few set of selected variables out of all the variables without analyzing the relationship between them.

Dimensionality reduction can be done directly by principal component analysis (PCA) or by first using K-means or other clustering techniques to partition the data into a fixed number of clusters and then projecting the data matrix into the space spanned by the centroid of the clusters (Karypis and Han 2000). Another method similar to PCA is factor analysis. This method is preferred when the goal is to describe the variance of a group of correlated variables with a single variable. This is equivalent to clustering groups of linearly related variables around a small number of latent variables called factors (Spearman 1904).

A disadvantage of dimensionality reduction techniques like PCA and factor analysis is that we cannot explicitly define the relationship between the latent (unobserved) variables and the observed variables. Also information is lost by using a few principal components instead of all components. There is also the problem of choosing how many latent variables or number of principal components. To solve the problems encountered by feature selection and dimensionality reduction, a method for variable reduction where all variables are retained is needed.

Variable clustering is a technique for reducing the number of variables to smaller set of clusters where the variables in a cluster are similar to other variables in the cluster and are dissimilar to other variables in another cluster. The similarity measure used here is that of correlation or other measures with the same meaning like mutual information. The advantage is that all variables are available for use after the clustering process. For example, the procedures of feature selection can be applied to the clustered variables by selecting variables from the clusters instead of individually (Bühlmann et al. 2013).

In this research, we introduce the idea of clustering variables using fuzzy equivalence relation. We also discuss the potential of the proposed model as both a feature selection and dimensionality reduction preprocessing tool. The contribution of this paper is to provide an alternative method to variable clustering and also give new interpretations to the clustered variables and discuss their relationship with PCA, factor analysis, feature selection and high dimensional data clustering.

# 2 Literature Review

Although clustering variables is not as popular as data clustering, some advances have been made in this area. The first approach to variable clustering used principal component regression (Kendall 1957). Other variable clustering methods have been developed with regression analysis in mind. This process leads to an algorithm that performs variable clustering and model fitting at the same time. Examples of such algorithms are the one proposed by Dettling and Bühlmann (2004) and the OSCAR method developed by Bondell and Reich (2008).

Hastie et al. (2000) proposed the 'gene shaving' technique which uses principal components analysis to find a group of highly correlated variables. This method was applied to cluster genes and can either be unsupervised or supervised with a response variable. Hastie et al. (2001) also proposed the 'tree harvesting' method to select groups of predictive variables formed by hierarchical clustering in a supervised learning scheme.

For unsupervised variable clustering, other methods have been developed. PROC VARCLUS is an unsupervised variable clustering algorithm developed by SAS Institute Inc. This algorithm uses an iterative approach to find groups of variables that are as correlated as possible among themselves and as uncorrelated as possible with variables in other clusters. The algorithm begins with all variables in one cluster. A cluster is then chosen to split into two clusters by using the first two principal components and assigning the variables to the component with which they have the highest correlation (Nelson 2001).

Vigneau et al. (2001) used a method similar to PROC VARCLUS for variable clustering. In their approach, the goal was to maximize the correlation between the variables and their cluster centroid. The algorithm is similar to K-means and requires the user to specify the number of clusters. The authors also suggested a hierarchical variable clustering algorithm based on the same criteria to help with the selection of the number of clusters.

Palla et al. (2012) developed the Dirichlet Process Variable Clustering (DPVC). They clustered the variables by partitioning the observed dimensions by the Chinese Restaurant Process. DPVC exhibits the usual advantages over other methods because it is probabilistic and non-parametric. It can handle missing data, learn the appropriate number of clusters from data, and avoid overfitting.

Bühlmann et al. (2013) proposed a bottom-up agglomerative variable clustering algorithm based on canonical correlation. The algorithm starts with all variables in different clusters and successively merges the two clusters with the highest canonical correlation.

To our knowledge, fuzzy equivalence relation has not been applied to cluster variables. In this paper, we partition the variables into clusters using fuzzy equivalence relations. Our approach is clearly linked to the data clustering application of fuzzy equivalence relations (Klir and Yuan 1995). All techniques used are the same as in data clustering. The only difference is that the distance or similarity measure used is the correlation among the variables. As noted earlier, we provide some new insight to the interpretation of the clustered variables by giving example of how the method can be used both as a standalone tool for unsupervised clustering of variables or as a preprocessing tool for variable selection in regression analysis, factor analysis, principal component analysis and clustering in high dimensional space. We consider that these new interpretations will be useful for other variable clustering methods too. The advantages of the proposed model over existing variable clustering methods are:

- Apart from using the correlation measures of the variables, the fuzzy proximity matrix in our proposed method can also be constructed using expert opinion on the relationship between the variables. This is very useful for variables that are not quantifiable or with missing, incomplete or unreliable data. The other methods mentioned above do not possess this flexibility, as they rely on the correlation calculations or other estimations made from the available data.
- There is no need to specify the number of clusters. The method produces a hierarchical cluster of all variables, if all α-cuts are used, and can also partition variables into an appropriate number of clusters, by choosing an appropriate α-cut most suitable for a particular application.
- The α-cut value used to cluster the variable represents the least correlation strength of all variables in the cluster. All variables therefore are equivalent at some level chosen by the user. This makes choosing one variable from the cluster or using the centroid of the variables as a cluster representative meaningful. Other existing methods do not have this property as they only find variables with the highest correlation to the centroid of the variables in the cluster or to other latent variable found by PCA.

## 3 Fuzzy Equivalence Relations

In the following we give some well-known notions and definitions.

**Definition 1.** Let $X$ be a universal set. Every function of the form $A: X \to [0, 1]$ is called a fuzzy set or a fuzzy subset of $X$ and $\mu_A(x)$ is the membership degree of $x$ in the fuzzy set $A$.

**Definition 2.** The α-cut of the fuzzy set $A$ is defined as the crisp set:

$$^\alpha A = \{x \in R: \mu_A(x) \geq a\}, \, a \in (0, 1]$$

**Definition 3.** Let $X$, $Y$ be universal sets. Then $R = \{[(x, y), \mu_R(x, y)] | (x, y) \in X \times Y\}$ is called a fuzzy relation on $X \times Y$.

**Definition 4.** Let $R \subset X \times Y$ and $S \subset Y \times Z$ be two fuzzy relations. The max-min composition $R \circ S$ is defined as:

$$R \circ S = \left\{ \left[ (x, z), \max_y \left\{ \min_{x,z} \{ \mu_R(x, y), \mu_S(y, z) \} \right\} \right] | x \in X, \ y \in Y, \ z \in Z \right\}$$

**Definition 5.** A fuzzy relation $R$ on $X \times X$ is called a fuzzy equivalence relation if it satisfies the following conditions:

1. Reflexive, that is $\mu_R(x, x) = 1, \forall x \in X$.
2. Symmetric, that is $\mu_R(x, y) = \mu_R(y, x), \ \forall x, y \in X$.
3. Transitive, that is $\mu_R(x, z) \geq \max_y \left\{ \min_{x,z} \{ \mu_R(x, y), \mu_R(y, z) \} \right\} \forall x, y, z \in X$.

**Definition 6.** A fuzzy relation on $X \times X$ is called a fuzzy compatibility relation if it is reflexive and symmetric.

**Definition 7.** The transitive closure $R_T$ of the fuzzy relation $R$ is the relation that is transitive, contains $R$ and has the smallest possible membership grades.

**Definition 8.** Given a fuzzy compatibility relation $R$ on $X \times X$, the transitive closure $R_T$ can be calculated (Klir and Yuan 1995) by using the algorithm below:

1. $R' = R \cup (R \circ R)$.
2. If $R' \neq R$, make $R = R'$ and go to step 1.
3. Stop when $R' = R_T$.

The type of composition and set union in step 1 must be compatible with the definition of transitivity used. The max-min transitive closure corresponds to using the max-min composition and max operator for the set union.

**Definition 9.** The $\alpha$-cut matrix of the fuzzy relation $R$ is defined as:

$$R_\alpha = \left\{ \begin{array}{l} ((x, y), \mu_{R_\alpha}(x, y)) | \mu_{R_\alpha}(x, y) = 1, \text{ if } \mu_{R_\alpha}(x, y) \geq \alpha, \\ \mu_{R_\alpha}(x, y) = 0, \text{ if } \mu_{R_\alpha}(x, y) < \alpha, \ (x, y) \in X \times Y, \ \alpha \in [0, 1] \end{array} \right\}$$

# 4 Proposed Method for Clustering Variables and Factor Analysis Based on Fuzzy Equivalence Relation

Suppose we have $N$ variables to use to model a system, where $N$ is considerably large. We form an $N \times N$ relation matrix $R$ of the degree of 'closeness' between the variables and we express this degree of closeness in the interval [0, 1]. The proximity matrix of the variables is reflexive and symmetric and hence it is a fuzzy compatibility relation. The transitive closure $R_T$ of the fuzzy compatibility relation is then calculated using the algorithm in Definition 8. The new relation is now reflexive, symmetric and transitive and hence a fuzzy equivalence relation. The variables can then be clustered by choosing appropriate α-cuts. For an $M \times N$ data matrix $X$, where the rows represent data samples and the columns represent variables, the variable clustering procedure is summarized below.

---

**Algorithm 1 – Variable clustering**
**Step 1** – Calculate the pairwise Pearson correlation (proximity) matrix of all variables.

**Step 2** – Convert the matrix to fuzzy compatibility relation $R$ by taking the absolute values of all entries. This is to make sure all values lie in the closed interval [0, 1].

**Step 3** – Find the transitive closure $R_T$ of the fuzzy compatibility relation $R$ using Definition 8.

**Hierarchical clustering** – Find all feasible clusters by using the α-cut matrices (Definition 9) of all unique values in $R_T$.

**Partition clustering** – Use prior knowledge about the variables or a cluster validity index to choose a suitable α-cut value that gives the best number of clusters.

---

Note that step 1 and step 2 above can be omitted if the fuzzy compatibility matrix is formed using expert knowledge about the variables. Also step 2 can be omitted if all values in $R$ are in the interval [0, 1]. Finally any proximity metric which is reflexive and symmetric and whose values lie in the closed interval [0, 1] can be used.

## 4.1 Empirical Example – Variable Clustering

We illustrate the variable clustering method using variables from a published regression analysis of the annual railway passenger demand of Greece (Profillidis and Botzoris 2005; Adjenughwure et al. 2013). There were a total of 12 independent variables in the analysis: average rail passenger travel distance, unit cost of rail transport, car ownership index, number of buses working in interurban routes, total bus vehicle-km travelled in interurban routes, average bus vehicle-km travelled in interurban routes, average bus passenger travel distance, unit cost of non-rail transport, the ratio of unit cost of bus transport to the unit cost of rail transport, cost of petrol, per capita Gross Domestic Product of Greece, and a variable which was represents habitual inertia and constraints on supply.

The goal is to cluster the variables unsupervised and then use the results of the clustered variables for further analysis. To enhance the readers understanding of the method, we demonstrate a simple example with 7 independent variables.

**Step1 –**
**Correlation matrix:**

$$R = \begin{bmatrix} 1 & -0.68 & 0.73 & 0.77 & 0.74 & 0.58 & 0.60 \\ -0.68 & 1 & -0.68 & -0.85 & -0.76 & -0.57 & -0.55 \\ 0.73 & -0.68 & 1 & 0.72 & 0.57 & 0.38 & 0.95 \\ 0.77 & -0.85 & 0.72 & 1 & 0.77 & 0.51 & 0.59 \\ 0.74 & -0.76 & 0.57 & 0.77 & 1 & 0.94 & 0.43 \\ 0.58 & -0.57 & 0.38 & 0.51 & 0.94 & 1 & 0.26 \\ 0.60 & -0.55 & 0.95 & 0.59 & 0.43 & 0.26 & 1 \end{bmatrix}$$

**Step2 –**
**Absolute value of**
**correlation matrix:**

$$R = \begin{bmatrix} 1 & 0.68 & 0.73 & 0.77 & 0.74 & 0.58 & 0.60 \\ 0.68 & 1 & 0.68 & 0.85 & 0.76 & 0.57 & 0.55 \\ 0.73 & 0.68 & 1 & 0.72 & 0.57 & 0.38 & 0.95 \\ 0.77 & 0.85 & 0.72 & 1 & 0.77 & 0.51 & 0.59 \\ 0.74 & 0.76 & 0.57 & 0.77 & 1 & 0.94 & 0.43 \\ 0.58 & 0.57 & 0.38 & 0.51 & 0.94 & 1 & 0.26 \\ 0.60 & 0.55 & 0.95 & 0.59 & 0.43 & 0.26 & 1 \end{bmatrix}$$

**Step3 –**
**Transitive closure of R:**

$$R_T = \begin{bmatrix} 1 & 0.77 & 0.73 & 0.77 & 0.77 & 0.77 & 0.73 \\ 0.77 & 1 & 0.73 & 0.85 & 0.77 & 0.77 & 0.73 \\ 0.73 & 0.73 & 1 & 0.73 & 0.73 & 0.73 & 0.95 \\ 0.77 & 0.85 & 0.73 & 1 & 0.77 & 0.77 & 0.73 \\ 0.77 & 0.77 & 0.73 & 0,77 & 1 & 0.94 & 0.73 \\ 0.77 & 0.77 & 0.73 & 0.77 & 0.94 & 1 & 0.73 \\ 0.73 & 0.73 & 0.95 & 0.73 & 0.73 & 0.73 & 1 \end{bmatrix}$$

**Example of partition**
**clustering** ($\alpha$ – cut matrix
of $R_T$ for $\alpha = 0.8$):

$$^{\alpha}R_T = \begin{bmatrix} 1 & 0 & 0 & 0 & 0 & 0 & 0 \\ 0 & 1 & 0 & 1 & 0 & 0 & 0 \\ 0 & 0 & 1 & 0 & 0 & 0 & 1 \\ 0 & 1 & 0 & 1 & 0 & 0 & 0 \\ 0 & 0 & 0 & 0 & 1 & 1 & 0 \\ 0 & 0 & 0 & 0 & 1 & 1 & 0 \\ 0 & 0 & 1 & 0 & 0 & 0 & 1 \end{bmatrix}$$

Each row or column corresponds to a variable. If any rows are equal, then they belong to the same cluster. From the matrix above, we find the clusters $\{1\}$, $\{2, 4\}$, $\{3, 7\}$, $\{5, 6\}$.

**Example of hierarchical clustering:** The unique value of $R_T$ are 0.73, 0.77, 0.85, 0.94, 0.95. We form the $\alpha$-cut matrix of $R_T$ for each of these values and obtain the clusters below:

$$\{1, 2, 3, 4, 5, 6, 7\} \qquad \alpha = 0.73$$
$$\{1, 2, 4, 5, 6\}, \{3, 7\} \qquad \alpha = 0.77$$
$$\{1\}, \{2, 4\}, \{5, 6\}, \{3, 7\} \qquad \alpha = 0.85$$
$$\{1\}, \{2\}, \{4\}, \{5, 6\}, \{3, 7\} \qquad \alpha = 0.94$$
$$\{1\}, \{2\}, \{4\}, \{5\}, \{6\}, \{3, 7\} \qquad \alpha = 0.95$$
$$\{1\}, \{2\}, \{4\}, \{5\}, \{6\}, \{3\}, \{7\} \qquad \alpha = 1.00$$

The clusters produced for the 12 variables by the Algorithm 1 for $\alpha = 0.85$ is shown below:

$$\{1,3,7,8,11\}, \{2,4\}, \ \{5,6\}, \{9\}, \ \{10\}, \ \{12\}$$

As the researchers, now we have the clustered variables. We can select variables from the groups instead of individually as stated earlier. We can choose one or more variables from each group using the same feature selection criteria that we would use if the variables were not clustered. For instance, we can apply the variable selection technique to the first cluster. So instead of selecting among 12 variables $\{1, 2, 3, 4, 5, 6, 7, 8, 9, 10, 11, 12\}$, we reduce the problem to selecting among 5 variables $\{1, 3, 7, 8, 11\}$. This is then followed by selection among 2 variables $\{2, 4\}$ and $\{5, 6\}$.

This technique has two significant advantages: firstly, the computation cost of the feature selection algorithm is greatly reduced, and secondly, the grouping helps to reduce the problem of omitting some relevant variables since the set to choose from has been reduced. Note that the simplest way to select the variables is to pick one or more variable from each group which has the highest correlation with the dependent variable. This is of course not the best way as there are many other factors to be considered when selecting the variables. Luckily, there are many variable selection techniques any of which can be applied successively to the clustered variables to get the final set of variables. There are also available techniques specially made for selecting among clustered variables like the one used in PROC VARCLUS (Nelson 2001), or the one proposed by Bühlmann et al. (2013).

In our example of regression analysis with statistical criteria, the appropriate model was calibrated following the technique 'general-to-specific model selection' (Hendry 2000). The feature selection on all 12 variables with the dependent variables, gave the variables $\{2\}$, $\{3\}$, $\{9\}$, $\{10\}$, $\{11\}$, $\{12\}$ as the best calibration subset (Profillidis and Botzoris 2005; Adjenughwure et al. 2013). This is equivalent to choosing variables $\{3\}$ and $\{11\}$ from the first cluster, choosing variable $\{2\}$ from the second cluster, variables $\{9\}$, $\{10\}$ and $\{12\}$ from the single clusters and none from the cluster $\{5, 6\}$. We note that the criterion or method used for selecting the variables after clustering is left to the user. Our main aim for the paper is to propose an efficient method to quickly cluster variables unsupervised before the feature selection process begins.

Apart from the advantages previously listed, the proposed method is easier to implement compared to other variable clustering methods currently available. The only parameter to choose is $\alpha$. A very small $\alpha$ will give fewer clusters with weakly correlated variables while a very large $\alpha$ will result in many clusters with highly correlated variables. If the goal is mainly to reduce the computational cost of feature selection algorithms, then use a small $\alpha$ to reduce the variables into few clusters and perform feature selection on each cluster. On the other hand, if we want to find or eliminate groups of highly correlated variables, we can use a bigger $\alpha$. Note that the smallest $\alpha$ is the minimum value of the transitive closure matrix while the largest $\alpha$ is 1.

## 4.2 Empirical Example – Factor Analysis

We will use the proposed variable clustering based on fuzzy equivalence relation to perform factor analysis on the quality of service of public transport in Greece. The commuters' perception on service quality offered by the public transport system of the city of Thessaloniki (Greece's second-largest city) was recently measured by using customer satisfaction survey (nineteen closed-ended questions, five-point Likert scale answers, 450 respondents) and an exploratory factor analysis was performed to determine the principal components of service quality, (Delinasios 2014).

So, there are nineteen variables (the nineteen closed-ended questions) and our goal is to represent all these variables by using latent variables called factors. The procedure for factor analysis is almost the same as variable clustering with the only difference being that a single variable cannot form a cluster. Any cluster with at least two variables is called factor. The algorithm for factor analysis using fuzzy equivalent relation is described below.

---

**Algorithm 2 – Factor analysis**

**Step 1** – Perform hierarchical variable clustering using the Algorithm 1 for variable clustering.

**Step 2** (classical factor analysis) – Select the α-cut level with the highest number of factors and choose the smallest α-cut level with the same number factors as the selected.

**Step 3** (variable assignment) – For each single non-assigned variable, use the correlation matrix to find the variable with which it has the highest correlation and assign the variable to the appropriate factor.

**Step 4** (hierarchical factor analysis) – Repeat step 2 and step 3 for all unique number of factors to form a hierarchical type of factor analysis.

---

### Step 1 – Hierarchical variable clustering:

$\{1,2,3,4,5,6,7,8,9,10,11,12,13,14,15,16,17,18,19\}$     $\alpha = 0.34$

$\{1,2,3,4,5,6,7,8,9,10,11,12,13,16,17,18\}, \{14,15,19\}$     $\alpha = 0.40$

$\{1,2,3,4,16\}, \{5,6,7,8,9,10,11,12,13,17,18\}, \{14,15,19\}$     $\alpha = 0.48$

$\{1,2,3,16\}, \{4\}, \{5,6,7,8,9,10,11,12,13,17,18\}, \{14,15,19\}$     $\alpha = 0.49$

$\{1,2,3,16\}, \{4\}, \{5,6,7,8,9,10,17\}, \{11,12,13,18\}, \{14,15,19\}$     $\alpha = 0.56$

$\{1,2,3,16\}, \{4\}, \{5,7,8,9,17\}, \{6\}, \{10\}, \{11,12,13,18\}, \{14,15,19\}$     $\alpha = 0.57$

..........

$\{1\}, \{2\}, \{3\}, \{4\}, \{5\}, \{6\}, \{8\}, \{10\}, \{16\}, \{7,9,17\}, \{11,12,13,18\}, \{14,15,19\}$   $\alpha = 0.61$

..........

$\{1\}, \{2\}, \{3\}, \{4\}, \{5\}, \{6\}, \{7\}, \{8\}, \{9\}, \{10\}, \dots, \{15\}, \{16\}, \{17\}, \{18\}, \{19\}$     $\alpha = 1.00$

**Step 2 – Classical factor analysis:** Remember that a factor is a group of two or more variables. From step 1, the highest number of factors is 4. There are many $\alpha$-cut levels with four factors. We choose the smallest which is $\alpha = 0.56$.

$$\{1, 2, 3, 16\}, \{4\}, \{5, 6, 7, 8, 9, 10, 17\}, \{11, 12, 13, 18\}, \{14, 15, 19\} \quad \alpha = 0.56$$

**Step 3 – Variable assignment:** Only the 4th variable was separated. Using the correlation matrix, we find the variable to which $\{4\}$ has the highest correlation. This is the $\{16\}$, so we put $\{4\}$ in the same group as $\{16\}$ and the final result is:

$$\{1, 2, 3, 4, 16\}, \{5, 6, 7, 8, 9, 10, 17\}, \{11, 12, 13, 18\}, \{14, 15, 19\}$$

This is exactly the factor analysis result with four factors as published by Delinasios (2014).

**Step 4 – Hierarchical factor analysis:** We only need to find the unique number of factors between the highest and the lowest number of factors from the $\alpha$-cut levels. Since the highest number of factors is 4 and the lowest number of factors is 1. We repeat step 2 and step 3 of Algorithm 2 for number of factors 2 and 3. The smallest $\alpha$-cut levels for 2 factors is $\alpha = 0.40$, while for 3 factors, $\alpha = 0.48$. The hierarchical factor analysis is presented below, (Fig. 1):

| | | |
|---|---|---|
| $\{1, 2, 3, 4, 5, 6, 7, 8, 9, 10, 11, 12, 13, 14, 15, 16, 17, 18, 19\}$ | 1 factor, | $\alpha = 0.34$ |
| $\{1, 2, 3, 4, 5, 6, 7, 8, 9, 10, 11, 12, 13, 16, 17, 18\}, \{14, 15, 19\}$ | 2 factors, | $\alpha = 0.40$ |
| $\{1, 2, 3, 4, 16\}, \{5, 6, 7, 8, 9, 10, 11, 12, 13, 17, 18\}, \{14, 15, 19\}$ | 3 factors, | $\alpha = 0.48$ |
| $\{1, 2, 3, 4, 16\}, \{5, 6, 7, 8, 9, 10, 17\}, \{11, 12, 13, 18\}, \{14, 15, 19\}$ | 4 factors, | $\alpha = 0.56$ |

These are exactly the factor analysis results with 1, 2, 3, and 4 factors as produced by SPSS. Note that in step 2 of Algorithm 2, any number of factors can be selected as long as there is a corresponding $\alpha$-cut level with the selected number of factors. As an advantage, *since* the $\alpha$-cut levels are ordered, step 2 is equivalent to choosing the first $\alpha$-cut level which has the same number of factors as the desired.

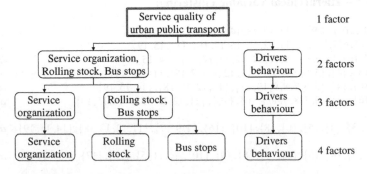

**Fig. 1** The hierarchical factor analysis based on fuzzy equivalence relation

# 5   Some Interpretations of the Clustered Variables

Although the proposed algorithms are quite different from classical factor analysis and dimensionality reduction techniques, they share some similarities. Each cluster can be viewed as a principal component or a latent variable (factor) formed using the variables in that cluster. For example, we can form 'factors', 'latent variables' or 'principal components' by taking the average of the variables in each cluster (Vigneau et al. 2001). Therefore, the clustered variables can be used as a pre-processing tool to select the appropriate number of latent variables or principal components for the purpose of factor analysis or principal components analysis.

We note that PCA can also be performed separately on each cluster. To do this, we form the data matrix with rows as examples and columns as variables using only the variables in the cluster. We can then choose one or more components from the PCA of that cluster as the cluster representative. We underline that this last advantage of variable clustering has not been explored in details and it would be interesting to compare the components produced by the clustered variables with that of direct PCA using all variables.

The clustered variables can also be useful for clustering high dimensional data. This can be done either by doing data clustering using only variables in the same group (a subspace of the 'closest' variables) or using one variable from each variable cluster(subspace of the 'farthest' variables).

# 6   Conclusions

A variable clustering method based on fuzzy equivalence relation has been proposed. We have applied the method to cluster variables from a regression analysis. The results show that the method can be used to reduce the number of variables used for modeling a system and thus is an alternative to other variable clustering algorithm currently available. We have also modified the proposed method and used it for factor analysis. The results show that the method yields similar results to classical factor analysis.

The proposed method has some advantages over existing methods, as it does not require specification of number of clusters or of number of factors. Additionally, it can be implemented using expert opinion about the relationship of the variables. For regression purpose, the clustered variables are considered equivalent at some chosen level hence selecting one variable from each cluster or taking their average is justified. Finally, some interpretations of clustered variables have been offered to help the reader to better understand and use the results of variable clustering for other applications.

**Acknowledgments** This research was financially supported through the university scholarship for doctoral students (KE 81516) in the section of Mathematics and Informatics, Department of Civil Engineering. The scholarship is awarded by the Research Committee of the Democritus University of Thrace.

# References

Adjenughwure, K., Botzoris, G., Papadopoulos, B.: Neural, fuzzy and econometric techniques for the calibration of transport demand models. J. Math. Sci. Eng. Appl. **7**, 385–403 (2013)

Beyer, K., Goldstein, J., Ramakrishnan, R., Shaft, U.: When is 'nearest neighbor' meaningful? In: 7th International Conference on Database Theory, pp. 217–235 (1999)

Bondell, H.D., Reich, B.J.: Simultaneous regression shrinkage, variable selection, and supervised clustering of predictors with OSCAR. Biometrics **64**, 115–123 (2008)

Bühlmann, P., Rütimann, P., van de Geer, S., Zhang, C.-H.: Correlated variables in regression: clustering and sparse estimation. J. Stat. Plann. Infer. **143**, 1835–1858 (2013)

Delinasios, N.: Parameterization of the Perceived by Users Quality of Service of Thessaloniki Urban Transport Organization. Democritus University of Thrace, Xanthi (2014)

Dettling, M., Bühlmann, P.: Finding predictive gene groups from microarray data. J. Multivar. Anal. **90**, 106–131 (2004)

Guyon, I., Elisseeff, A.: An introduction to variable and feature selection. J. Mach. Learn. Res. **3**, 1157–1182 (2003)

Hastie, T., Tibshirani, R., Botstein, D., Brown, P.: Supervised harvesting of expression trees. Genome Biol. **2**, 1–12 (2001)

Hastie, T., Tibshirani, R., Eisen, M., Alizadeh, A., Levy, R., Staudt, L., Chan, W.C., Botstein, D., Brown, P.: `Gene shaving' as a method for identifyingdistinct sets of genes with similar expression patterns. Genome Biol. **1**, 1–21 (2000)

Hendry, D.F.: The success of general-to-specific model selection. In: Hendry, D.F. (ed.) Econometrics: Alchemy or Science? Essays in Econometric Methodology, pp. 467–490. Oxford University Press, Oxford (2000)

Karypis, G., Han, E.-H.: Fast supervised dimensionality reduction algorithm with applications to document categorization & retrieval. In: 9th Conference on Information and Knowledge Management, pp. 12–19 (2000)

Kendall, M.: A Course in Multivariate Analysis. Griffin, London (1957)

Klir, G.J., Yuan, B.: Fuzzy Sets and Fuzzy Logic Theory and Application. Prentice Hall PTR, New Jersey (1995)

Nelson, B.D.: Variable reduction for modeling using PROC VARCLUS. In: 26th Annual SAS® Users Group International Conference, pp. 261–263 (2001)

Palla, K., Knowles, D., Ghahramani, Z.:A nonparametric variable clustering model. In: 26th Annual Conference on Neural Information Processing Systems, pp. 2987–2996 (2012)

Profillidis, V., Botzoris, G.: A comparative analysis of the forecasting ability of classical econometric and fuzzy models. Fuzzy Econ. Rev. **10**, 35–46 (2005)

Sanche, R., Lonergan, K.: Variable reduction for predictive modeling with clustering. In: Casualty Actuarial Society Winter Forum, pp. 89–100 (2006)

Spearman, C.: `General intelligence`, objectively determined and measured. Am. J. Psychol. **15**, 201–292 (1904)

Tibshirani, R.: Regression shrinkage and selection via the Lasso. J. Roy. Stat. Soc. **58**, 267–288 (1996)

Vigneau, E., Qannari, E.M., Punter, P.H., Knoops, S.: Segmentation of a panel of consumers using clustering of variables around latent directions of preference. Food Qual. Prefer. **12**, 359–363 (2001)

# Fuzzy EOQ Inventory Model With and Without Production as an Enterprise Improvement Strategy

Federico González-Santoyo, Beatriz Flores,
Anna M. Gil-Lafuente and Juan J. Flores

**Abstract** This work presents a theoretical extension to the inventory model EOQ with and without production, representing all variables as fuzzy quantities. The model is compared against the classical EOQ model with and without production. In this comparison, crisp and fuzzy data were used, and the results and conclusions were contrasted. We present the advantages of the fuzzy theory vs. classical theory in decision-making in the enterprise.

**Keywords** Fuzzy · Inventory · EOQ · Decision Making · Enterprise · Planning

## 1 Introduction

In the production dynamics of a company, the production and inventory management requires flawless strategic planning (González S.F. et al. 2010, 2011, 2013). This planning must include product demand forecast, optimal use of the plant capacity, and optimization of human resources, manufacture and acquisition times and amounts.

F. González-Santoyo (✉) · B. Flores
Facultad de Contaduría y Ciencias Administrativas, Universidad Michoacana, Morelia, Mexico
e-mail: fsantoyo@umich.mx

B. Flores
e-mail: betyf@umich.mx

A.M. Gil-Lafuente
Universidad de Barcelona, Barcelona, Spain
e-mail: amgil@ub.edu

J.J. Flores
Facultad de Ingeniería Electrica, Universidad Michoacana, Morelia, Mexico
e-mail: juanf@umich.mx

© Springer International Publishing Switzerland 2015
J. Gil-Aluja et al. (eds.), *Scientific Methods for the Treatment of Uncertainty in Social Sciences*, Advances in Intelligent Systems and Computing 377,
DOI 10.1007/978-3-319-19704-3_19

231

Kaufmann A. and Gil Aluja J. (1986) define the production process as the central nucleus of the production process. The enterprise's activity revolves around this nucleus, demanding raw material and finished products supply. That makes necessary the design of an efficient material delivery program. Otherwise, the plant may become inactive due to the lack of raw material. This situation leads to high cost levels, produced by operating the plant at levels below its capacity.

Managers keep raw material and finished product stocks, which represent static assets. These assets could be used in other productive activities. This situation arise for the following reasons:

- Productive activity makes impossible to maintain a given stock level.
- Uncertainty in future demand leads to keeping a minimum inventory level.
- Speculations arise when a sudden increase in prices is expected, or there is a high possibility of sales increase in the future.

Inventory control (Narasimhan S. et al. 1996) is a critical aspect of successful management. When keeping inventories is costly, companies cannot have high stock volumes.

To minimize the stocks, the company must execute a flawless planning to match the offer and demand levels, seeking the condition where the stock amount be minimal. Inventory is an amount of stored materials to be used in production or to satisfy the consumers demand (Schoeder R.G. 1992). Basic decisions to be made in stock management, among others, are:

- When to order?
- How much to order?

To answer these questions we need to know the behavior of the company's expected demand for the period of time under analysis, annual stock cost (**h**), generally a percentage of the item cost, the item cost (C), and service costs (S).

Stock management is one of the most important managerial functions, since it demands assets and if not performed properly, it can delay delivery of products to consumers. Optimal stock management has impact on production, marketing, and finance. The operational components found in stock management are:

- Financial. Seeks to keep low inventory levels, to maintain low costs.
- Marketing. Seeks to keep high inventory levels, to assure supply and sales.
- Operating. Seeks to keep adequate stock levels to guarantee an efficient production and homogeneous usage levels.

A company needs strong stock management systems to balance the above requirements, whose ideal stock levels conflict. This leads to seeking an optimal stock level, which allows the company to satisfy the market needs using the least possible amount of financial resources.

In a stock system, there exists uncertainty in the offer-demand behavior, and in the time required to complete the process until products reach consumers.

The problem addressed in this paper is the determination of how much to order to maintain a minimum stock level and still be able to face an uncertain demand. We also determine the time to order, when the company is or is not in production.

This paper is organized as follows: Section 1 provides an introduction; Section 2 provides background knowledge on stock costs; Section 3 explains demand behavior; Sections 4 and 5 present the classical EOQ model with ad without production, respectively; Section 6 presents a case analysis; Section 7 presents the results; Sections 8 and 9 present the conclusions and recommendations.

# 2 Stock Costs

The structure of inventory costs includes the following types of costs:

i. **Item cost.** The cost of purchasing and/or producing the stock items. Generally expressed as the unit cost times the stock capacity.
ii. **Ordering cost, preparation, or waste.** This cost is related to the purchase of a group or lot of items. This cost does not depend on the number of items.
iii. **Inventory cost.** This cost is related with storing items for a period of time. This cost is usually expressed as a percentage of the item value per unit of time.

Inventory costs normally have three components:

i. **Equity cost.** This cost arises when items are stored and equity is not available for other purposes. This cost represents the cost of not performing other investments.
ii. **Storage cost.** This cost includes components that vary with space, insurance, and taxes.
iii. **Obsolescence, damage, and waste cost.** These costs are assigned to items that age or expire, the higher the risk to become unusable, the greater the cost rate. The cost of items that expire are added to aging costs. For instance, in some grocery items, the loss costs include stolen items and damage related to maintaining them in stock.
iv. **Out of stock cost.** This kind of cost reflects the consequences of running out of stock for each product in the inventory. It includes raw materials, finished products, etc. The lack of an item brings causes the loss of an opportunity to produce or sell a product.

# 3 Demand Behavior

Future demand in an enterprise can be classified according to what we know about it (Kaufmann and Gil Aluja 1986; González Santoyo and Flores Romero 2002):

- When the company knows exactly the demand's behavior with time. This fact represents deterministic or certain demand.
- When the company does not know exactly how demand behaves. This situation represents a probabilistic or stochastic behavior.
- When the company does not know the future levels of demand, but takes advantage of a set of experts. Knowledge in uncertain and the reasoning framework to be used in fuzzy logic.

# 4 Classical Economic Order(ing) Quantity (EOQ)

F.W. Harris developed this methodology in 1915, and it is still in use for inventory management when demand is an independent variable.

The basic assumptions of the model are:

- The demand rate is known and constant along time.
- Delivery time is constant and zero.
- Since demand and delivery are instantaneous, there is no stocking.
- Materials are bought or produced in groups or lots, and place in stock.
- Unit cost per item is constant and there is no discount for bulk sales.
- The cost to place an order is **k** monetary units.
- An item unit cost is **c**.
- The unit storage cost is **h**.
- There is no interaction between products.

According to these assumptions, stock behaves as shown in Fig. 1.

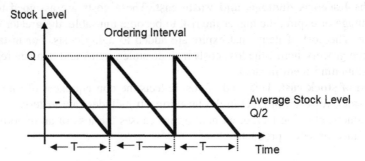

**Fig. 1** EOQ model

where Q is the ordering amount (in number of units), d is demand (in number of units/time), K is the fixed cost, c is the cost per unit ($/unit), and h is the storage cost per unit = i %(c).

Figure 2 shows the behaviour of cost; as **Q** increases, the purchasing cost decreases, since we place less orders per year. At the same time, the stocking cost increases, since the stock level increases. Therefore, purchasing and stocking costs compensate, one decreases while the other one increases. To determine the value of **Q**, that minimizes **CP(Q)** we compute the partial derivative of **CP(Q)** and solve for **Q** when it is zero. The cost per period is given by Eq. (1).

$$CT(Q) = k + cQ + h\left(\frac{Q}{2}\right)T \tag{1}$$

The optimal cost is given by Eq. (2).

$$CP(Q) = \lim_{n \to \infty}\left[\frac{n\,CT(Q)}{n\,T}\right] = \frac{CT(Q)}{T} = \frac{k + cQ + h\left(\frac{Q}{2}\right)T}{T} = \frac{k\,d}{Q} + c\,d + \frac{h\,Q}{2} \tag{2}$$

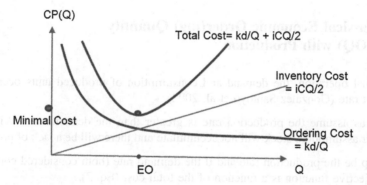

**Fig. 2** Cost behavior

But

$$T = \frac{Q}{d} \tag{3}$$

$$\frac{\partial\,CP(Q)}{\partial\,q} = 0 = -\frac{k\,d}{Q^2} + \frac{h}{2} = 0$$

$$\frac{k\,d}{Q^2} = \frac{h}{2}$$

$$Q^2 = \frac{2\,k\,d}{h}$$

Assuming $C_p = k$ and $h = C_h$, we obtain Eq. (4).

$$Q = \sqrt{\frac{2\,k\,d}{h}} = \sqrt{\frac{2\,C_p d}{C_h}} \qquad (4)$$

Q represents the ordering size that minimizes the stock average operation cost. Q is generally computed per year, but any time unit can be used. To determine the time required for the stock to reach zero, we use Eq. (5).

$$T = \frac{Q}{d} = \sqrt{\frac{2\,C_p}{C_h d}} = \frac{1}{N} \qquad (5)$$

The stock optimal average cost can be computed using Eq. (6).

$$CP(Q) = \frac{kd}{Q} + cd + \left(\frac{hQ}{2}\right) \qquad (6)$$

# 5   Classical Economic Order(ing) Quantity (EOQ) with Production

In normal operation, the demand and consumption of produced units occur at a constant rate (González Santoyo et al. 2002).

- Let us assume the production rate is greater than the demand rate. With any other assumption, stock will not accumulate and there will be a lack of products.

Let **p** be the production rate and **d** the demand rate (both considered constant). The objective function is a function of the total cost (Eq. 7).

$$CIT = Ordering\ Cost + Maintenance\ Cost \qquad (7)$$

The ordering cost is given by Eq. (8).

$$C_p \left[\frac{d}{Q}\right] \qquad (8)$$

The interpretation of the ordering cost while producing is known as startup. This cost includes man-hours worked, material, and production loss cost (incurred while getting the production system ready for operation).

- It is a fixed cost for each production lot, independent of the number of items being produced.

- Startup downtimes are integrated to the production plan development costs for each item, ordering formulation, all paperwork needed to prepare machinery and equipment, and the order flow control along the company's process.
- Maintenace cost is the unit cost to keep equipment running, times the mean stock level.

Since the production of the ordered amount (**Q**) takes place over a period of time defined by the production rate (**p**) and the parts enter the stock at the production rate, given a consumption rate, we obtain the inventory behavior shown in Fig. 3. The maximum and mean inventory levels are a function of the lot size, de production rate (**p**) and the demand rate (**d**) (Guiffrida 2010).

To determine the mean inventory level ($I_p$), since items are being received and consumed simultaneously, we first compute the time ($t_p$) required to produce the amount (**Q**). See Eq. (9).

$$t_p = \frac{Q}{p} \tag{9}$$

where $t_p$ is the time required to produce the ordered amount Q, given the supply rate p.

The maximum Inventory level is given by Eq. (10).

$$I_{max} = t_p(p - d) = (p - d)\left(\frac{Q}{p}\right) = Q\left(1 - \frac{d}{p}\right) \tag{10}$$

where (p-d) is the stocking rate and $t_p$ is the replenishing time (e assume p > d). Between replenishing times, stock decreases at a demand rate d.

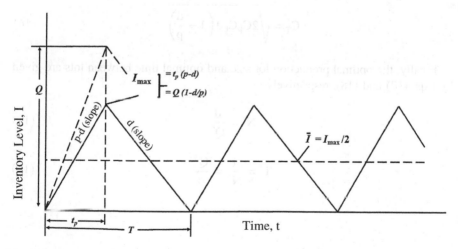

**Fig. 3** Inventory model with production

To compute the total inventory cost, we need to express the maximum stock level in terms of the ordering amount (Gallagher 1982). The mean inventory level is given by Eq. (11).

$$I_p = \frac{t_p(p-d)}{2} = \frac{I_{max}}{2} \tag{11}$$

Substituting Eq. (9) in Eq. (11), we obtain Eq. (12).

$$I_p = \left[\frac{Q}{2}\right]\left[1 - \frac{d}{p}\right] \tag{12}$$

The annual maintenance cost $(\mathbf{CM_a})$ and the total cost $(\mathbf{C_T})$ are given by Eqs. (13) and (14), respectively.

$$CM_a = C_h\left(\frac{Q}{2}\right)\left[1 - \frac{d}{p}\right] \tag{13}$$

$$C_T = C_p\left(\frac{d}{Q}\right) + C_h\left(\frac{Q}{2}\right)\left(1 - \frac{d}{p}\right) \tag{14}$$

Since the ordering amount is given by Eq. (15), the total cost can be expressed as in Eq. (16).

$$Q^* = \sqrt{\frac{2C_p d}{C_h\left[1 - \frac{d}{p}\right]}} \tag{15}$$

$$C_T^* = \sqrt{2C_p C_h d\left(1 - \frac{d}{p}\right)} \tag{16}$$

Finally, the optimal production lot size and optimal time between lots are given by Eqs. (17) and (18), respectively.

$$N^* = \frac{d}{Q^*} \tag{17}$$

$$T^* = \frac{1}{N^*} = \frac{Q^*}{d} \tag{18}$$

## 6　Case Analysis

To illustrate the application of the (EOQ) model, with and without production, both under certainty and uncertainty, we will use the case described in this section.

The company El Zapato Dorado, is a world-class company that ships shoes world wide from León, Guanajuato, Mexico. According to its records, the total stock is 10,000 pairs of shoes. The mean cost per pair is $12.00, so the total inventory cost is $1,200,000.00. The equity cost is estimated as an annual rate of 5 %, taxes, insurance, damages, wastes, and storage management costs are also 5 %.

The most requested shoes in the market are type 1. A marketing research and statistics indicate that last year 10 orders of 1,000 pairs were placed per period (5 weeks), at a cost of $20.00 per pair. The manufacturer guarantees that each order is delivered in 3 days, which has been accomplished so far. The average demand is 200 pairs per week.

The company takes 30 min to process an order. The cost per order is $16.00 per hour. Other costs include office supplies, mailing, telephone, clerical work, and transportation amount to $1.00 per order. Given this, the total cost of ordering is $17.00.

The company faces the choice of keeping a small stock and order frequently, or keep a large stock and order infrequently. The first choice may produce excessive ordering costs, while the second one would imply a higher stocking cost. So, we need to obtain an optimal ordering amount, minimizing stocking costs and still satisfying all market requirements.

As $Q$ grows, the stock management costs grow. This implies that the annual number of orders decreases non-linearly, tending to zero, asymptotically.

In the model that includes production, we assume the company installs a new production plant next to the main storage. Let us assume the plant's capacity (production rate) $p$=15,000 pairs per year, $C_h = $2.00, and $C_p = $9.00.

To analyze the system under uncertainty, we use a fuzzy logic model, using triangular fuzzy numbers. Following the information provided by a panel of experts, using the Delphi, we have: $\tilde{d} = (9\ 500,\ 10\ 000,\ 10\ 500)$, $\tilde{C}_p = (8.5,\ 9,\ 9.5)$, and $\tilde{C}_h = (1.5,\ 2,\ 2.5)$. The analysis is performed using a 11-value scale for fuzzy linguistic terms. To each $\alpha$-cut $[0 \leq \alpha_k \leq 1]$, corresponds a confidence interval $[r_k^\alpha, s_k^\alpha]$ that can be expressed as a function of $\alpha_k$ (see Eq. 19).

$$[r_k^\alpha, s_k^\alpha] = [r + (m - r)\alpha_k, s - (s - m)\alpha_k] \tag{19}$$

## 7　Results

Table 1 compares the results of the EOQ model without production in the classical and the fuzzy versions.

**Table 1** Results of the classical and fuzzy EOQ models without production

| Classical EOQ | Fuzzy EOQ |
|---|---|
| Q = 300 units | $\tilde{Q} = (254, \mathbf{300}, 364)$ |
| CIT = $600.00 | $\widetilde{CIT} = (492.18, 600, 706.22)$ |
| | $\approx (492, \mathbf{600}, 706)$ |
| N = 33.33 ≈ 33 orders per year | $\tilde{N} = (26, 33.33, 41.33)$ |
| T = 0.030 per year | $\tilde{T} = (0.0241, \mathbf{0.030}, 0.0384)$ |

Table 2 compares the results of the EOQ model with production in the classical and the fuzzy versions.

**Table 2** Results of the classical and fuzzy EOQ models with production

| Classical EOQ with production | Fuzzy EOQ with production |
|---|---|
| Q = 522 units | $\tilde{Q} = (409, \mathbf{520}, 694)$ |
| CIT = $346.41 | $\widetilde{CIT} = (258.52, \mathbf{346.44}, 539.39)$ |
| N = 19.14 orders per year | $\tilde{N} = (13.68, \mathbf{19.23}, 25.67)$ |
| T = 0.052 per year | $\tilde{T} = (0.038, \mathbf{0.052}, 0.073)$ |
| = 18.2 days | |

# 8 Conclusions

From the analysis performed, we conclude that it is necessary to increase the used market capability, taking into account a variation on the demand level of (9,500, 10,000, 10,500) pairs of shoes and the variation of costs (fixed and inventory management unit cost) $\tilde{C}_p = (8.5, 9, 9.5)_p$ and $\tilde{C}_h = (1.5, 2, 2.5)$. For a company with the adequate conditions to produce directly the demand requirements, or seek external providers to satisfy ts needs and turn into a marketer of products that designs and outsource. According to the obtained results, it is convenient for the company El Zapato Dorado to adopt the EOQ system with production. This conclusion is derived from the fact that when it operates with production, the annual operation cost (CIT) almost doubles for non-production conditions. Similarly, although the inventory level with production almost doubles the non-production scenario, it allows increasing the sales level and starting new markets. These new markets will allow the company to place that stock and increase its profit level.

# 9  Recommendations

Taking the conclusions as reference, we recommend to include fuzzy logic to the inventory operation analysis. We recommend the deployment of the EOQ model with production. This model will provide a competitive advantage in decision making when used under uncertainty. This is a consequence from the fact that classical theory hides information that the fuzzy theory reveals. Using this approach, we include higher quality information to the analysis scenario, which allows us to direct the strategic planning of the company, which leads to better financial results and an advantage in the market position for the company.

# References

Gallagher, C.H.: Métodos cuantitativos para la toma de decisiones en la administración. Mc. Graw Hill, México (1982)

González Santoyo, F., Flores Romero, B., Gil Lafuente, A.M., Juan, F.: Uncertain optimal inventory as a strategy for enterprise global positioning. AMSE. Chania- Grecia (2013)

González Santoyo, F., Flores Romero, B., Gil Lafuente, A.M.: Modelos y teorías para la evaluación de inversiones empresariales. FeGoSa-Ingeniería Administrativa S.A. de C.V., UMSNH, IAIDRES, Morelia México (2010)

González Santoyo, F., Flores Romero, B.: Teoría de Inventarios en la empresa (notas de seminario). Doctorado en Economía y Empresa. Universitat Rovira i Virgili, España (2002)

González Santoyo, F., Flores Romero, B., Gil Lafuente, A.M.: Procesos para la toma de decisiones en un entorno globalizado. Editorial Universitaria Ramón Areces, España (2011)

Guiffrida, A.: Fuzzy Inventory Models. In: Jaber, M.Y. (ed.) Inventory Management: Non-Classical Views. Chapter 8. CRC. Press, FL, Boca Raton, pp. 173–190

Kaufmann, A, Gil Aluja, J.: Introducción de la teoría de subconjuntos borrosos a la gestión de las empresas. Velograf S.A, España (1986)

Kaufmann, A, Gil Aluja, J, Terceño, G.A.: Matemáticas para la economía y la gestión de empresas. Foro Científico, Barcelona- España (1994)

Moskowitz, H., Wright Gordon, P.: Investigación de Operaciones. Prentice Hall, México (1982)

Narasimhan, S., Mc. Leavey, D.W., Billington, P.: Planeación de la producción y control de inventarios. Prentice Hall, México (1996)

Schoeder Roger, G.: Administración de operaciones. Toma de decisiones en la función de operaciones. Mc. Graw Hill, México (1992)

# Part V
# Modelling and Simulation Techniques

# A Bibliometric Overview of Financial Studies

José M. Merigó, Jian-Bo Yang and Dong-Ling Xu

**Abstract** Academic research in modern finance has been developing over the last decades. Many important contributions have been published in the main journals of the field. This paper analyzes scholarly research in finance by using bibliometric indicators. The main results are summarized in three fundamental issues. First, the citation structure in finance is presented. Next, the paper studies the influence of financial journals by using a wide range of indicators including publications, citations and the $h$-index. The paper ends with an overview of the most influential papers. In general, the results are in accordance with the expectations where the Journal of Finance, the Journal of Financial Economics and the Review of Financial Studies are the most popular journals and the USA is clearly the dominant country in finance.

**Keywords** Bibliometrics · Web of science · Journal rankings · Finance

## 1 Introduction

Bibliometric studies are very common in the literature. They analyze quantitatively the bibliographic material. Due to the strong development of computers and internet, today it is a very popular field to assess the state of the art of a research area because the information is easily available to any scientific institution. The definition of bibliometrics has brought many discussions in the literature. A very common definition is the one provided by Broadus (1987) that clearly defined the topic considering future

J.M. Merigó (✉)
Department of Management Control and Information Systems, University of Chile, Av. Diagonal Paraguay 257, 8030015 Santiago, Chile
e-mail: jmerigo@fen.uchile.cl

J.M. Merigó · J.-B Yang · D. L Xu
Manchester Business School, University of Manchester, Manchester M156PB, UK
e-mail: jian-bo.yang@mbs.ac.uk

D.-L Xu
e-mail: ling.xu@mbs.ac.uk

© Springer International Publishing Switzerland 2015
J. Gil-Aluja et al. (eds.), *Scientific Methods for the Treatment of Uncertainty in Social Sciences*, Advances in Intelligent Systems and Computing 377,
DOI 10.1007/978-3-319-19704-3_20

developments due to modern technologies. He stated that bibliometrics is "the quantitative study of physical published units, or of bibliographic units, or of surrogates of either". Recently, several studies (Alonso et al. 2009) have provided deeper understandings of the concept including the work of Bar-Ilan (2008) that connected it with a more general framework that included scientometrics and informetrics.

The aim of this paper is to present a bibliometric analysis of the most fundamental research developed in finance since the creation of JF. For doing so, it is used the Web of Science (WoS) as the database for collecting the information because it is usually regarded as the most influential one for scientific research. Thus, this study provides a modern approach for analyzing the state of the art in finance from a bibliometric perspective. Its main advantage is that it analyzes the information considering a wide range of variables that permits to detect the weaknesses and strengths of a journal or a paper. Moreover, it provides an updated list of results that contributes to the knowledge in this field following previous works in this direction by Alexander and Mabry (1994), Chan et al. (2002), Chung et al. (2001) and Currie and Pandher (2011). The analysis is divided in two parts that can be classified in journal analysis and the most cited papers since 1946.

The paper focuses on the most relevant journals. It analyzes those journals with a strong financial orientation indexed in WoS that are usually found in the subject category of "business finance". By doing so, it is assumed that only the most relevant journals in financial research are included because the aim of WoS is only to include those journals that accomplishes with high quality standards including a rigorous peer-review process and regular publication of issues without delays. Several indicators are used in order to analyze their quality (Merigó et al. 2015) including the number of papers and citations, the $h$-index and the impact factor provided by WoS – Journal Citation Reports (JCR). The results are in accordance with the common knowledge and previous studies (Chan et al. 2013), being JF, JFE and RFS the most influential journals.

An analysis of the most cited papers in financial research is also presented. This part of the work studies the 50 most cited papers in finance according to the results found in WoS. Thus, it is possible to get an overview of those papers that have a strongest impact in the scientific community. The results are more or less in accordance with the common knowledge where many of the well-known papers appear highly ranked in the list. For doing this, this paper is organized as follows. Section 2 presents a journal ranking in finance according to some bibliometric indicators. Section 3 studies the most cited papers in finance of all time. Section 4 summarizes the main results and conclusions of the paper.

## 2   Citation Structure in Finance

In April 2013, there were 39,440 papers in WoS in the 50 financial journals available in the WoS category of Business, Finance. If only articles, notes and reviews are considered, the number is reduced to 32,087. The global $h$-index of finance according to the 50 selected journals is 257. That is, 257 papers of the whole set of 39,440 papers have received at least 257 citations.

An important issue when analyzing the publication and citation structure is to consider the number of papers that have surpassed a citation threshold. This indicates the level of citation that most of the papers receive and permits to identify the number of citations that the top papers usually receive. Table 1 presents the general

**Table 1** General citation structure in finance according to WoS

| | ≥500 | ≥100 | ≥50 | ≥10 | ≥5 | ≥1 | Total |
|---|---|---|---|---|---|---|---|
| 1983 | 1 | 26 | 67 | 181 | 254 | 355 | 389 |
| 1984 | 2 | 24 | 62 | 193 | 248 | 371 | 438 |
| 1985 | 3 | 29 | 57 | 192 | 261 | 374 | 434 |
| 1986 | 3 | 30 | 69 | 215 | 298 | 406 | 466 |
| 1987 | 3 | 27 | 66 | 207 | 268 | 378 | 431 |
| 1988 | 7 | 40 | 80 | 204 | 285 | 406 | 447 |
| 1989 | 3 | 40 | 88 | 249 | 318 | 437 | 496 |
| 1990 | 1 | 46 | 96 | 266 | 342 | 464 | 520 |
| 1991 | 2 | 43 | 86 | 257 | 321 | 449 | 502 |
| 1992 | 3 | 43 | 84 | 260 | 323 | 436 | 480 |
| 1993 | 6 | 39 | 104 | 287 | 366 | 468 | 519 |
| 1994 | 2 | 33 | 73 | 242 | 319 | 424 | 467 |
| 1995 | 2 | 49 | 106 | 313 | 406 | 523 | 574 |
| 1996 | 2 | 49 | 112 | 319 | 401 | 519 | 558 |
| 1997 | 6 | 54 | 121 | 332 | 425 | 553 | 588 |
| 1998 | 2 | 46 | 111 | 322 | 412 | 527 | 608 |
| 1999 | 2 | 55 | 110 | 347 | 433 | 554 | 635 |
| 2000 | 2 | 66 | 134 | 340 | 446 | 565 | 677 |
| 2001 | 0 | 47 | 114 | 386 | 498 | 644 | 706 |
| 2002 | 1 | 54 | 144 | 425 | 538 | 695 | 790 |
| 2003 | 0 | 33 | 104 | 436 | 577 | 774 | 852 |
| 2004 | 0 | 24 | 100 | 497 | 665 | 874 | 954 |
| 2005 | 0 | 24 | 102 | 505 | 694 | 981 | 1073 |
| 2006 | 0 | 14 | 61 | 499 | 757 | 1086 | 1194 |
| 2007 | 0 | 11 | 43 | 429 | 685 | 1092 | 1245 |
| 2008 | 0 | 5 | 25 | 394 | 732 | 1402 | 1632 |
| 2009 | 1 | 3 | 10 | 365 | 731 | 1577 | 1955 |
| 2010 | 0 | 0 | 2 | 167 | 477 | 1435 | 1997 |
| 2011 | 0 | 0 | 0 | 40 | 175 | 1151 | 2127 |
| 2012 | 0 | 0 | 0 | 2 | 13 | 482 | 2050 |
| Total | 54 | 954 | 2331 | 8871 | 12668 | 20402 | 25804 |
| % | 0.21 % | 3.70 % | 9.03 % | 34.38 % | 49.09 % | 79.07 % | 100 % |

Abbreviations: ≥500, ≥100, ≥50, ≥10, ≥5, ≥1 = number of papers with more than 500, 100, 50, 10, 5 and 1 citations; % = Percentage of papers

citation structure considering several citation thresholds and developing an annual analysis since 1983.

Only 54 papers have received at least 500 citations between 1983 and 2012 which represents 0.2 % of all the papers. About 9 % of the papers receive more than 50 citations. Focusing on the last 10 years, the citation level is very low because more time is needed in order to consolidate a huge number of citations. Only one paper has received more than 500 citations and a very low number of the papers have already surpassed the 50 citation threshold. However, it is expected that many of these papers will receive a lot of citations in the future because currently they are new papers that still have not reached a consolidated position in the scientific community.

## 3 Journal Rankings

This section presents a journal ranking according to the data available in WoS. The journals are ranked according to the $h$-index although many other indicators are included in order to get a complete picture of each of them. The reason for using the $h$-index is because it combines publications and citations in the same measure (Olheten et al. 2011). However, it has some limitations when dealing with journals with significant differences in the number of publications. Usually, a higher number of papers bring a higher $h$-index independently of the quality of the papers. Although it is more efficient that the total number of papers and citations, it still cannot totally control this issue. The other alternative is to consider the ratio citation/papers or the impact factor but the problem here is that very small journals may get higher positions although not many people take them into account. Table 2 presents the journal ranking. Although the ranking is established according to the $h$-index, the rest of the indicators give a complete view of each of the journals. Note that in the case of tie in the $h$-index, it is selected first the journal with the lowest number of papers because this indicates that with a lower number of papers it has been able to reach a higher $h$-index.

JF and JFE are clearly the most influential journals in finance. Next, it is found RFS that is growing a lot during the last years and JFQA. Note that JB ended publication in 2006 but if included in the ranking would appear in the fourth position before JFQA. More or less, the results are in accordance with the common knowledge being the most popular journals in the first positions (Alexander and Marbry 1994; Chan et al. 2013). However, some deviations are found due to the particular nature of the $h$-index that requires a consolidation process throughout time. Therefore, young journals that have been recently included in WoS appear in lower positions. In some cases, these can be seen as a deviation since some of these journals should probably appear in a better position including RF and EFM. But in general, the results seem to be logical.

**Table 2** Most influential financial journals according to WoS

| R | Name | H | TC | TP | C/P | Y | Vol. | IF | IF5 | T50 |
|---|---|---|---|---|---|---|---|---|---|---|
| 1 | JF | 187 | 195677 | 4768 | 41 | 1946 | 1 | 4.333 | 6.185 | 21 |
| 2 | JFE | 169 | 136090 | 1973 | 69 | 1976 | 3 | 3.424 | 5.087 | 20 |
| 3 | RFS | 103 | 42501 | 1266 | 34 | 1990 | 3 | 3.256 | 5.367 | 1 |
| – | JB[1] | 93 | 44204 | 2298 | 19 | 1928 | 1 | – | – | 4 |
| 4 | JFQA | 72 | 29278 | 1957 | 15 | 1966 | 1 | 1.636 | 2.130 | 0 |
| 5 | JBF | 69 | 36733 | 3372 | 11 | 1980 | 4 | 1.287 | 1.721 | 0 |
| 6 | JMC | 67 | 28262 | 2112 | 13 | 1976 | 8 | 1.104 | 1.700 | 2 |
| 7 | JIMF | 55 | 16961 | 1523 | 11 | 1983 | 2 | 0.858 | 1.434 | 0 |
| 8 | FM | 42 | 11089 | 1283 | 9 | 1972 | 1 | 1.330 | 1.568 | 0 |
| 9 | MF | 40 | 7432 | 429 | 17 | 1997 | 7 | 1.000 | 1.463 | 1 |
| 10 | JFI | 35 | 4686 | 345 | 14 | 1995 | 4 | 2.208 | 2.460 | 0 |
| 11 | JFM | 34 | 9929 | 1547 | 6 | 1981 | 1 | 0.782 | 0.855 | 0 |
| 12 | JPM | 33 | 5682 | 1445 | 4 | 1984 | 10 | 0.525 | 0.562 | 0 |
| 13 | JCF | 32 | 4693 | 536 | 9 | 2001 | 7 | 1.035 | 1.774 | 0 |
| 14 | FS | 26 | 2986 | 299 | 10 | 2002 | 6 | 1.212 | 1.597 | 0 |
| 15 | QF | 25 | 3384 | 796 | 4 | 2001 | 1 | 0.824 | 0.957 | 0 |
| 16 | FAJ | 24 | 2529 | 447 | 6 | 2001 | 57 | 0.952 | 0.959 | 0 |
| 17 | JFSR | 23 | 2177 | 366 | 6 | 2008 | 33 | 1.176 | – | 0 |
| 18 | IJFE | 20 | 1917 | 393 | 5 | 1997 | 2 | 0.784 | 0.776 | 0 |
| 19 | JFMk | 18 | 1998 | 223 | 9 | 2002 | 5 | 1.093 | 1.505 | 0 |
| 20 | FRBSL | 16 | 1011 | 244 | 4 | 2004 | 86 | 0.640 | 0.748 | 0 |
| 21 | EFM | 16 | 1201 | 271 | 4 | 2005 | 11 | 0.738 | 1.431 | 0 |
| 22 | RF | 11 | 485 | 144 | 3 | 2008 | 12 | 1.440 | – | 0 |
| 23 | JOR | 10 | 333 | 102 | 3 | 2006 | 1 | 0.182 | 0.427 | 0 |
| 24 | JFEC | 10 | 422 | 128 | 3 | 2007 | 5 | 0.976 | 1.580 | 0 |
| 25 | JEF | 10 | 643 | 290 | 2 | 2008 | 15 | 0.934 | – | 1 |

[1]JB ended publication in 2006

Abbreviations: R = Rank; H = $h$-index; TC and TP = Total citations and papers; Y = Year when the journal was included in WoS; Vol. = First volume included in WoS; IF and IF5 = 2 and 5-Year Impact Factor 2011; T50 = Number of papers in the Top 50 list shown in Table 3

Journal abbreviations: JF = J. Finance; JFE = J. Financial Economics; RFS = Review of Financial Studies; JB = J. Business; JFQA = J. Financial and Quantitative Analysis; JBF = J. Banking and Finance; JMC = J. Money, Credit and Banking; JIMF = J. Int. Money and Finance; FM = Financial Management; MF = Mathematical Finance; JFI = J. Financial Intermediation; JFM = J. Futures Markets; JPM = J. Portfolio Management; JCF = J. Corporate Finance; FS = Finance and Stochastics; QF = Quantitative Finance; FAJ = Financial Analysts J.; JFSR = J. Financial Services Research; IJFE = Int. J. Finance & Economics; JFMk = J. Financial Markets; FRBSL = Federal Reserve Bank of St Louis; EFM = European Financial Management; RF = Review of Finance (before European Finance Review); JOR = J. Operational Risk; JFEC = J. Financial Econometrics; JEF = J. Empirical Finance

## 4 The Most Influential Papers in Finance

Many papers have made fundamental contributions to the financial literature. Some of them have even led to the Nobel Prize in economics. This section tries to identify the most influential ones by analyzing the 50 most cited papers of all time according to WoS. This measure aims at identifying the influence and popularity that a paper has reached in the financial literature. However, several limitations may occur due to the specific research considered in each paper that may attract more researchers and citations than other very good papers but with less use in the scientific community. In general terms, it is assumed that the most cited studies represent the majority of the key papers in the financial literature although some exceptions may appear. Table 3 presents a list with the 50 most cited articles in finance.

The most cited papers are very well-known in the scientific community. Some of them have led to the Nobel Prize in economics including the paper by Markowitz,

**Table 3** The 50 most cited papers in finance according to WoS

| R | J | TC | Title | Author/s | Year |
|---|---|---|---|---|---|
| 1 | JFE | 7264 | Theory of firm: managerial behaviour, agency costs and ownership structure | MC Jensen, WH Meckling | 1976 |
| 2 | JF | 3185 | Portfolio selection | H Markowitz | 1952 |
| 3 | JF | 2630 | Capital asset process: a theory of market equilibrium under conditions of risk | WF Sharpe | 1964 |
| 4 | JF | 2335 | Efficient capital markets: review of theory and empirical work | EF Fama | 1970 |
| 5 | JFE | 2231 | Common risk factors in the returns on stocks and bonds | EF Fama, KR French | 1993 |
| 6 | JFE | 2095 | Corporate financing and investment decisions when firms have information that investors do not have | SC Myers, NS Majluf | 1984 |
| 7 | JF | 1691 | The cross-section of expected stock returns | EF Fama, KR French | 1992 |
| 8 | JF | 1621 | Counter speculation, auctions, and competitive sealed tenders | W Vickrey | 1961 |
| 9 | JFE | 1559 | Determinants of corporate borrowing | SC Myers | 1977 |
| 10 | JF | 1515 | A survey of corporate governance | A Shleifer, RW Vishny | 1997 |
| 11 | JB | 1488 | The behaviour of stock-market prices | EF Fama | 1965 |
| 12 | JB | 1460 | The variation of certain speculative prices | B Mandelbrot | 1963 |
| 13 | JF | 1428 | Legal determinants of external finance | R La Porta, F Lopez De Silanes, A Shleifer et al. | 1997 |
| 14 | JF | 1382 | Corporate ownership around the world | R La Porta, F Lopez De Silanes, A Shleifer | 1999 |
| 15 | JF | 1277 | Pricing of corporate debt: risk structure of interest rates | RC Merton | 1974 |
| 16 | MF | 1209 | Coherent measures of risk | P Artzner, F Delbaen et al. | 1999 |

(continued)

**Table 3** (continued)

| R | J | TC | Title | Author/s | Year |
|---|---|----|-------|----------|------|
| 17 | JF | 1173 | Financial ratios, discriminant analysis and prediction of corporate bankruptcy | EI Altman | 1968 |
| 18 | JFE | 1167 | Management ownership and market valuation: an empirical analysis | R Morck, A Shleifer, RW Vishny | 1988 |
| 19 | RFS | 1121 | A closed form solution for options with stochastic volatility with applications to bond and currency options | SL Heston | 1993 |
| 20 | JFE | 1117 | Equilibrium characterization of term structure | O Vasicek | 1977 |
| 21 | JFE | 1110 | Option pricing: simplified approach | JC Cox, SA Ross, M Rubinstein | 1979 |
| 22 | JF | 1086 | On persistence in mutual fund performance | MM Carhart | 1997 |
| 23 | JF | 1060 | The modern industrial revolution, exit, and the failure of internal control systems | MC Jensen | 1993 |
| 24 | JF | 998 | On the relation between the expected value and the volatility of the nominal excess returns on stocks | LR Glosten, R Jagannathan, DE Runkle | 1993 |
| 25 | JFE | 993 | Using daily stock returns: the case of event studies | SJ Brown, JB Warner | 1985 |
| 26 | JF | 940 | Returns to buying winners and selling losers: implications for stock market efficiency | N Jegadeesh, S Titman | 1993 |
| 27 | JFE | 913 | Option pricing when underlying stock returns are discontinuous | RC Merton | 1976 |
| 28 | JFE | 886 | The market for corporate control: the scientific evidence | MC Jensen, RS Ruback | 1983 |
| 29 | JF | 873 | Informational asymmetries, financial structure, and financial intermediation | HE Leland, DH Pyle | 1977 |
| 30 | JMC | 843 | A general equilibrium approach to monetary theory | J Tobin | 1969 |
| 31 | JF | 842 | The pricing of options on assets with stochastic volatilities | J Hull, A White | 1987 |
| 32 | JFE | 836 | Bid, ask and transaction prices in a specialist market with heterogeneously informed traders | LR Glosten, PR Milgrom | 1985 |
| 33 | JF | 802 | Efficient capital markets. 2 | EF Fama | 1991 |
| 34 | JF | 791 | Does the stock market overreact | WFM Debondt, R Thaler | 1985 |
| 35 | JF | 747 | Multifactor explanations of asset pricing anomalies | EF Fama, KR French | 1996 |
| 36 | JFE | 729 | Estimating betas from nonsynchronous data | M Scholes, J Williams | 1977 |
| 37 | JFE | 726 | Valuation of options for alternative stochastic processes | JC Cox, SA Ross | 1976 |
| 38 | JMC | 724 | Postwar US business cycles: an empirical investigation | RJ Hodrick, EC Prescott | 1997 |
| 39 | JB | 723 | Dividend policy, growth, and the valuation of shares | MH Miller, F Modigliani | 1961 |

(continued)

**Table 3** (continued)

| R | J | TC | Title | Author/s | Year |
|---|---|---|---|---|---|
| 40 | JF | 698 | The capital structure puzzle | SC Myers | 1984 |
| 41 | JFE | 691 | Outside directors and CEO turnover | MS Weisbach | 1988 |
| 42 | JFE | 684 | Critique of asset pricing theory tests 1: Past and potential testability of theory | R Roll | 1977 |
| 43 | JFE | 678 | Higher market valuation of companies with a small board of directors | D Yermack | 1996 |
| 44 | JEF | 675 | A long memory property of stock market returns and a new model | Z Ding, CWJ Granger, RF Engle | 1993 |
| 45 | JF | 667 | Problems in selection of security of portfolios: performance of mutual funds in period 1945–1964 - 1 | MC Jensen | 1968 |
| 46 | JFE | 663 | Industry costs of equity | EF Fama, KR French | 1997 |
| 47 | JFE | 663 | Expected stock returns and volatility | KR French, GW Schwert, RF Stambaugh | 1987 |
| 48 | JFE | 658 | The separation of ownership and control in East Asian Corporations | S Claessens, S Djankov, LHP Lang | 2000 |
| 49 | JFE | 651 | Additional evidence on equity ownership and corporate value | JJ McConnell, H Servaes | 1990 |
| 50 | JB | 641 | Economic forces and the stock-market | NF Chen, R Roll, SA Ross | 1986 |

Sharpe, Vickrey and Merton. JF and JFE dominate the list with more than 80 % of the papers. Note that the last column of Table 2 shows the number of papers that each journal has in the top 50. Note that the results shown in Table 3 are in accordance with previous studies (Alexander and Marbry 1994; Chung et al. 2001; Arnold et al. 2003) although important deviations are found due to the evolution of financial research during the last years and the strong growth seen in WoS that has increased a lot the number of citations.

Regarding authors in the list, it is worth noting that Eugene F. Fama and Andrei Shleifer have published four of the top 50 papers. Michael C. Jensen also obtain remarkable results having three papers in the top 50. Kenneth R. French. Rafael La Porta, Florencio López De Silanes, Robert Merton, Stewart C. Myers and Robert W. Vishny have two papers each. Another interesting issue is that most of the papers come from American institutions and authors. Non-English speaking countries have a very low presence in the list. Currently, it seems that they are increasing their number of publications and citations but still far away from the USA.

Note that many key papers in finance have been published in a more general economic journal. Among others, it is found the famous paper of Black and Scholes (1973) published in the Journal of Political Economy (JPE) about the pricing of options that gave Myron Scholes the Nobel Prize in economics. Observe that this paper, if ranked in the list, would get the second position with 5253 citations. Table 4 presents an additional list of highly cited papers in finance that were not published in journals strictly dedicated to finance.

**Table 4** Other highly cited papers published in other journals

| J | TC | Title | Author/s | Y |
|---|---|---|---|---|
| JPE | 5253 | Pricing of options and corporate liabilities | F Black, M Scholes | 1973 |
| AER | 2720 | Agency costs of free cash flow, corporate-finance, and takeovers | MC Jensen | 1986 |
| JPE | 2400 | Law and finance | R La Porta, F Lopez-de-Silanes, A Shleifer et al. | 1998 |
| JLE | 2203 | Separation of ownership and control | EF Fama, MC Jensen | 1983 |
| BJE | 2178 | Theory of rational option pricing | RC Merton | 1973 |
| AER | 2038 | The cost of capital, corporation finance and the theory of investment | F Modigliani, MH Miller | 1958 |
| JPE | 1693 | Agency problems and the theory of the firm | EF Fama | 1980 |
| RESt | 1631 | The valuation of risk assets and the selection of risky investments in stock portfolios ... | J Lintner | 1965 |
| ECMT | 1619 | A theory of the term structure of interest rates | JC Cox, JE Ingersoll, SA Ross | 1985 |
| JPE | 1546 | Risk, return and equilibrium: Empirical tests | EF Fama, JD MacBeth | 1973 |
| ECMT | 1366 | Continuous auctions and insider trading | AS Kyle | 1985 |
| RES | 1202 | Financial intermediation and delegated monitoring | DW Diamond | 1984 |
| ECMT | 1179 | Intertemporal capital asset pricing model | RC Merton | 1973 |
| JPE | 1147 | Bank runs, deposit insurance, and liquidity | DW Diamond, PH Dybvig | 1983 |
| JECM | 1041 | ARCH modelling in finance: a review of the theory and empirical evidence | T Bollerslev, RY Chou, KF Kroner | 1992 |
| JPE | 1021 | The structure of corporate ownership: Causes and consequences | H Demsetz, K Lehn | 1985 |
| JPE | 993 | Large shareholder and corporate control | A Shleifer, RW Vishny | 1986 |
| JET | 954 | Martingales and arbitrage in multiperiod securities market | JM Harrison, DM Kreps | 1979 |
| JET | 932 | Arbitrage theory of capital asset pricing | SA Ross | 1976 |
| AER | 844 | Financial dependence and growth | RG Rajan, L Zingales | 1998 |

Note that the requirement to be included in this list is to have received at least 840 citations and be within the scope of finance. Abbreviations are available in Tables 2 and 3 except for JPE = J. Political Economy; AER = American Economic Review; JLE = J. Law & Economics; BJE = Bell J. Economics; RESt = Review of Economics and Statistics; ECMT = Econometrica; RES = Review of Economic Studies; JECM = J. Econometrics; JET = J. Economic Theory

Apart from JPE, it is worth noting that American Economic Review and Econometrica are those journals that have also published many leading papers in financial research.

# 5 Conclusions

A general bibliometric overview of scholarly research in finance has been presented. Several fundamental issues have been considered including a journal analysis and the most influential papers in the field. A major result found in the paper is that the USA clearly dominates the field have the leading papers and authors in finance. Moreover, they are responsible for publishing the leading journals. The information has been collected through WoS that is usually regarded as the main database for academic research. The results are in accordance with the common knowledge being the most popular research ranked in the first positions.

This paper has provided a general bibliometric overview of financial research over the last decades. Although the results are in accordance with the common knowledge, it is worth noting that some important limitations may produce changes on the results shown in the paper. Therefore, the paper aims to be informative rather than trying to provide some general strict rankings.

**Acknowledgments** Support from the European Commission through the project PIEF-GA-2011-300062 is gratefully acknowledged.

# References

Alexander Jr, J.C., Mabry, R.H.: Relative significance of journals, authors, and articles cited in financial research. J. Fin. **49**, 697–712 (1994)

Alonso, S., Cabrerizo, F.J., Herrera-Viedma, E., Herrera, F.: H-index: A review focused on its variants, computation and standarization for different scientific fields. J. Inf. **3**, 273–289 (2009)

Arnold, T., Butler, A.W., Crack, T.F., Altintig, A.: Impact: What influences finance research? J. Bus. **76**, 343–361 (2003)

Bar-Ilan, J.: Informetrics at the beginning of the 21st century—a review. J. Inf. **2**, 1–52 (2008)

Black, F., Scholes, M.: Pricing of options and corporate liabilities. J. Polit. Econ. **81**, 637–654 (1973)

Broadus, R.N.: Toward a definition of "Bibliometrics". Scientometrics **12**, 373–379 (1987)

Chan, K.C., Chang, C.H., Chang, Y.: Ranking of finance journals: some Google scholar citation perspectives. J. Empir. Fin. **21**, 241–250 (2013)

Chan, K.C., Chen, C.R., Steiner, T.L.: Production in the finance literature, institutional reputation, and labor mobility in academia: A global perspective. Fin. Manage. **31**, 131–156 (2002)

Chung, K.H., Cox, R.A.K., Mitchell, J.B.: Citation patterns in the finance literature. Fin. Manage. **30**, 99–118 (2001)

Currie, R.R., Pandher, G.S.: Finance journal rankings and tiers: An active scholar assessment methodology. J. Bank. Fin. **35**, 7–20 (2011)

Merigó, J.M., Gil-Lafuente, A.M., Yager, R.R.: An overview of fuzzy research with bibliometric indicators. Appl. Soft Comput. **27**, 420–433 (2015)

Olheten, E., Theoharakis, V., Travlos, N.G.: Faculty perceptions and readership patterns of finance journals: a global view. J. Fin. Quant. Anal. **40**, 223–239 (2011)

# A Theoretical Approach to Endogenous Development Traps in an Evolutionary Economic System

Silvia London and Fernando Tohmé

**Abstract** The representation of evolving economies can be formally represented through evolutionary self-organized systems (ESO), a cellular automata model with endogenous rules of change. Another possibility is to consider economic evolution as the result of the nested application of rules of changes on certain structures we call economic systems. Both approaches show disadvantages: ESO systems are too general and arbitrary, while the application of rules to rules lacks in most cases an effective characterization. In order to get the best out of both worlds we define here a notion of economic ESO systems. We show that the class of these systems is equivalent to a subclass of economic systems. The systems in this subclass can be effectively represented, with the extra bonus that the crucial role we assume knowledge plays in the mechanism of economic evolution becomes explicit. We claim that these systems are particularly fit for representing the notion of "poverty trap". In fact, they arise in an economic system that is unable to surpass a critical boundary.

**Keywords** Evolution · Self-organized criticality · Effective representations · Model theory

## 1 Introduction

An economic system can be fully described by means of three "parameters": its institutional structure, its technology and the agent's preferences. A formal framework intended to represent the evolution of an economy involves the coupled dynamical paths of those parameters.

S. London (✉) · F. Tohmé
Departamento de Economía, Investigaciones Económicas y Sociales del Sur, Universidad Nacional del Sur, CONICET, 12 de Octubre y San Juan, 7mo Piso, 8000 Bahía Blanca, Argentina
e-mail: slondon@uns.edu.ar

F. Tohmé
e-mail: ftohme@criba.edu.ar

© Springer International Publishing Switzerland 2015
J. Gil-Aluja et al. (eds.), *Scientific Methods for the Treatment of Uncertainty in Social Sciences*, Advances in Intelligent Systems and Computing 377, DOI 10.1007/978-3-319-19704-3_21

A large body of literature has been devoted to analyze the reasons for the success of some economies versus the failure of others. Neoclassical growth theory has been unable to explain those differences among economies. In turn, other theories have provided more cogent arguments on this question, ranging from endogenous growth theory (Lucas 2002) to the dynamics of distributions (Durlauf and Quah 1999) and the institutional approach (Azariadis and Stachurski 2005). In the latter, the institutional setting of an economy is deemed as a deep parameter, influencing the performance of the economy, which is also affected by changes in preferences and technology. In this way, a poverty trap, "... any self-reinforcing mechanism which causes poverty to persist" (op. cit.), can be interpreted as a situation in which the parameters of a poor economy do not change endogenously, even under exogenous shocks.

The first step towards the full characterization of poverty traps is to describe the evolutionary processes that may affect the economy. A conceptual tool that can be applied to do this is the model of self-organized critical (SOC) systems. This model is quite adequate to represent economies that change and adjust in time.[1] Per Bak (1996, in London 2008) emphasizes the analogy with the behavior of sandpiles. That is, real world economic systems are analogous to structures made out of sand, since they are discrete, plastic and generate endogenous perturbations. Moreover, according to this author, economies seem to go from one punctuated equilibrium to another (technically: from a meta-stable state to another): long periods of gradual change are followed by short bursts of radical change. The idea is that local changes are rather frequent while global changes are sporadic. In contrast to other self-organizing systems, when a SOC system is externally perturbed it evolves towards a stable critical state, which shows long range spatial, and temporal correlations. Once in that state, a SOC system generates variations of all sizes and frequencies, which, seen from the outside, appear as meta-stable states of the system.

Despite their advantages, a fundamental feature of SOC systems makes them inappropriate for the formalization of evolving economies: the system always runs under the same set of rules. An economy in evolution can hardly be conceived in that way. To overcome this difficulty we introduce a more general model of process where rules change in time, Evolutionary Self-organized (ESO) systems. In these systems the change of rules is endogenous to the process: changes of rules arise as a result of the internal dynamics instead as an external imposition.

On the other hand, ESO systems show a serious drawback in order to represent economic systems where the intentionality of agents plays a fundamental role. In this case it becomes important to specify how the evolution of the system as a whole affects the characteristics of the agents. In turn, those characteristics define the behavior of the system. We obtain, therefore, a loop between the system and the

---

[1]For instance, Arenas et al. (2000) present a specification of technological evolution based on the local interaction of a finite number of agents in a low dimensional space. This is the typical case of a SOC system.

decision-making units. Any description of the evolution of an economy requires a specification of how this loop arises. The idea discussed in the New-Institutional literature is that the link between the system and the agents is given by the aggregate (social) knowledge.

North (1993) defines Institutions as the norms of a society. They are designed constraints on human interaction and behavior, shaping the incentives in various kinds of interactions (economic, political or social nature). They also define the boundaries of choices available to individuals. Their origin is the need to reduce the uncertainty due to the lack of full information. Institutions arise to provide guidance as well as to limit the arbitrariness in such contexts.

It is useful to differentiate institutions from organizations. While both provide structure to human interactions, the latter require more homogeneity in their constituent bodies, and implement results induced by the institutions. In turn, they modify their parent institutions, becoming the main source of institutional change.

In fact, institutional change is the result of the combined use of information and the power exerted. In particular, as clearly exhibited in certain models where all the relevant parameters (institutions, technology, and characteristics of the agents) change in time: the evolutionary path of an economy is a consequence of how the economic agents make use of aggregate knowledge (Tohmé and London 1998). The entire set of possibilities for an economy can be represented as a tree of possible histories, each one conditioned by its associated sequence of states of knowledge. A problem is that albeit being recursively defined, this tree is not effective, i.e. there is no algorithmic procedure generating the actual history of the economy.

In this paper, we present an effective characterization of economic evolution combining the algorithmic nature of ESO systems with the intuitions developed about the role that knowledge plays in economic evolution. It will be shown that social knowledge can be embedded in an ESO representation of economic systems as an extra parameter. This simple procedure has the advantage of preserving all the theorems valid for general ESO systems and at the same time it provides a more economic-like representation. A theorem of soundness will be provided for this representation and the lack of a completeness result will also be discussed. The only difference worth mentioning is that instead of having a probabilistic foundation, it can be softened by means of fuzzy gradations. This is not a big departure, in the sense that, as pointed out by Kosko (1990), the former can be seen as a case of the latter. Another important feature of this model is that we focus on the existence of poverty traps. We characterize them as systems in which endogenous thresholds to change exist, such that shocks are unable to move the system away from a (low) stationary state.

In the next section we give an overview of ESO systems and their properties. In Sect 3 we give a formal characterization of evolving economies. Then we show how this characterization can be reduced to an ESO system using a simple specification of the aggregate knowledge.

## 2  ESO Systems

An ESO system (ESOs) is a sequence of structures that change in time due to the action of external shocks (London and Tohmé 2006). More precisely:

**Definition 1:** An ESO system, S is $E = (T, R, M)$, where $T = (T_0, T_1, \ldots)$ is a sequence of topologies, $R = (R_0, R_1, \ldots)$ is a sequence of uniform rules, and M is a set of meta-rules

A topology $T_i$ is $T_i = (S_i, S_{oi}, C_i)$, where $S_i$ is a finite set of sites, $S_{oi}$ is the initial state of those sites, and $C_i$ is the structure of the connections among them. Each s in $S_i$, can be in a state $|s|$, a numerical value drawn from $\{0, 1, \ldots, n_i\}$. Given a site s and the values of its neighborhood in time $t$, $|Ns|_t$, the value of the site in $t + 1$ results from applying the rule $R_i$: $|s|_{t+1} = R_i(|Ns|_t, |s|_t)$. This means that the state of the site depends both on the state of the site as well as on the state of its neighbors. Each $R_i$ must be a non-linear function in order to avoid the trivial asymptotic result where all sites reach the same value.

A shock $D_s$ is represented by the random variation of the state of a site. The size of the shock must be evaluated before applying a rule $R_i$. If that size exceeds the maximum feasible value, the meta-rule transforms both the topology and the rules. More precisely, if the shocks in time t are in range, that is, $|s| + D_s < n_t$, then $T_{t+1} = T_t$, and $R_{t+1} = R_t$. Otherwise, any meta-rule $M_K$ is selected at random, and $(T_{t+1}, R_{t+1}) = M_K(T_t, R_t)$: the meta-rule generates a new topology and a new set of rules. We add an requirement of parsimony: $s_t$ y $s_{t+1}$ cannot be disjoint.

If shocks are always in range, the system behaves like the Game of Life, which has been proven to be a SOC system. The non-linearity of the rules generates $1/f$ noise, fractal spatial structures and universal computability (London and Tohmé 2006). It follows trivially that:

**Proposition 1:** if for each t, $|s| + D_s < n_t$, then E is a SOC system.

**Proof:** if for every t, $|s| + D_s < n_t$, $T_{t+1} = T_t$, and $R_{t+1} = R_t$. That means that E can be represented as $E(T_0, R_0)$ since the initial topology and rules are permanent. The set of meta-rules M is irrelevant, since these rules are never activated. As R0 is non-linear, E is equivalent to the Game of Life: each site's state depends non-linearly on the state of its neighbors and the range of values is discrete, with more than two possible values. Therefore, since the Game of Life is a SOC system it follows that E is also a SOC.                                                                    □

When shocks are out of range, the entire system "mutates", adding new sites, new connections or new rules. Ad-hoc measures must be defined in order to characterize the degree of change. One possibility is to consider the degree of homogeneity among sites. It suffices to take into account the mean value $s^* = Mean(|s|)$, and the standard deviation $\sigma(st)$ ($\sigma t$, for short) of values in the system. It follows: that:

**Proposition 2:** if E is a SOC system, the sequence $(\sigma_0, \sigma_1, \ldots)$ constitutes in average a 1/f noise.

**Proof:** For any site $s$, the sequence of its values in time, $s = (|s|_0, |s|_1 \ldots)$ must be a 1/f noise since E is by hypothesis a SOC system. That means that its Fourier transform $F(s)$, which gives the **spectral density** of the time series, must be $F(s) \approx \lambda^{-\alpha}$. with $0 < \lambda < 2$, where $\lambda \in [0, \infty]$ is the frequency (see Mandelbrot 2002). Since $F(\cdot)$ is a linear operator the transform of the sequence of average values, $s^* = s_0^*, s_1^* \ldots)$, $F(s^*)$ is the average of the transforms of the individual sites. Call $\alpha_h$ the highest value of the $\alpha$ parameter (for site $i_h$) while $\alpha_l$ is the lowest value among all the sites. Then $\alpha_l < -Log(F(s^*))/Log\lambda < \alpha_h$ . This means that $F(s^*)$ cannot behave as a power law with a parameter outside the interval $[a_l, a_n] \subset [0, 2]$. I.e. it is, in average, a 1/f noise. Then since the standard deviation $\sigma$ is such that $-2s^* < \sigma < 2s^*$ and a power law is preserved under a multiplication by a constant, the sequence of standard deviations, $(\sigma_0, \sigma_1, \ldots)$ must be also, in average, a 1/f noise. $\qquad \square$

Therefore, to identify the meta-stable states of an ESO system (a sequence of SOCs) is equivalent to find ranges of robustness for the parameter of its spectral density. That is, given the history of an ESO system it is possible to find time intervals in which the value of the characteristic parameter of that distribution can be clearly distinguished from its values in other periods. It can be argued that each of those distinguished periods is a stage of meta-stability.

Evolution is characterized by the presence of punctuated equilibria. This means that evolution proceeds by sudden jumps followed by long periods of stability. Stability means here that the basic structure remains basically unchanged. In this setting, ESOs are good representations of evolutionary phenomena if evolutionary stability is identified with meta-stability. A consequence of representing evolution by means of ESOs is the following:

**Theorem 1:** an ESO system never reaches a stationary state.

**Proof:** Trivial. Since the sequence of standard deviations $(\sigma_0, \sigma_1, \ldots)$ is a 1/f noise, there cannot exist a $t^*$ such that for each $t > t^*$, $\sigma_{t+1} = \sigma_t$ even if the number of sites remains fixed. $\qquad \square$

This result shows why it is practically impossible to predict the behavior of evolutionary phenomena. However, we can study these processes by means of computational simulation. ESOs facilitate this task since, as a particular type of cellular automata, they have an effective specification.

# 3   Economic Systems in Evolution[2]

ESOs constitute a general framework that can be used to represent systems in evolution. Their characterization can be applied to establish how systems go from periods of stability to other, interrupted by short periods of sudden change. In periods of sudden change the critical parameter of the system mutates, leading to the apparition of new rules and connections among sites. The jump to the new status quo is given by the random selection of a new meta-rule. But for real world economies it is not reasonable to postulate that. It seems rather clear that the intentionality of the agents has to play a role in the choice of meta-rules: we must consider the effects that increases in the level of knowledge have on the stability of an economy.

Other drawbacks of ESOs stem from two features that are, in a certain sense, opposite. One is that they are too general and abstract, being not useful to understand the intrinsic mechanisms that lead to changes. In the study of real world cases, the functional structures involved must to be explicit. The other feature that diminishes the usefulness of ESOs is that they are two demanding in informational terms, requiring a precise description of the probability distributions involved. We address here these two aspects by, on one hand, incorporating specific and concrete items in the description of the system, and on the other replacing probabilistic characterizations by fuzzy gradations.

An important step towards a more concrete specification: we reconsider the interpretation of changes in the meta-rules of the system, seeing them as institutional ones. The degree of development achieved by the system depends on the complexity of interactions entertained.

A characterization of economic system, which improves over the drawbacks of ESO systems, is:

**Definition 2:**  an economic system $\varepsilon$, in period $t$, is

$$\varepsilon^t = \langle I, X^t, O^t, K^t, Q^t, R^t, cd^t, w^t, c^t, p^t \rangle$$

where: (a) I is a fixed number of agents (b) $X^t$, is a compact subset in a Euclidean space Rl. It represents the set of feasible resource allocations (c) $O^t \in \prod_{i \in I} O_i$ is a vector of characteristics of the agents. $K^t \in \prod_{i \in I} O_i$ describes the information held by the agents. For each i, $O_i^t$ is the space of her possible types, and $K_i^t$ is a variable representing the possible value of $\varepsilon^t$, subject to a fuzzy distribution. To $O^t \times K^t \times p^t$ corresponds a $cd^t \in X^t$. Each $O_i^t$ provides a description of the preferences and endowments of agent $i$ while $k_i^t$ describes the model that $i$ has about the economy, $cd^t$ is the bundle that $i$ desires to consume in period $t$ according to his type and beliefs. That is, $cd^t$ is i's demand function, (d) Q t is a compact correspondence, representing the productive sector of the economy in such a way that: $x^t = (w^{t+1} + c^{t+1}) \in Q^t(w^t, p^t) \subseteq X^{t+1}$. Given a vector of inputs, $w^t$ and their current

---

[2]We follow closely London 2008.

prices pt, Q t (wt, pt) is a set of possible outputs in the next period (obviously, only one element can be realized in $t+1$), (e) $R^t$ is a vectorial operator representing the socio-political structure of the economy, that for, $x^t \in X^t$, $R^t(cd^t, x^t) = (w^t, c^t, p^t)$. Given the output and the desired consumption, the rules indicate which will be the prices in the next period and how the product will be shared among agents and productive units, and (f) $w^t \in X^t$ are the inputs for the productive structure, $C^t$ is the bundle of commodities effectively assigned to consumption and $p^t \in [0, 1]$ is the normalized vector of prices.

In this framework, when will transitions occur? There are two possibilities. The first case is when changes are incremental, i.e. changes that arise due to the normal behavior of the system without modifying its fundamental parameters (growth theory analyzes this type of changes). The second case involves evolutionary (or non-smooth) changes, which arise from changes in the fundamental parameters of the economy. These changes may range from institutional crises to technological breakthroughs. Conditions that may lead to these evolutionary jumps are in a certain way related to changes in the state of knowledge in the economy:

Changes in the characteristics of the agents, due to a better knowledge of the economy. The following operator represents this fact: $\phi 1 : K \to O$

Changes in the space of commodities: a previously unknown commodity enters in the set of possible choices of the agents. This can be represented as: $\phi 2 : O \to X$, then: $\phi 2 \ o \ \phi 1 : K \to X$

Changes in the productive structure, following the change in the space of commodities: $\phi 3 : X \to Q$, and therefore: $\phi 3 \ o \ \phi 2 \ o \ \phi 1 : K \to Q$

Changes in the political structure, given that the agents decide on which "rules of the game" they want to change: $\phi 4 : O \to R$, then: $\phi 4 \ o \ \phi 1 : K \to R$

Knowledge propagates among the relevant parameters through the characteristics of the agents. We assume that the operators act in such a way that a change in $K^t$ defines the values of the parameters in $t + 1$. This is very demanding, but we will keep this assumption, since it introduces little change in the main results we obtain and simplify notoriously the presentation.

Knowledge affects the performance of the economy through its organizations. As long as they are willing to apply resources to induce changes, they will act successfully (London 2008).

Unless the state of knowledge remains stationary, all the parameters will be in permanent change. This is not independent of the particular rules of the system. In general, rules tend to avoid sudden changes: institutions are mainly concerned with providing stability to the system. On the other hand, it is also true that the state of knowledge changes from one period to the next. Its effects will depend on the nature of the relation between institutions and organizations.

To make this interaction among knowledge and rules precise we define a learning operator:

**Definition 3:** the learning operator $\gamma$ is such that: $\gamma: \{\xi^{t-1}\} \to K$ where $\xi^{t-1}$ is the set of $\varepsilon$'s. possible histories until $t-1$. Given a generic $\xi^{t-1}$, the knowledge in t, $K^t$, is $\gamma(\xi^{t-1})$.

The change of structure obtains combining learning with the influence of knowledge on the parameters of the system:

**Definition 4:** the operator of endogenous structural change is:

$$\Phi_0 \equiv \langle \phi_1, \phi_2 \circ \phi_1, \phi_3 \circ \phi_2 \circ \phi_1, \phi_4 \circ \phi_1 \rangle \circ \gamma: \{\xi^{t-1}\} \to X^t \times O^t \times Q^t \times R^t$$

This indicates that the previous history of the economy limits the selection of feasible future values of the parameters. Therefore:

**Definition 5:** an evolutionary transition from $t$ to $t+1$ occurs if

$$X^t \times O^t \times Q^t \times R^t \neq \Phi_0(\xi^t).$$

A problem that arises in this characterization is that the structural change operator might also change. That is, the system may not be autonomous. To solve this difficulty we distinguish between two types of operators. First, we consider operators of type I (OP1), which modify the structure of the economy according to its previous history. Each $\Phi_0 \in$ OP1 has the form $\Phi_0(\xi^{t-1}) \to X^t \times O^t \times Q^t \times R^t$. The set of type II operators (OP2), instead, can be partitioned in a hierarchy of sets $\{OP_{k+1}\}$, where each $\Phi_{k+1} \in OP_{k+1}$ is such that $\Phi_{k+1}: OP_k \to OP_K$. In fact, at each stage t the operator that can be applied can be seen as a nested application of operators with indexes less than $t^2$. In other words, each operator can be put in terms of a modification of the operator of type I applied in the first period: $\Phi_0^t = \rho \circ \Phi_0^{t-1}$, where $\rho: OP1 \to OP1$.

Moreover: $\Phi_0^t \equiv \Gamma^t \circ \Phi_0^{t-1}$, where $\Gamma^t \in \{\Phi_0^t \circ \Phi_0^{t-1} \ldots \circ \Phi_0^1$, for $\Phi_0^i \in OP_i = 1\infty \subseteq OP2\}$

We can see, then, that each operator of structural change can be defined in terms of the action of OP2 on $\Phi_0^t$. It follows that:

**Theorem 2:** (OP2, o) constitutes a semigroup.

**Proof:** OP2 is the set $\{\rho: \rho: OP1 \to OP1\}$. If follows trivially that (OP2, o) verifies the following properties (required to qualify as a semigroup):

- (closure under o) For $\rho1, \rho2 \in OP2, \rho1 \, o \, \rho2 \in OP2$: immediate from $\rho1, \rho2: OP1 \to OP1$ that is, domains and codomains coincide.
- (neutral element) There exists $I \in OP2$ such that for each $\rho \in OP2, I \, o \, \rho = \rho \, o \, I$. It is evident that the identity operator $I_d: OP1 \to OP1$ coincides with I.                     □

This result can be combined with the following definition:

**Definition 6:** a cascade is a one parameter semi-group $\Gamma$ acting on a set $M$: for each $t$, there exists a $\rho^t: M \to M$ such that given $x(0) = X_0$, then $x(t) = \rho^t x_0$.

In discrete terms, a cascade is the equivalent to a flow in continuous time and it defines a dynamic system. It can be proven that OP2 acting on OP1 constitutes a cascade (London 2008); evolutionary transitions can be seen as non-deterministic consequences of the history of the economy. Proposition 3 shows this:

**Proposition 3:** for each $\xi$ t there exists at least one $\Phi_0^t$.

**Proof:** Trivial. $\xi^t$ is the history of $\varepsilon$ until period $t$. According to the definition of cascade, for each $t$ there exists at least one $\rho^t \in OP2$ such that $\Phi_0^t = \rho^t o \Phi$. □

Each sequence $\{\Phi_0^k\}_{k \geq t0}$ obtained as a result of the action of OP2 on $\Phi_0^t$ is an orbit in OP1. On the other hand, an orbit in OP1 determines a trajectory in $X^t \times O^t \times Q^t \times R^t$ (i.e. a time path for the economy). This is equivalent to say that:

**Proposition 4:** there exists a $\theta: \{\xi^t\}_{t=to}^x \times \{\varepsilon^t\}_{t=to}^x$, such that $\theta(., OP2 \, o \, \Phi_0^{t0}(.))$ acting on $\{\xi^t\}_{t=to}^x$ is a cascade.

**Proof:** From the definition of economy: for each t, $X^t \times O^t \times Q^t \times R^t$ defines a $\xi^t$. In functional terms, $\xi^t = \varepsilon(X^t \times O^t \times Q^t \times R^t)$. This defines a $\sigma: \{\xi^t\}_{t=to}^x \times \{\varepsilon^t\}_{t=to}^x \to \{\varepsilon^t\}_{t=to}^x, \xi^{t+1} = \sigma(\xi^t, \varepsilon(X^{t+1} \times O^{t+1} \times Q^{t+1} \times R^{t+1}))$.

On the other hand, $OP2 \, o \, \Phi_0^{t0} = \{\rho^t o \, \Phi_0^{t0} : \rho^t \in OP2\}$ is such that for each $\rho* \in OP2 \, o \, \Phi_0^{t0} \rho*: \{\xi^t\}_{t=to}^x \to \{\varepsilon^t\}_{t=to}^x$. It is trivial to show that $\sigma(., OP2 \, o \, \Phi_0^{t0}(.))$ is a semigroup. In particular, $Id^l(.; Id^2(.))$, where $Id^l \equiv Id\{\varepsilon^t\}_{t=to}^x$ constitutes the neutral element. $Id^l$ existence is ensured by the existence of a $\rho \in OP2$ such that $\rho o \, \Phi_0^{t0}(\xi^t) = (X^t \times O^t \times Q^t \times R^t)$.

Finally, taking 0 as $x_0$, we can define recursively $\xi^{t+1} = \sigma(0, \varepsilon^{to})$, since given $t$, there exists a $\rho \in OP2$ such that $\xi^t = \sigma(\xi^{t-1}, \varepsilon \, o \, \rho^t o \, \Phi_0^{t0}(\varepsilon^t))$. □

This cascade defines a sequence of meta-stable states of the economy. We show that an economy can be seen as a dynamic system submitted to external shocks that only induce evolutionary changes when the internal conditions are the appropriate. The external shocks are here identified with shocks of knowledge.

# 4 Economics ESO Systems

The notion of economic system given in the previous section defines the rules of structural change in terms of the histories of the system. Despite being conceptually simple, this characterization is not algorithmically definable. If the temporal horizon is infinite or random the entire set of histories of the economy is non-denumerable. This characterization lacks of the property of being effective. This is in sharp contrast with ESO systems, which as simple automata are by definition equivalent to partial recursive functions.

On the other hand, we already discussed the limitations of pure ESO system as representation of economic evolution. We will here tailor the definition of ESO systems, incorporating the main elements that define economic systems in evolution. We have then:

**Definition 7:** an economic ESO system (EESO) is $E = (T, R, M)$, where $T = (T_0, T_1, \ldots)$ is a sequence of topologies, $R = (R_0, R_1 \ldots)$ is a sequence of uniform rules, and M is a set of meta-rules. A topology $T_t$ is $T_t = (S_t, C_t)$, where $S_t$ is a finite set of sites in period $t$ while $C_t$ is the structure of connections among them. Given $i = 1, \ldots, n^*, n+1, \ldots, I$, a fixed number of agents, $i < n_t$ are the indexes that represent the consumers while $n_t < i < I$ corresponds to the productive units. The sets in this partition are denoted $S_c^t$ and $S_p^t$ respectively. The states of the sites are given by:

- For each $s_i^t \in S_c^t$, her state is a vector $|s_i^t| = <X^t, O_i^t, K_i^t>$. To each $O_i^t \times K_i^t \times p^t$ corresponds a $c_i^t$, $i$'s demand.
- For each $s_j^t \in S_p^t$, its state is $|s_j^t| = <X^t, Q_j^t, w_j^t>$., such that given its inputs $w_j^t$ and the current prices $p^t$, next period's production is $x_j^{t+1} = Q_j^t(w_j^t, p^t)$
- The institutional rule $R_t$ determines $(w_t, c_t, p_t)$.
- A shock in t will affect the value of $K^t$.

It is necessary to show how shocks affect the value of knowledge and how meta-rules act on the system. The specification of K is as follows:

**Definition 8** $K_i^T = Expected_i(E^{T+1} / \{E^t\}_{t=0}^t)$, where the forecasted state of the EESO system in period $T + 1$, given the sequence of previous states until period T, is computed using any standard forecast algorithm.[3] This value is individual for each site $s_j^T \in S_C^T$. A shock on the vector $K^T$ consists in the information to one of the sites, $s_i^T$, that the state of the system in $T + 1$ will be $E^* \neq K_i^T$. Given the meta-rule $M^T$, if E* differs less than a fuzzy degree $\alpha T$ than specified by $K_i^T$ we say that the shock is in range. Otherwise a new meta-rule $M^{T+1}$ applies (predefined by $M^T$) and a new topology $T^{T+1}$ and a new set of rules $R^{T+1}$ becomes active.

According to this, the set of meta-rules is self-referential, since each meta-rule prescribes which meta-rule has to be activated according to the size of shocks (being itself if the shock is small). This property of being self-referential presents meta-theoretical questions about the appropriateness of these systems as representation of economic systems in evolution.

There are two main meta-theoretic properties that can be satisfied by a representation (with respect to the represented object). One is soundness (every assertion

---

[3]A "soft" computation approach is presented in Weiss and Kulikowski (1990).

true in the representation corresponds to an assertion true in the represented object). The other is its converse, completeness, i.e. that every assertion true in the represented object corresponds to an assertion true in the representation (see Boolos and Jeffrey 1974). In the case of economic ESOs as representation of economic systems in evolution, soundness follows immediately:

**Proposition 5:** given an EESO system E, there exists an economic system $\varepsilon$ such that for each t, the values of $X^t, O^t, Q^t, R^t, c_d^t, w_t, c_t$ and $p_t$ correspond to the values of a state $\varepsilon^t$.

**Proof:** in each period $T$ the meta-rule $M^T$ can be seen as a rule of type $OP_t$ and the transition towards $M^{T+1}$ results from applying a rule $\rho: OP_T \rightarrow OP_T$. Given the characterization of the topology and the rules of the system is follows trivially that there exists a $\varepsilon^t$ with the same values of the EESO system.                    □

This result is based on considering that the shock of knowledge is external to the system. But it is easy to imagine situations in which the internal dynamics of knowledge leads to sudden changes in the behavior of the system. This kind of situations, which is quite easy to formalize inside an economic system in evolution, cannot be represented in an EESO system.

Nevertheless, only the appropriate internal conditions ensure the possibility of evolutionary changes. The institutions are instrumental in facilitating these kinds of transitions. Furthermore, they tend to be "smooth", i.e. without discontinuities, since the more resilient institutions are those more deeply entrenched in the society. In turn, the demand for changes arises from organizations and individuals that expect to more than compensate the costs of the changes. But these mechanisms operate through both the individual and social knowledge in the economy. This is consistent with the point of view of New-Institutionalism, which shows that a source of institutional change is the variation in relative prices, frequently endogenously determined by innovations or changes of taste. In any case, they can be traced back to changes in knowledge, which mediates in all these interactions.

## 4.1 Development Traps and Unsuccessful EESO Systems

To understand the process of institutional change it is useful to think of an initial situation of equilibrium, in which no agent finds advantageous to devote resources to modify the agreements that hold together the society. Then, a modification in the structure of relative prices generates incentives for rewriting original contracts. But these, in turn, cannot be modified unless some higher order rules are in turn changed. This pressure on the whole structure ends up with institutional changes of various types. When constraints are intended to be removed resources, time and cultural erosion have to be applied to accomplish the task of changing them.

The demand for institutional changes arises from the perception of potential gains for individuals and organizations. If the costs involved in the change are less than those profits, the change will be effected. But these costs are related to the efficiency of the political markets in which the changes are to be negotiated. These costs are particularly high in developing economies (London 2008).

This may provide an interesting interpretation of the critical boundaries on the size of shocks: they are related to idiosyncratic aspects of the economy. Poverty traps arise when those boundaries cannot be surpassed. Therefore, the economic system tends to remain in a fixed state, failing to be a full-blown EESO system.

Formally:

**Definition 9:** Consider an EESO system E at period T in which a shock on the vector $K^T$ is the information to one of the sites, $s_i^T$, that the state of the system in T + 1 will be $E^* \neq K_i^T$. If the fuzzy degree of the proposition [$E^*$ is at less than a $\alpha^T$ fuzzy degree from the specification of $K_i^T | M^T$) is close to 1, independently of $E^*$, we say the E is in a poverty trap.

Then,

**Proposition 6:** If E is in a poverty trap it is not evolutionary.

**Proof:** By Proposition 5, an EESO system E at period T corresponds to an economic system at state $\epsilon^t$. If E is in a poverty trap, no matter the shock that E receives, there will be no transition. But since a change happens if $E^*$ is at more than a $\alpha^T$ degree of $K_i^T$ it must happen that for all $E^*$, statement [$E^*$ is at less than a $\alpha^T$ fuzzy degree from the specification of $K_i^T | M^T$] has a degree close to 1. This means that E has reached a stationary state. But according to Proposition 2 this implies that it is not an ESO system.                                                                    □

In turn, since the definition of poverty traps is conditioned by the actual set of meta-rules ($M^T$]) this means that the critical boundaries (the $\alpha^T$ degree of $K_i^T$) are endogenous to the state at T. Formally:

**Proposition 7:** For E in a poverty trap, the critical boundary that defines it is endogenous. That is $\alpha^T(M^T)$.

**Proof:** Suppose that $\alpha^T$ is independent of $M^T$. Then, consider the fuzzy distribution of $E^*$, given $M^T$. Then, if ($E^*|M^T$) has degree 1, [$E^*$ is at less than a $\alpha^T$ fuzzy degree from the specification of $K_i^T | M^T$] must get a degree strictly less than 1. Contradiction, since E is in a poverty trap.                                                                    □

## 5   Discussion

We presented a formal schema that can be applied to the representation of the concept of economic evolution. This formalism incorporates the best features of ESOs (self-organization and computability) and of economic systems in evolution (the role of knowledge and intentionality in economic evolution).

Each component lacks certain features provided by the other. So, ESOs are too general, but economic systems provide them the relevant economic notions. Economic systems, in turn, lack a computational characterization, which is provided by the automata-like behavior of ESO systems.

The resulting notion of economic ESO system incorporates the best of both approaches. It captures all non-linear economic phenomena: irreversibility, path-dependence and uncertainty. The later is associated with the inability of the agents to predict the size of the shocks on the system. Another interesting feature incorporated in economic ESOs is the role played by knowledge and its interaction with the meta-rules. The resilience to the action of the meta-rules, which leads to a stationary state, is interpreted as the trademark of poverty traps.

We have shown that the current characterization of economic ESOs is sound and only needs to expand the properties of its components. We think that one way of doing this is by means of the incorporation of reasoning into each site that corresponds to an intentional agent. It is matter of future work to analyze how this should be done, although some hints are already present in the growing literature on knowledge in games and decision making.

# References

Arenas, A., Díaz-Guilera, A., Llas, M., Perez Conrad, J., Vega-Redondo, F.: A Self-organized model of technological progress. In: CEF 2000 proceedings (2000)

Azariadis, C., Stachurski, J.: Poverty traps. In: Aghion, P., Durlauf, S. (eds.) Handbook of Economic Growth. Elsevier, Amsterdam (2005)

Boolos, G., Jeffrey, R.: Computability and Logic. Cambridge University Press, Cambridge (1974)

Kosko, B.: Fuzziness vs. probability. Int. J. Gen. Syst. **17**, 211–240 (1990)

London, S.: Instituciones, Sistemas Complejos, y la Teoría Económica Evolucionista. In: Jardón Urrieta, J. (ed.) Evolucionismo Económico, Instituciones y Sistemas Complejos Adaptativos. Porrúa, México (2008)

Lucas, R.: Lectures on Economic Growth. Harvard University Press, Cambridge (2002)

Mandelbrot, B.: Fractional Brownian motions, fractional noises and applications. SIAM Revi. **10**, 422–437 (2002)

North, D.: Institutions and Economic Theory. Am. Econ. **36**, 3–6 (1993)

Tohmé, F., London, S.: A Mathematical Representation of Economic Evolution. Math. Comput. Model. **27**(8), 29–40 (1998)

# ABC, A Viable Algorithm for the Political Districting Problem

Eric-Alfredo Rincón-García, Miguel-Ángel Gutiérrez-Andrade, Sergio-Gerardo de-los-Cobos-Silva, Pedro Lara-Velázquez, Roman-Anselmo Mora-Gutiérrez and Antonin Ponsich

**Abstract** Since 2004, the Federal districting processes have been carried out using a Simulated Annealing based algorithm. However, in 2014, for the local districting of the state of México, a traditional Simulated Annealing technique and an Artificial Bee Colony based algorithm were proposed. Both algorithms used a weight aggregation function to manage the multi-objective nature of the problem, but the population based technique produced better solutions. In this paper, the same techniques are applied to six Mexican states, in order to compare the performance of both algorithms. Results show that the Artificial Bee Colony based algorithm is a viable option for this kind of problems.

**Keywords** Districting · Artificial bee colony · Simulated annealing · Heuristic

E.-A. Rincón-García (✉) · R.-A. Mora-Gutiérrez · A. Ponsich
Departamento de Sistemas, Universidad Autónoma Metropolitana - Azcapotzalco,
Av. San Pablo 180. Col, Reynosa Tamaulipas C.P 02200, Mexico D.F.
e-mail: rigaeral@correo.azc.uam.mx

R.-A. Mora-Gutiérrez
c-mail: mgra@correo.azc.uam.mx

A. Ponsich
e-mail: aspo@correo.azc.uam.mx

M.-Á. Gutiérrez-Andrade · S.-G. de-los-Cobos-Silva · P. Lara-Velázquez
Departamento de Ingeniería Eléctrica, Universidad Autónoma Metropolitana - Iztapalapa,
Av. San Rafael Atlixco 186. Col, Vicentina. del Iztapalapa C.P 09340, Mexico D.F.
e-mail: gamma@xanum.uam.mx

S.-G. de-los-Cobos-Silva
e-mail: cobos@xanum.uam.mx

P. Lara-Velázquez
e-mail: plara@xanum.uam.mx

© Springer International Publishing Switzerland 2015                           269
J. Gil-Aluja et al. (eds.), *Scientific Methods for the Treatment of Uncertainty
in Social Sciences*, Advances in Intelligent Systems and Computing 377,
DOI 10.1007/978-3-319-19704-3_22

# 1 Introduction

The zone design problem arises from the need of aggregating small geographical units (GUs) into regions, in such a way that one (or more) objective function(s) is (are) optimized and some constraints are satisfied. The constraints can include, for example, the construction of connected zones, with the same amount of population, clients, mass media, public services, etc. (Bernabé et al. 2012; Rincón-García et al. 2010; Duque et al. 2007; Kalcsics et al. 2005). The zone design is used in diverse problems like school redistricting (Ferland and Guenette 1990; Caro et al. 2001; DesJardins et al. 2006), police district (Dell'Amico et al. 2002), service and maintenance zones (Tavares-Pereira et al. 2007), sales territory (Zoltners and Shina 2005) and land use (Williams 2002).

The design of electoral zones or electoral districting is a well-known case, due to its influence in the results of electoral processes and its computational complexity, which has been shown to be NP-Hard (Altman 1997). In this framework, the GUs are grouped into a predetermined number of zones or districts, and democracy must be guaranteed through the satisfaction of restrictions that are imposed by law (Mehrotra et al. 1998; Ricca et al. 2011). In particular, some generally proposed criteria are population equality, to ensure the "one man one vote" principle; compactness, to avoid any unfair manipulation of the border or shape of electoral zones for political purposes, and contiguity, to prevent from designing fragmented districts.

Nowadays, different meta-heuristics have been reported in specialized literature to produce automated districting plans. The most common techniques include local search algorithms such as Simulated Annealing (SA), Tabu Search and Old Bachelor Acceptance (Macmillan 2001; Bozkaya et al. 2003; Ricca and Simeone 2008, Rincón-García et al. 2013). Although some Genetic (Bacao et al. 2005), Evolutionary (Chung-I 2011) and Swarm Intelligence (Rincón-García et al. 2012) algorithms have been applied.

In Mexico, the Federal Electoral Institute[1] (IFE), for the federal districting processes in 2004 and 2013, used a SA based algorithm. However, in 2014, for the local districting problem of the State of Mexico, two automated districting algorithms were proposed: a Simulated Annealing based algorithm, and, for the first time in Mexico, a population based technique was used: an Artificial Bee Colony (ABC) algorithm.

The primary purpose of this paper is to compare the performance of these algorithms in different instances. To address this issue, we provide a description of the problem in Sect. 2 a brief overview of the inner working mode of the SA and ABC algorithms is presented in Sect. 3. Some computational results are detailed in Sect. 4. Finally, conclusions and perspectives for future work are drawn in Sect. 5.

---

[1]The National Electoral Institute, INE, since April 4, 2014.

## 2 Problem Description

As mentioned previously, population equality and compactness are important principles that should be promoted in the design of electoral districts. For this reason, the objective functions should guide the search towards regular shaped districts with approximately the same amount of population.

In order to promote the principle "one man one vote", different measures to quantify the population equality have been proposed, for example the difference (or ratio) between the most and the least populated zone, or the sum of the absolute values of the difference between the average and the real number of inhabitants in each zone. In particular, in 2014, the Electoral Institute of the State of Mexico (IEEM) proposed the following measure:

$$C_1(P) = \sum_{s \in S} \left(1 - \frac{P_s}{(P_T/n)}\right)^2 \left(\frac{1}{0.15}\right)^2 \tag{1}$$

where $P = \{Z_1, Z_2, \ldots, Z_n\}$ is a districting plan. Each district $Z_s$ is defined through a set of binary variables $x_{is}$ such that $x_{is} = 1$ if the $i$-th GU belongs to district $s$ and $x_{is} = 0$ otherwise. $P_T$ is the population of the considered state, $P_s$ is the population of district $s$, 0.15 is the maximum percentage of deviation allowed for the state, $S = \{1, 2, 3, \ldots, n\}$, $n$ is the number of electoral districts that must be generated in the state. Thus, the lower the cost $C_1$, the better the population equality of a solution. Indeed, the perfect population equality is achieved when all districts have the same number of inhabitants, and in this case the measure assigns a value of zero to $C_1$.

Regarding the assessment of district compactness, several measures have been proposed, as observed in (Niemi et al. 1990). The IEEM introduced a metric that can be easily computed, in order to improve the runtime performance. This measure of compactness compares the perimeter of each district with that of a square having the same area.

$$C_2(P) = \sum_{s \in S} \left(1 - \frac{PC_s}{\sqrt{AC_s}} - 1\right) \tag{2}$$

where $PC_s$ and $AC_s$ are the perimeter and the area of the considered district $s$, respectively. Thus, districts with a good compactness will have a compactness value close to 0.

Finally, in order to handle the multi-objective nature of the problem, the IEEM used a weight aggregation function strategy.

$$\textit{Minimize } f(P) = \lambda_1 C_1(P) + \lambda_2 C_2(P) \tag{3}$$

where, $\lambda_1$, $\lambda_2$, are weighting factors that set the relative importance of equality population and compactness criteria.

This problem formulation therefore seeks for a districting plan that represents the best balance between population equality and compactness, a balance obviously biased by the weighting factors. In 2014 the IEEM proposed $\lambda_1 = 1$, and $\lambda_2 = 0.5$. However, these factors can be modified due to political agreements, thus it is impossible to predict the factors that will be used in future processes. Therefore it has been decided, in the framework of the present work, to produce a set of efficient solutions, using different weighting factors for the computational experiments.

## 3 Heuristic Algorithms

Since the design of electoral zones is an NP-Hard problem, automated heuristic algorithms are an appropriate strategy to design electoral districting plans. In Mexico, since 2004, a SA strategy has been applied to carry out the federal districting processes. However, new heuristic techniques have proven to get better solutions in these kinds of problems. Whereby the IEEM decided to propose a new strategy, based on Artificial Bee Colony, for districting the state of Mexico. This strategy also includes, nevertheless, a traditional SA algorithm, in order to produce solutions of a quality at least as good as those proposed in last processes.

In this section we provide a brief description of the Simulated Annealing and Artificial Bee Colony strategies used by the IEEM.

### 3.1 Simulated Annealing

Simulated Annealing is a metaheuristic introduced by Kirkpatrick in (Kirkpatrick et al. 1983). The SA algorithm starts with an initial solution $P$ and generates, in each iteration, a random neighbor solution $Q$. If this neighbor improves the current value of the objective function, it is accepted as the current solution. If the neighbor solution does not improve the objective value, then it is accepted as the current solution according to a probability $\eta$ based on the Metropolis criterion:

$$\eta = \exp\left(\frac{f(P) - f(Q)}{T}\right) \tag{4}$$

where $f(P)$ and $f(Q)$ represent the objective value of the current and neighbor solutions, respectively. $T$ is a parameter called temperature, which is controlled through a cooling schedule that defines the temperature decrease and the (finite) number of iterations for each temperature value.

## 3.2 Simulated Annealing Adaptation

A classical implementation of SA was proposed by IEEM, with a geometric decreasing cooling schedule. The initial solution is created using the following strategy. All GUs are labeled as available. The algorithm then selects randomly $n$ GUs, assigns them to different zones and labels them as not available. At this moment, each zone has therefore only one GU. Finally, each zone is iteratively extended by adding an available GU having a frontier with the zone in its current shape. Every time a GU is incorporated to a zone, it is labeled as not available in order to avoid the construction of overlapping zones. The latter step is performed until all the GUs are labeled as not available. This process ensures that the initial solution consists of $n$ connected zones that include all GUs. Note that SA and ABC use the same procedure to create initial solutions.

Regarding now the construction of a neighbor solution, a random zone is chosen and a GU in this zone is moved to a neighbor zone. Therefore, the neighbor solution is identical to the current one, except that one GU is reassigned to an adjacent zone. The new solution is evaluated and accepted or rejected according to the Metropolis criterion. If the neighbor solution is rejected, another GU is randomly selected; this process is repeated until the temperature reaches a predefined lower bound.

## 3.3 Artificial Bee Colony

Artificial Bee Colony (ABC) is a bio-inspired metaheuristic, originally proposed by Karaboga and based on the natural behaviour of honey bees for finding food resources (Karaboga and Basturk 2007). Artificial bees are classified into three groups of bees: employed, onlookers, and scouts.

In the ABC algorithm, each solution to the problem under consideration is called a food source and represented by a $D$-dimensional real-valued vector. The fitness of a solution is associated to the amount of nectar in the food source. The algorithm cycle begins with an improvement phase, which is carried out by employed bees. For each solution $Y_i = [y_{i,1}, y_{i,2}, ..., y_{i,D}]$ in the population, a solution $Y_k = [y_{k,1}, y_{k,2}, ..., y_{k,D}]$ $(i \neq k)$ is randomly chosen to produce a new solution $V_i = [v_{i,1}, v_{i,2}, ..., v_{i,D}]$ according to the following equation:

$$V_{i,j} = y_{i,j} + r*(y_{i,j} - y_{k,j}) \tag{5}$$

where $j$ is an index randomly generated in $\{1, ..., D\}$ and $r$ is a uniformly distributed real random number in the range $(-1,1)$. Each produced solution $V_i$ is subsequently evaluated by the employed bees and passes through a greedy selection process: if the new food source has a nectar amount equal to or better than the employed bee's current food source, it replaces by the current solution; otherwise, the old food source is kept.

When the employed searching phase is over, each onlooker bee evaluates the nectar information provided by the employed bees and chooses a food source according to a probability $p_i$, computed on the basis of its nectar amount (i.e., the corresponding fitness):

$$p_i = \frac{fit_i}{\sum\limits_{j=1}^{M} fit_j} \tag{6}$$

where $fit_i$ is the fitness value of the food source $Y_i$ and $M$ is the number of food sources. Once the food source is chosen, the onlooker tries to improve the corresponding solution generating a new one, in the same way the employed bees do, i.e. through Eq. (5). The new solution is then evaluated by the onlooker bee, its quality is compared with that of the current one and the best solution is selected applying a greedy selection process.

If a solution cannot be further improved through a predetermined number of trials, the solution is abandoned and a scout bee produces a new random solution. The termination criterion is, typically, a fixed number of computed iterations of the previously described cycle.

## 3.4 Artificial Bee Colony Adaptation

The ABC heuristic was originally designed for continuous optimization problems, and cannot directly be used for discrete cases. Thus, IEEM's algorithm included some modifications to the ABC method based on a recombination strategy.

First, $M$ food sources are generated using the strategy described in Sect. 3.2. The number of onlooker and employed bees is set equal to the number of food sources, and exactly one employed bee is assigned to each food source.

According to Eq. (5), employed and onlooker bees generate new solutions by combining two food sources. However, after some experiments, the performance of the algorithm was improved when employed and onlooker bees used different strategies to explore the solution space. Each employed bee must apply a local search, similar to the strategy used by SA described in Sect. 3.2, while onlooker bees use a recombination technique inspired in Eq. (5).

First, each employed bee, $i$, modifies its food source, $P_i$, using the following strategy. A random zone is chosen and a GU in this zone is moved to a neighbour zone in such way that the new solution, $V_i$. If the new solution $V_i$ has a nectar amount better than or equal to that of $P_i$, $V_i$ replaces $P_i$ and becomes a new food source exploited by the hive. In other case, $V_i$ is rejected and $P_i$ is preserved.

As soon as the employed bees' process has been completed, each onlooker bee chooses two solutions. The first solution, $P_1$, is selected depending on the probability given by (6), where the fitness value of the food source is given by (3).

The second solution, $P_2$, is randomly selected from the food sources exploited by the hive. A new food source, $V_1$, is produced through a recombination technique described straightforward.

A GU $k$ is randomly selected. Thus, there is a zone $Z_i \varepsilon P_1$ and a zone $Z_j \varepsilon P_2$ such that $k \varepsilon Z_i \cap Z_j$. Let us now consider the following sets:

$$H_1 = \{l: x_{li} = 0, x_{lj} = 1\} \tag{7}$$

$$H_2 = \{l: x_{li} = 1, x_{lj} = 0\} \tag{8}$$

Then a GU in $H_1$ is inserted into $Z_i$, and a GU in $H_2$ are extracted from $Z_i$, and inserted into any randomly chosen zone contiguous to $Z_i$.

Note that these moves can produce a disconnection in zone $Z_i$, so that a repair process must be applied. The number of connected components in $Z_i$ is counted after all the moves previously described. If the number of connected components equals 1, then the zone is connected. Otherwise, the algorithm defines the connected component that includes GU $k$ (i.e., the GU used within the above-described recombination strategy) as zone $Z_i$, subsequently, the remaining components are assigned to other adjacent zones.

## 4 Computational Experiments

The two algorithms described in the previous sections were tested on six Mexican states: Distrito Federal, Guerrero, Jalisco, Mexico, Michoacán and Nuevo León. For each state, the number of inhabitants, sections and zones to be created are presented in Table 1. We must remark that some sections are usually grouped in order to decrease the number of variables, and to reduce the complexity of the districting process. However, we decided that the performance of both algorithms should be tested in difficult problems, thus for these computational experiments the sections were kept separated.

In agreement with the Federal requirements stipulated for México's 2006 and 2013 elections, a maximum percentage of deviation (in terms of population in each district) d = 15 % was considered.

**Table 1** Data for the six tackled instances

| State | Inhabitants | Geographic units | Number of zones |
|---|---|---|---|
| Distrito Federal | 8,605,239 | 5,535 | 27 |
| Guerrero | 3,079,649 | 2,784 | 9 |
| Jalisco | 6,322,002 | 3,326 | 19 |
| México | 13,096,686 | 5,930 | 40 |
| Michoacán | 3,985,667 | 2,677 | 12 |
| Nuevo León | 3,834,141 | 2,135 | 12 |

Regarding the settings of parameters shared by the considered techniques, the set $L = \{\lambda_1, \lambda_2\}$ of weighting factors is defined as follows: $\lambda_1 = \{0.99, 0.98, 0.97, ...,$ $0.02, 0.01\}$, while $\lambda_2$ is set to $\lambda_2 = 1 - \lambda_1$. Besides, in order to deal with the stochastic effect inherent to heuristic techniques, 990 independent executions were performed for each algorithm on each of the 6 instances (10 runs for each weight couple $(\lambda_1, \lambda_2)$). Each run produces a single solution and the resulting 990 solutions are subsequently filtered through a Pareto sorting procedure, which identifies the final non-dominated solutions.

The approximated Pareto fronts obtained by the algorithms, in Jalisco and Michoacán, are illustrated in Fig. 1. These fronts were obtained combining the solutions produced by the various executions of each algorithm. From the graphical observation of the sets of non-dominated solutions, it seems that ABC provides better solutions, in terms of convergence. However, these observations need to be confirmed by a performance metric.

Due to the multi-objective nature of the problem, we decided to compare the algorithms using the front coverage measure.

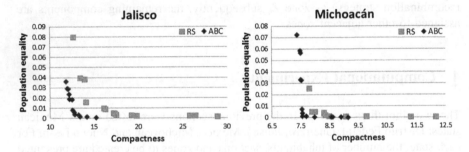

**Fig. 1** Non-dominated solutions for the tested algorithms

The front coverage $C(A_1; A_2)$ is a binary metric that computes the ratio of efficient solutions produced by one algorithm $A_2$ dominated by or equal to at least one efficient solution produced by another competing algorithm $A_1$. Note that, commonly, $C(A_1; A_2) \neq C(A_2; A_1)$, so that both values must be computed. The expression of $C(A_1; A_2)$ is provided straightforward:

$$C(A_1; A_2) = \frac{|s_2 \in PF_2; \exists s_1 \in PF_1 : s_1 \succ s_2|}{|PF_2|} \tag{20}$$

where $PF_1$ and $PF_2$ are the approximated Pareto sets obtained by algorithms $A_1$ and $A_2$ respectively. If $C(A_1; A_2)$ is equal to 1, then all the efficient solutions produced by $A_2$ are dominated by efficient solutions produced by $A_1$.

The front coverage metric results are presented in Table 2 for all instances. The ABC algorithm seems to be the best option since its solutions dominate in a great extent those produced by SA; simultaneously, a low percentage of its solutions are dominated. For example, all SA solutions are dominated in Jalisco,

México, Michoacán and Nuevo León, while only the 38.89 % of ABC solutions are dominated in Distrito Federal.

**Table 2** Front coverage metric

| Distrito Federal | | SA | ABC | México | | SA | ABC |
|---|---|---|---|---|---|---|---|
| A1 | SA | – | 0.3889 | A1 | SA | – | 0.0 |
| | ABC | 0.6364 | – | | ABC | 1.0 | – |
| Guerrero | | SA | ABC | Michoacán | | SA | ABC |
| A1 | SA | – | 0.0 | A1 | SA | – | 0.0 |
| | ABC | 0.7273 | – | | ABC | 1.0 | – |
| Jalisco | | SA | ABC | Nuevo León | | SA | ABC |
| A1 | SA | – | 0.0 | A1 | SA | – | 0.0 |
| | ABC | 1.0 | – | | ABC | 1.0 | – |

# 5 Conclusions

In this paper, the performance level of an ABC based algorithm, previously developed for the districting process of the state of México in 2014, is studied. In addition, a SA based technique, traditionally used by IFE, is employed as a reference to evaluate the population-based algorithm. The respective performances of both algorithms were evaluated in terms of the quality of the approximated Pareto front and of efficiency. The computational experiments proved that the ABC algorithm produces better quality efficient solutions than its counterpart.

Despite the computational experiments were carried out over six instances, it is likely that the overall performance level of the ABC algorithm can be meaningfully generalized to this formulation of the districting problem.

Finally, it is clear that this work can be enhanced focusing on some global guidelines such as implementing multiobjective versions of these algorithms to exploit the characteristics of this bi-dimensional problem. These guidelines provide perspectives for future work.

# References

Altman, M.: Is automation the answer: the computational complexity of automated redistricting. Rutgers Comput. Law Technol. J. **23**, 81–141 (1997)

Dacao, F., Lobo, V., Painho, M.: Applying genetic algorithms to zone design. Soft. Comput. **9**, 341–348 (2005)

Bernabé, B., Coello-Coello, C.A., Osorio-Lama, M.: A multiobjective approach for the heuristic optimization of compactness and homogeneity in the optimal zoning. J. Appl.Res. Technol. **10** (3), 447–457 (2012)

Bozkaya, B., Erkut, E., Laporte, G.: A tabu search heuristic and adaptive memory procedure for political districting. Eur. J. Oper. Res. **144**, 12–26 (2003)

Caro, F., Shirabe, T., Guignard, M., Weintraub, A.: School redistricting: embedding GIS tools with integer programming. J. Oper. Res. Soc. **55**, 836–849 (2001)

Chung-I, C.: A Knowledge-based Evolution Algorithm approach to political districting problem. Comput. Phys. Commun. **182**(1), 209–212 (2011)

Dell'Amico, S., Wang, S., Batta, R., Rump, C.: A simulated annealing approach to police district design. Comput. Oper. Res. **29**, 667–684 (2002)

DesJardins, M., Bulka, B., Carr, R., Jordan, E., Rheingans, P.: Heuristic search and information visualization methods for school redistricting. AI Mag. **28**(3), 59–72 (2006)

Duque, J.C., Ramos, R., Suriñach, J.: Supervised regionalization methods: a survey. Int. Reg. Sci. Rev. **30**(3), 195–220 (2007)

Ferland, F., Guenette, G.: Decision support system for the school districting problem. Oper. Res. **38**, 15–21 (1990)

Kalcsics, J., Nickel, S., Schröder, M.: Towards a unified territorial design approach: applications, algorithms and GIS integration. TOP **13**(1), 1–74 (2005)

Karaboga, D., Basturk, B.: A powerful and efficient algorithm for numerical function optimization: artificial bee colony (ABC) algorithm. J. Glob. Optim. **39**, 459–471 (2007)

Kirkpatrick, S., Gellat, C.D., Vecchi, M.P.: Optimization by simulated annealing. Science **220**, 671–680 (1983)

Macmillan, W.: Redistricting in a GIS environment: an optimization algorithm using switching-points. J. Geogr. Syst. **3**, 167–180 (2001)

Mehrotra, A., Johnson, E.L., Nemhauser, G.L.: An optimization based heuristic for political districting. Manage. Sci. **44**(8), 1100–1114 (1998)

Niemi, R.G., Grofman, B., Carlucci, C., Hofeller, T.: Measuring compactness and the role of a compactness standard in a test for partisan and racial gerrymandering. J. Polit. **52**, 1155–1181 (1990)

Ricca, F., Simeone, B.: Local search algorithms for political districting. Eur. J. Oper. Res. **189**(3), 1409–1426 (2008)

Ricca, F., Scozzari, A., Simeone, B.: Political districting: From classical models to recent approaches. J. Oper. Res. **204**(1), 271–299 (2011)

Rincón-García, E.A., Gutiérrez-Andrade, M.A., Ramírez-Rodrígez J., Lara-Velázquez, P., de-los-Cobos-Silva, S.G.: Applying simulated annealing to design compact zones. Fuzzy Econ. Rev. **15**(2), 11–24 (2010)

Rincón-García, E.A., Gutiérrez-Andrade, M.A., de-los-Cobos-Silva, S.G., Lara-Velázquez, P., Mora-Gutiérrez, R.A., Ponsich, A.: A Discrete Particle Swarm Optimization Algorithm for Designing Electoral Zones. Methods for Decision Making in an Uncertain Environment. World Scientific Proceedings Series on Computer Engineering and Information Science, Reus, Spain, vol. 6, pp. 174–197 (2012)

Rincón-García, E.A., Gutiérrez-Andrade, M.A., de-los-Cobos-Silva, S.G., Lara-Velázquez, P., Mora-Gutiérrez, R.A., Ponsich, A.: A multiobjective algorithm for redistricting. J. Appl. Res. Technol. **11**(3), 324–330 (2013)

Shirabe, T.: A Model of contiguity for spatial unit allocation. Geogr. Anal. **37**, 2–16 (2005)

Tavares-Pereira, F., Rui, J., Mousseau, V., Roy, B.: Multiple criteria districting problems: The Public Transportation Network Pricing System of the Paris Region. Ann. Oper. Res. **154**, 69–92 (2007)

Williams, J.C.: A Zero-one programming model for contiguous land acquisition. Geogr. Anal. **34**(4), 330–349 (2002)

Zoltners, A.A., Sinha, P.: Sales territory design: thirty years of modeling and implementation. Market. Sci. **24**(3), 313–331 (2005)

# Asymmetric Uncertainty of Mortality and Longevity in the Spanish Population

Jorge M. Uribe, Helena Chuliá and Montserrat Guillén

**Abstract** Using data of specific mortality rates, discriminating between males and females, we estimate mortality and longevity risks for Spain in a period spanning from 1950 to 2012. We employ Dynamic Factor Models, fitted over the differences of the log-mortality rates to forecast mortality rates and we model the short-run dependence relationship in the data set by means of pair-copula constructions. We also compare the forecasting performance of our model with other alternatives in the literature, such as the well-known Lee-Carter Model. Finally, we provide estimations of risk measures such as VaR and Conditional-VaR for different hypothetical populations, which could be of great importance to assess the uncertainty faced by firms such as pension funds or insurance companies, operating in Spain. Our results indicate that mortality and longevity risks are asymmetric, especially in aged populations of males.

**Keywords** Longevity risk · Mortality risk · Dynamic factor models · Copulas

J.M. Uribe (✉)
Department of Economics, Universidad del Valle and University of Barcelona,
Calle 13 # 100-00, Cali, Colombia
e-mail: jorge.uribe@correounivalle.edu.co

H. Chuliá · M. Guillén
Department of Econometrics, University of Barcelona, Av. Diagonal 690,
08034 Barcelona, Spain
e-mail: hchulia@ub.edu

M. Guillén
e-mail: mguillen@ub.edu

© Springer International Publishing Switzerland 2015
J. Gil-Aluja et al. (eds.), *Scientific Methods for the Treatment of Uncertainty in Social Sciences*, Advances in Intelligent Systems and Computing 377,
DOI 10.1007/978-3-319-19704-3_23

# 1  Introduction

The key issue about modeling longevity and mortality risks arises because of the stochastic nature of mortality rates. This fact has been recognized by regulators and extensively explored by the academic literature in recent times (Cairns et al. 2011; Continuous Mortality Investigation 2004, 2005, 2013; Hollmann et al. 2000). Within the set of available alternatives to approach this random nature, factor models constitute a very attractive tool, given the low frequency of the mortality data and the relative high number of specific mortality rates to be forecasted. Specific mortality rates, which discriminate between age and sex, are arguably more appropriate to estimate mortality projections than aggregate rates, since there is great variability in the mortality dynamics thorough different ages and cohorts.

Factor models were initially proposed in the actuarial literature to model and project mortality rates by Lee and Carter (1992). The Lee-Carter model is a single-factor model, in which the factor is a stochastic trend shared by all the specific mortality rates. The model has been subject to several criticisms (Dushi et al. 2010; Mitchell et al. 2013), but it still remains as a plausible alternative within the academia. Indeed, it has become a 'workhorse' within the actuarial field, together with its extension to Poisson-log bilinear projections proposed by Brouhns and Denuit (2002).

In this paper we make use of a methodology to estimate longevity and mortality risks that takes into account advances in two different fronts, point estimation and scenarios construction. The first part implies the use of Dynamic Factor Models (DFM) to forecast mortality rates, and the second part the calculation of risk measures using pair-copula constructions. In this way, we provide a robust alternative to measure longevity and mortality risks by using distortion risk measures, such as Value at Risk, or Tail-Value at Risk (i.e. Conditional VaR). Everything is illustrated for hypothetical populations of males and females within different ranges of age.

The methodology is applied to forecast mortality rates and to estimate risk measures of longevity for the Spanish population that can be used as benchmarks by annuity or pension schemes, using data from 1950 to 2012 provided by Human Mortality Database. Our modeling strategy directly approaches the dependence structure in the construction of scenarios for mortality rates projections. The estimation of VaRs and Tail-VaRs could be useful to assess suitable capital requirements for the operation of firms exposed to products linked to mortality or longevity risks. Our strategy yields considerable gains in model fitting and therefore improves the forecasting performance of the model, allowing us to provide more accurate estimations of risk than those previously available in the literature.

# 2 Methodology

## 2.1 Factor Models

Following Bai and Ng (2008), let $N$ be the number of cross-sectional units and $T$ the number of time series observations. In our case, we have 220 cross-sectional units (mortality rates for ages from 0 to 109+ years, for males or females). If we consider males and females separately then $N = 110$. For $i = 1, \ldots, N$ and $t = 1, \ldots, T$. The Static Factor Model (SFM) is defined as:

$$x_{it} = C_{it} + e_{it}, \tag{1}$$

where $C_{it} = \lambda_i F_t$ is referred as the common component of the model, $e_{it}$ is an idiosyncratic error and $\lambda_i$ are factor loadings. Loadings are coefficients that relate the corresponding $r$ static common factors $F_t$ to the specific mortality rate in the $i$-th specific mortality category. If we define $X_t = (x_{1t}, x_{2t}, \ldots, x_{Nt})'$ and $\Lambda = (\lambda_1, \ldots, \lambda_N)'$, in vector form, for each period, we have:

$$\underset{(N \times 1)}{X_t} = \underset{(N \times r)(r \times 1)}{\Lambda \; F_t} + \underset{(N \times 1)}{e_t}, \tag{2}$$

where $e_t = (e_{1t}, e_{2t}, \ldots, e_{Nt})'$. Notice that even when the model specifies a static relationship between $x_{it}$ and $F_t$, the latter can be described by a dynamic vector process, as follows:

$$A(L)F_t = u_t, \tag{3}$$

where $A(L)$ is a polynomial of the lag operator. The SFM is implemented in several studies in order to forecast mortality rates in Spain, for instance by Alonso (2008). Note that if we set $L = 1$ and $r = 1$ we are facing the traditional model proposed by Lee and Carter (1992).

In the general case, $F_t$ could contain stationary and non-stationary factors (Peña and Poncela 2006; Bai and Ng 2004). Nevertheless, for empirical applications it is convenient to restrict the attention to the cases where either all the factors in $F_t$ are stationary or all of them evolve according to the same stochastic trends (by assumption). This is especially true for models with very large $N$, where traditional cointegration tests based on Vector-Autoregressive (VAR) representations of the original series are not suitable. In this study we prefer to fulfill the stationarity assumption, by differentiating the series before the actual estimation of the factor model. In this way, we avoid (or at least minimize) the possible arising of spurious coefficients in our empirical forecasting equations.

The static model above can be compared with the Dynamic Factor Model (DFM), defined as:

$$x_{it} = \lambda_i(L)f_t + e_{it}, \tag{4}$$

where $\lambda_i(L) = (1 - \lambda_{i1}L -, \ldots, - \lambda_{is}L^S)$ is a vector of dynamic factor loadings of length $s$. In case that we allow $s$ to be infinite we are in presence of the Generalized Dynamic Factor Model (GDFM) originally proposed by Forni and Reichlin (1998) and Forni et al. (2000). It was till Forni et al. (2005) that it could be used to forecasting purposes. It is a generalization of the DFM (Stock and Watson 2002) since it allows for a richer dynamic structure in the factors and it does not assume mutual orthogonality of the idiosyncratic components $e_{it}$. Regardless the dimension of $s$, $f_t$ evolves according to:

$$f_t = C(L)\varepsilon_t, \tag{5}$$

in this case $\varepsilon_t$ are *iid* errors with the same dimension than $f_t$. We denote it with the letter $q$.

Lastly, within the factor models typology, we have that when the idiosyncratic disturbances are allowed to be weekly correlated, we are talking about an approximate factor model. On the contrary, when they are not, it becomes an exact factor model. We use the former approach here.

We can rewrite the model in (4) as a SFM by redefining the vector of factors and factors' loads (Bai and Ng 2008), and hence we can present it as follows in a general way:

$$\underset{(N \times T)}{X} = \underset{(N \times r)(r \times T)}{\Lambda F} + \underset{(N \times T)}{e}, \tag{6}$$

where $X = (X_1, \ldots, X_N)$ and $F = (F_1, \ldots, F_T)$. $F$ and $\Lambda$ not separately identifiable and therefore some restrictions have to be imposed to uniquely fix them. The estimation of the factors is based on the method of Principal Components, which guarantees the fulfillment of the identification restrictions by construction.

Stock and Watson (2006) propose to construct the h-step ahead forecast by projecting $x_{it+h}$ onto the estimated factors lagged $h$ periods in the following way:

$$X_{t+h} = \beta F_t + e_{t+h}. \tag{7}$$

In the present context, unknown factors can be replaced by their (consistent) estimations $\hat{F}_t$ and hence linear projections are enough to recover the coefficients in (7). Direct forecast can be less efficient than iterated forecast, but it is also more robust facing model misspecification. Alternatively, in the context of the GDFM the forecasting equations will take the form:

$$X_{t+h} = \left[ \Gamma_h^C \hat{Z}' \left( \hat{Z} \hat{\Gamma}_0 \hat{Z}' \right)^{-1} \right] (\hat{Z} X_t) + e_{t+h}, \tag{8}$$

$\hat{C}_t$ is the estimation of the common component, and $\Gamma_0^C$ and $\hat{\Gamma}_0$ are contemporaneous- covariance matrices of the common components and the $x'$ s, respectively. The first matrix is estimated based on spectral density methods. $\hat{Z}$ are generalized eigenvectors and, therefore, $\hat{Z}X_t$ are the generalized principal components.

In our empirical application we ignore the forecasting of the idiosyncratic components, hence we concentrate in the estimation of (7) and (8), using only the information provided by the common factors, leaving the dependence relationship to affect only the simulated scenarios of mortality and longevity. We model the dependence relationship between the differences of the mortality rates for different ages, using copulas. We follow the strategy proposed by Aas et al. (2009). These authors show that multivariate data, which exhibit involved patterns of dependence in the tails, can be modeled using a cascade of pair-copula, acting on two variables at a time.

## 3　Data and Results

Our data set is composed by annual mortality rates for males and females in Spain from 1950 to 2012. The data for 0 to 101 years were taken from the web page of Human Mortality Database. The data from 102 to 109+ years were calculated using a linear extrapolation taken as a starting point the mortality rate at 101 years and a mortality rate equal to 1 at 110 years. This strategy assumes than nobody survives after 110 years. The reason to approach the higher rates in this way is the great variability of the data for ages above 101 years. We also have added a 'stochastic noisy' variation to the deterministic trend in the older categories, to avoid perfect dependence between the series in the last fragment of the sample.

We select the number of static and dynamic factors following the criteria proposed by Bai and Ng (2002, 2007). We consider six different artificial populations: males and females from 0 to 109+ years; males and females from 18 to 64 years and, males and females from 65 to 109+ years. The total population in each case was set to 30,000 people.

Our main results consist on the comparisons of the accuracy of different forecasting models and the estimation of longevity and mortality risks for the Spanish population. In general, the GDFM and the DFM fitted over the differences of the log-mortality rates, perform better than the other models in forecasting, regardless the forecasting horizon. The worst model is the DFM fitted on the levels of log mortality rates, followed closely by the Lee-Carter Model. We confirm the finding by Mitchell et al. (2013), namely that the models 'in differences' outperform the models in 'levels' in forecasting. Given that we discriminate between sex and ages, and the fact that we utilize annual mortality rates and an automatic lag-selection in the dynamic model, this first result is important in its own right. It provides a tool for practitioners, who are interested in measuring and managing mortality and longevity risks, which in turns depend on the forecasting accuracy of the underlying employed model. This result can be related to the fact that the models operating on the log-rates (in levels) directly impose a common stochastic trend to describe the dynamics of every mortality rate. This long-run common variation can be false, given the high number of series (more than 100 for each sex). Even if some subsets of the variables are actually cointegrated (they shared the same common trend), some others will certainly not and therefore, imposing a common trend can derive in spurious estimation of the factor loads and the forecasting coefficients.

We also calculate TVaR and VaR measures one-year ahead using copulas. First we use the empirical distribution of the GDFM residuals to construct the pseudo-sample, necessary to estimate the copula. Second, we considered different empirical copulas, which summarize alternative possible dependence structures in the data in a flexible way. We compare copulas: Gaussian, Clayton, BB6, Survival-Joe, Rotated-Clayton (180 and 270°), which demonstrated to be the more appropriated in a preliminary estimation step. We selected the best copula among the candidates, through AIC criterion, and used it to construct the 'multivariate dependence tree'.

In Tables 1 and 2 we report the Tail-VaR and VaR at the 0.5 % (left tail) and the same statistics at the 99.5 percentiles (right tail), for populations of males and females, respectively. This confidence level is standard in the practice of the insurance companies and pension funds and it is recommend within the regulatory framework (for example in Solvency I and II). When we look at the left-tail-VaR or TVaR, we are asking about 'longevity risk'. Conversely, at the right tail we are referring to 'mortality risk'. In both cases risk has to be understood as a significant dispersion from the expected number of deaths, forecasted with the GDFM, which is the best model at hand, together with the DFM.

**Table 1** Longevity and mortality risks for portfolios of size 30,000. Males

|         | TVaR | VaR  | ED   | VaR  | TVaR |
|---------|------|------|------|------|------|
| 0–109+  | 250  | 251  | 257  | 264  | 265  |
| 18–64   | 74   | 74   | 76   | 78   | 78   |
| 65–109+ | 1341 | 1346 | 1371 | 1393 | 1395 |

**Table 2** Longevity and mortality risks for portfolios of size 30,000. Females

|         | TVaR | VaR  | ED   | VaR  | TVaR |
|---------|------|------|------|------|------|
| 0–109+  | 233  | 234  | 241  | 248  | 249  |
| 18–64   | 34   | 34   | 35   | 36   | 36   |
| 65–109+ | 1083 | 1085 | 1108 | 1129 | 1132 |

The results of the forecasting exercise can be read in several ways. Suppose a portfolio of 30,000 males between 0 and 109+ years that behaves according to the population structure of the Spanish males in 2012. Then the expected number of events (deaths) in one year is 257. Similarly, we expect to see 241 events in a population composed only by woman, under the same conditions. Instead, in a population of 30,000 males between 18 and 64 years, we expect to observe 76 deaths in one year, and 35 events in the case of females at the same ages. Lastly, and more important, if we were to construct a population-portfolio compose by males between 65 and 109+ years, we would expect to observe in one year 1,371 events of death. Under the same circumstances with female affiliates we would expect to observe 1,108 events.

In terms of longevity risk we find that, as expected, longevity risk increases as the age advances. In other words, the oldest part of the population does not only show higher mortality rates, as expected, but its projection also presents greater variability. Hence, forecasting for ages above 65 years is subject to considerably bigger uncertainty, which in turn, has to be assimilated to a greater level of risk. For example, the right-tail-VaR for males between 0 and 109+ years is 251 and the left-tail-VaR, 264. For woman these values are 234 and 248, respectively. These estimations provide useful insights for the operation of any insurance or pension's company. They tell us that it is possible to assert with a 99.5 % of statistical-confidence level, in one year, that no more than 264 persons (males) will die, or conversely, that no less than 251 will die, even when one expects 257 to die.

The calculations increase significantly for the upper ages, both for males and females. In a portfolio of males between 65 and 109+ years (the one with the higher longevity and mortality risks) we can expect with a 99.5 % of confidence level to observe $1,371 - 1,346 = 25$ persons surviving in one year above the projections provided by the best available model or $1,393 - 1,371 = 22$ people dying above such projection. If we instead use the TVaR's to make the same calculations, we would find that 24 people will survive above our expectations and 30 people could die above them.

These results evidence a relevant empirical finding of our study, related to the asymmetric nature of mortality and longevity risks. By using the TVaR at 99.5 %, longevity risk is 29 % higher than mortality risk for the portfolio of males (24 % in females). Moreover the longevity risk of the older population, only regarding males between 65 and 109+ years, is considerably higher than the mortality risk of the younger population between 18 and 64 years. Indeed, in order to make those risks equivalent, you would have to affiliate 15 times more people between 18 and 64 than those between 65 and 109+, expecting the longevity and mortality risks to compensate each other.

Lastly, we compare longevity risk in males and females about the same ages (but with different population structures, both of them resembling the Spanish population). We observe that longevity risk is bigger for males than females. This result is not contradictory with the very well documented fact that woman tend to live longer than males, indeed it is because of the female greater longevity that male rate variability is greater in the upper ages and therefore more difficult to forecast. Therefore, fixing the number of persons in each portfolio will produce more accurate results for woman and less uncertainty in the model projections.

It has to be noticed, however, that the models in differences perform worse than the models in levels, for short forecasting horizons at ages above 95 years (results not presented to save space). This finding has to do with the fact that the greater fitting of the models in differences comes at the expense of some degree of over-fitting for the older population, especially for males. This over-performance of the models in levels disappears as the forecast horizon increases and, indeed, for 10 years ahead it completely reverses. The intuition is that the imposition of the shared-stochastic trend deteriorates the forecasting in a cumulative fashion, and it results worse than the over-fitting disadvantage of the models in differences, for ages above 95 and a medium-term forecasting horizon.

## 4 Conclusions

In general lines, our results indicate that the DFM in differences performs better than the models in levels, at all forecasting horizons (between one and ten years). The worst model is usually the DFM in log-rates, followed closely by the Lee-Carter Model.

These results are important for the operation of an insurance company or a pension fund. Ours can be interpreted as a hypothetical exercise assuming a *numerarie* value for each mortality or longevity claim equal to 1. The mapping from mortality and longevity events to actual money is straightforward provided the specific operational values of a firm. Hence we prefer to present the results in terms of mortality or longevity events.

It should be notice that different from previous works in the literature, we do not (implicitly) assume a cointegration relationship operating on the 220 mortality rates of our empirical exercise. Therefore our forecasting exercise only includes the modeling of the stochastic volatility around the trend, equivalent to the second step in Plat's algorithm (Plat 2011). The simulation of the mortality trends can be achieve in our case, simply by projecting one-year independent random walks processes. We also consider a more general dependence structure compared to all other previous works, as we use dvine-copula model instead of local linear correlations estimators.

Our results also highlight the asymmetric nature of stochastic mortality and longevity risks, for different populations.

## References

Aas, K., Czado, C., Frigessi, A., Bakken, H.: Pair-copula constructions of multiple dependence. Insur. Math. Econ. **44**, 182–198 (2009)

Alonso, A.M.: Predicción de tablas de mortalidad dinámicas mediante un procedimiento bootstrapp. Fundación MAPFRE, Madrid (2008)

Bai, J., Ng, S.: Determining the number of factors in approximate factor models. Econometrica **70**, 191–221 (2002)

Bai, J., Ng, S.: A panic attack on unit roots and cointegration. Econometrica **72**, 1127–1177 (2004)

Bai, J., Ng, S.: Determining the number of primitive shocks in factor models. J. Bus. Econ. Stat. **25**, 52–60 (2007)

Bai, J., Ng, S.: Large dimensional factor analysis. Found. Trends. Econometr. **3**, 89–163 (2008)
Brouhns, N., Denuit, M.: Risque de longévité et rentes viagères II. Tables de mortalité prospectives pour la population belge. Belgian Actuarial Bull. **2**, 49–63 (2002)
Cairns, A., Blake, D., Dowd, K., Coughlan, G.D., Epstein, D., Khalaf-Allah, M.: Mortality density forecasts: An analysis of six stochastic mortality models. Insur. Math. Econ. **48**, 355–367 (2011)
Continuous Mortality Investigation: Projecting future mortality: A discussion paper. Working Paper 3. Institute and Faculty of Actuaries, UK (2004)
Continuous Mortality Investigation: Projecting future mortality: Towards a proposal for a stochastic methodology. Working Paper 15. Institute and Faculty of Actuaries, UK (2005)
Continuous Mortality Investigation: The CMI mortality projections model. Working Paper 63. Institute and Faculty of Actuaries, UK(2013)
Dushi, I., Friedberg, L., Webb, T.: The impact of aggregate mortality risk on defined benefit pension plans. J. Pension Econ. Fin. **9**, 481–503 (2010)
Forni, M., Hallin, M., Lippi, M., Reichlin, L.: The generalized dynamic-factor model: Identification and estimation. Rev. Econ. Stat. **82**, 540–554 (2000)
Forni, M., Hallin, M., Lippi, M., Reichlin, L.: The generalized dynamic factor model: One-sided estimation and forecasting. J. Am. Stat. Assoc. **100**, 830–840 (2005)
Forni, M., Reichlin, L.: Let's get real: Factor to analytical cycle approach dynamics disaggregated business. Rev. Econ. Stud. **65**, 453–473 (1998)
Hollmann, F.W., Mulder, T.J., Kallan, J.E.: Methodology and assumptions for the population projections of the United States: 1999 to 2100. Population Division U.S. Census Bureau Working Paper 38 (2000)
Lee, R.D., Carter, L.R.: Modeling and forecasting U.S. mortality. J. Am. Stat. Assoc. **87**, 659–671 (1992)
Mitchell, D., Brockett, P., Mendoza-Arriaga, R., Muthuraman, K.: Modeling and forecasting mortality rates. Insur. Math. Econ. **52**, 275–285 (2013)
Peña, D., Poncela, P.: Nonstationary dynamic factor analysis. J. Stat. Plan. Inf. **136**, 1237–1257 (2006)
Plat, R.: One -year value-at-risk for longevity and mortality. Insur. Math. Econ. **49**, 462–470 (2011)
Stock, J.H., Watson, M.W.: Forecasting using principal components from a large number of predictors. J. Am. Stat. Assoc. **97**, 1167–1179 (2002)
Stock, J.H., Watson, M.W.: Forecasting with many predictors. Handb. Econ. Forecast. **1**, 515–554 (2006)

# Joint Modeling of Health Care Usage and Longevity Uncertainty for an Insurance Portfolio

Xavier Piulachs, Ramon Alemany, Montserrat Guillén
and Carles Serrat

**Abstract** We study longevity and usage of medical resources of a sample of individuals aged 65 years or more who are covered by a private insurance policy. A longitudinal analysis is presented, where the annual cumulative number of medical coverage requests by each subject characterizes insurance intensity of care until death. We confirm that there is a significant correlation between the longitudinal data on usage level and the survival time processes. We obtain dynamic estimations of event probabilities and we exploit the potential of joint models for personalized survival curve adjustment.

**Keywords** Health insurance · Survival analysis · Longitudinal data

## 1 Introduction

The gradual development of medical science and technology leads to a larger number of years lived with disabilities, which in turn increases the demand of medical resources. This is a key challenge for health insurance companies who have to face additional costs in order to meet the care needs in the event of a large cohort

X. Piulachs (✉) · R. Alemany · M. Guillén
Department of Econometrics, Riskcenter, University of Barcelona, Av. Diagonal 690,
08034 Barcelona, Spain
e-mail: xavierpiulachs@hotmail.com

R. Alemany
e-mail: ralemany@ub.edu

M. Guillén
e-mail: mguillen@ub.edu

C. Serrat
Department of Applied Mathematics I, Universitat Politècnica de Catalunya, Av. Dr.
Marañón, 44-50, 08028 Barcelona, Spain
e-mail: carles.serrat@upc.edu

© Springer International Publishing Switzerland 2015
J. Gil-Aluja et al. (eds.), *Scientific Methods for the Treatment of Uncertainty
in Social Sciences*, Advances in Intelligent Systems and Computing 377,
DOI 10.1007/978-3-319-19704-3_24

of elderly people. Furthermore, it is well known that private insurance policy holders are generally supposed to have a higher socio-economic level compared to the rest of the population because they can afford private health coverage (Schoen et al. 2010). Consequently, the mortality tables of the general population may be biased for the insured customers and insurance companies estimate specific survival probabilities for their portfolios using standard actuarial methods (Yue and Huang 2011; Denuit 2009; Denuit and Frostig 2008). In practice, however, they disregard longitudinal information on their policy holders that is continuously being collected. Health insurance companies accumulate data on the intensity and the type of use of medical resources which can be extremely valuable to predict personalized survival probabilities and to quantify the risk of medical care demand of their clients above the expected values.

The aim of our study is to show how historical and follow-up records, which are in fact repeated measures of a longitudinal marker that counts the number of times that the policy holder has used the insurance policy coverage, can effectively predict personalized survival probabilities. Our proposed joint modeling approach, which is a powerful methodology used in statistics for medicine (Tsiatis et al. 1995; Rizopoulos 2012a, b; Rizopoulos and Lesaffre 2014), allows to examine the association between a given medical care usage trend and longevity prospects.

It is well known that medical usage intensity increases substantially at older ages (Blane et al. 2008) and end-of-life care expenditures are significantly larger than throughout life (Dao et al. 2014; Murphy 2012), but according to Bird et al. (2002) men's and women's health care experiences differ as they age. While increasing attention has been focused on gender differences in health status, prevalence of illnesses, and access to quality care among older adults, it is also needed deepen knowledge about differences in their health care in the last years of their lives. This is precisely what we study.

The dynamic personalized predictions that we are aiming at are based on both baseline subject's time-to-event covariates, recorded at the start of the study, and subject's longitudinal information measured at fixed time points within an observation window. Therefore, both the longitudinal and the survival information is part of a single statistical model, which allows: (i) to establish the degree of association between the value of the longitudinal variable and the time to event outcome, (ii) to estimate subject-specific survival probabilities based on personalized longitudinal outcomes and (iii) to update personalized survival estimations as additional longitudinal responses are collected. This can provide a comprehensive risk assessment of a health insurance portfolio using all available information.

To the best of our knowledge, no study has evaluated how health care usage and risk of death can be modelled jointly. The main results of our analysis are: (1) we confirm age and gender are main factors influencing changes in survival for a health insurance member, (2) we find evidence of a significant association between serial measurements of cumulative private insurance care usage and longevity, and (3) we obtain dynamic estimations of event probabilities by exploiting the potential of joint models. In summary, our contribution shows that an increase in health care usage intensity is negatively associated with survival, but its influence varies with usage accumulation and depending on other factors such as sex and age, as well as previous insurance conditions.

## 2 Data and Methods

The data correspond to information provided by a Spanish private health insurance mutual company, containing historical data which started being collected on January 1st, 2006 and ended on February 1st, 2014. In particular, our study is limited to 39,580 insurance policy holders (39.8 % men and 60.2 % women) who had reached the age of 65 before the observation period started.

Table 1 presents the definition of the variables that are used in the analysis. Two variables are central in our study. First, the longitudinal process which counts the number of times that the health insurance company has provided a service to the policy holder. The unit service can be a variety of possible coverage functions such as a doctor visit, a blood or an X-ray test, a prescribed therapy, a hospital stay and any other treatment that is established in the insurance contract. We do not distinguish between different types of services at this stage, but obviously the cumulated number of unit services provided over time to a given patient is strongly correlated with her health condition. Due to the right skewed shape exhibited by the longitudinal outcome, a logarithmic scale is applied (Verbeke and Molenberghs 2009). Second, we also consider the survival time, where the event of interest is death. Information is censored because the majority of individuals survive beyond the end of the study period. Some other cancel their insurance policy and therefore they quit the study automatically. These dropouts are considered at random, as they are generally due to personal reasons such as the decision not to renew the policy, or a change of company, which are supposed to introduce an independent censorship mechanism.

**Table 1** Variables in the private insurer dataset (2006–2014)

| Variable name | Definition |
|---|---|
| ID | Subject identifier: $i = 1,2,...,30580$ |
| SEX | Gender of the subject: $0$ = Male, $1$ = Female |
| OBSTIME | Age (years) in excess of 65 at each time point |
| CUM0 | Cumulative number of private health service usage units over the four years previous to entering the study |
| CUM | Cumulative number of private health service usage units at each observation time point |
| TIME | Final observation time (years), which may correspond to an event (death) or to a right-censored data |
| CENS | Censoring indicator (1) right-censored, (0) otherwise |

A private health service usage unit is a visit to a GP or a specialist, a hospital spell, a medical test, etc.

Let $y_i(t) = log.CUM_i(t) = log\{1 + CUM_i(t)\}$ be the response variable of the $i$-th subject, $i = 1, \ldots, n$, observed at time $t$, where $n$ is the total number of observed individuals in the sample. The outcome is linearly related to a set of $p$ explanatory covariates and $q$ random effects. In our application the first response is the number of private health usage service units after a logarithmic transformation.

In addition, let $m_i(t)$ denote the true underlying value of the longitudinal outcome, and $M_i(t) = \{m_i(s), 0 \leq s \leq t\}$ the complete longitudinal history up to time $t$. The joint modeling approach consists in defining: (i) a model for the marker trajectory, usually a mixed model; (ii) a model for the time-to-event, usually a proportional hazard model, and (iii) linking both models using a shared latent structure (Rizopoulos 2011).

## 2.1 Longitudinal Submodel: Random Intercept Model

The main goal of linear mixed effects models is to account for the special features of serial evaluations of outcomes over time, thus being able to establish a plausible model in order to describe the particular evolution of each subject included in a longitudinal study. The particular features of these models are that they work with unbalanced datasets (unequal number of follow-up measurements between subjects and varying times between repeated measurements of each subject), and that they can explicitly take into account that measurements from the same patient may be more correlated than measurements from different patients.

The model is specified as follows:

$$
\begin{cases}
log.CUM_i(t) = m_i(t) + \varepsilon_i(t) = \beta_0 + b_{i0} + \beta_1 t + \varepsilon_i(t) \\
\boldsymbol{\beta} = (\beta_0, \beta_1)^T \\
b_{i0} : N(0, \sigma_{b_0}^2) \\
\varepsilon_i(t) : N(0, \sigma^2)
\end{cases}
\tag{1}
$$

This model is straightforward. It just postulates that, besides individual random effects, a linear time trend governs the rate of increase of the number of accumulated service units provided to insurance policy holders. This seems plausible as we also expect that the older the policy holder the larger the rate at which the number of requested services increases.

## 2.2 Survival Submodel: PH Cox Model

The model provides the conditional hazard function $h_i(t|\boldsymbol{w}_i)$ at time $t$ of a subject's profile given by a set of $p$ time-independent explanatory covariates called baseline covariates.

We assume that survival time $T^*$, or *TIME* in our dataset, is a non-negative continuous random variable which represents the exact time until some specified event, which is death in our case. The survival model is specified through the hazard function as follows:

$$h_i(t|w_i) = h_0(t)exp\{\gamma log.CUM0_i\}, \tag{2}$$

where $h_0(t)$ is an unspecified and non-negative baseline hazard function, representing the hazard function where $w_i = 0$, and $w_i$ contains the information about the set of explanatory time-independent covariates that define the $i$-th subject's profile.

The Cox model (Cox 1972) is often called a proportional hazards (PH) model because, two individuals $i$ and $i'$ with respective covariate values $w_i$ and $w_{i'}$, have a constant hazard ratio, so their corresponding hazard functions are proportional to each other and do not change over time.

## 2.3 Joint Model for Longitudinal and Survival Data

To account for the fact that the longitudinal marker is an endogenous time-dependent covariate measured with error (Kalbfleisch and Prentice 2002) with respect to survival, it is assumed that the risk for an event depends on the true and unobserved value of the endogenous variable at time $t$, denoted by $m_i(t)$. The endogeneity problem is quite intuitive here. There is a latent factor causing a health deterioration, which in turn potentially implies an increase in the risk of death and more intensity of health care service use. So, survival and health care are strongly related each other, through this latent factor.

On the basis of the expressed considerations, the true and unobserved outcome at a specific time point $t$ can be modeled by joining the two above approaches (Rizopoulos 2012a, b):

$$h_i(t|M_i(t), w_i) = h_0(t)exp\{\gamma log.CUM0_i + \alpha(\beta_0 + b_{i0} + \beta_1 t)\} \tag{3}$$

The models presented in this section can be generalized to higher dimensions (Andrinopoulou et al. 2014). More information on joint modeling fitting can be found in Rizopoulos (2012a, b) and details about the R package implementation are given in Rizopoulos (2010). Further detail on the methods can be found in Hsieh et al. (2006). Applications to prostate cancer are studied in Proust-Lima and Taylor (2009), and Serrat et al. (2015). Further details on the current study are given in Piulachs et al. (2014).

## 3 Results and Prediction

The results for the private mutual insurer dataset are presented in Table 2 separately for men and women. The association parameter $\alpha$ is positive and significantly different from zero. This indicates that the larger the cumulative number of service units provided the larger the risk of death. This is consistent with intuition as an aggravated patient has a higher probability of death and as a consequence, he demands health care resources. At the same time, who demand health care services are certainly motivated by a deteriorated health condition and therefore expected survival time decreases. Note that a positive parameter factor in the hazard function means that the risk of death increases, whereas a negative parameter has the contrary effect.

**Table 2** Results of the joint model estimation in the private insurer dataset (2006–2014)

| Parameters | Men | | Women | |
|---|---|---|---|---|
| | Estimate | 95 % CI | Estimate | 95 % CI |
| $\beta_0$ | 2.140 | (2.100, 2.180) | 2.158 | (2.123, 2.192) |
| $\beta_1$ | 0.170 | (0.167, 0.172) | 0.157 | (0.155, 0.159) |
| $\sigma$ | 0.332 | (0.329, 0.334) | 0.314 | (0.312, 0.316) |
| $\sigma_{b_0}$ | 1.648 | (1.597, 1.698) | 1.791 | (1.748, 1.834) |
| $\gamma$ | −1.174 | (−1.306, −1.042) | −0.964 | (−1.042, −0.887) |
| $\alpha$ | **1.437** | (1.275, 1.598) | **1.273** | (1.179, 1.367) |

CI stands for confidence interval

We note that the association between the longitudinal process and the survival outcome is slightly higher for men (1.437) than for women (1.273), but the difference is not statistically significant. All other parameter estimates are similar for men and women except for estimated $\gamma$, which is not significantly different from zero for women. This means that pre-existing conditions, which are represented by *log.CUM*0 and which refer to the cumulated number of services received during the four years previous to the study, do influence negatively the hazard rate for men and women, even though slightly greater for men than for women. This result seems to indicate that a larger survival is expected for those who were using medical care more intensively than others. This result also means that if a patient accumulates a large number of services, but was able to survive to the starting date of the study, then he has a smaller hazard rate of death compared to another subject who has not accumulated as many services as him. This can be interpreted as the preventive effect or health care or the curative effect, which prove to be efficiently leading to longer life expectancy.

However, if suddenly a patient requires medical care, the number of accumulated services increases and therefore, since parameter estimates for $\beta_0$, $\beta_1$ and $\alpha$ are positive, that would synchronize with the hazard rate, which would increase and lead to a higher risk of death.

A log-unit increase in the cumulative number of visits entails a $exp(1.437) = 4.2$-fold increase in the men risk and $exp(1.273) = 3.6$-fold increase for women.

Some comments on $\sigma$ and $\sigma_{b_0}$ are needed. $\sigma_{b_0}$ can be interpreted as an inherent variability in the random effects of the longitudinal model while $\sigma$ is the variability of the model. Note that $\sigma_{b_0}$ is slightly larger for women than for men.

As an illustration, let us consider for instance the case of a woman 65 aged at study start point, for whom her cumulative service received during the four years prior to the study starting time point is known. In Fig. 1 we can observe how the model updates the predicted survival probabilities as new longitudinal information is collected. In 2010 the information provided by the survival curve indicates that life expectancy is large: However, after 2013 we see an increased cumulated number of services received (see the steep trend of cumulated dots on the right plot), which implies a survival curve that decreases more sharply than the one at the beginning of the study. Thus, at the end of 2013, predicted life expectancy diminishes with respect to the initial one. This methodology is a useful prognostic tool.

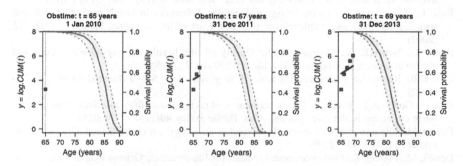

**Fig. 1** Dynamic survival probabilities for a woman aged 65 who is still alive at the end of the study, Source: Own calculations using the private insurer dataset and the JM package from freely available software R

## 4 Conclusions

From the analysis of our private insurance longitudinal data sample, we conclude that the observed cumulative number of health care service units provided is strongly as positively associated with the risk of death. The baseline cumulative number of health care service units provided to a patient has a protective effect. This is in line with evidence of a preventive affect. The joint modelling methodology allows to continuously updating the predictions of subject-specific survival probabilities, when new information on service usage comes along.

Further work is going to be pursued on the generalization of the statistical model to counting processes and to the implementation of multivariate longitudinal markers, as they seem very natural here. Indeed the number of medical care services needed can be categorized in big groups, those that are routine programmed actions and exceptional treatments, such as surgery or serious procedures.

One of the limitations of our study is the fact that all health services have the same importance in the longitudinal counter. Another practical issue is the fact that insurance customers switch between companies and new policy holders could enter the sample or leave the group motivated by health-related problems. However, we do not think that this was a problem in this particular dataset since our policyholders were above 65 years of age, and they have been in this mutual company for several years, because it is very infrequent to change the health insurance provider at this age.

# References

Andrinopoulou, E.-R., Rizopoulos, D., Takkenberg, J., Lesaffre, E.: Joint modeling of two longitudinal outcomes and competing risk data. Stat. Med. **33**(18), 3167–3178 (2014)

Bird, C., Shugarman, L., Lynn, J.: Age and gender differences in health care utilization and spending for medicare beneficiaries in their last years of life. J. Palliat. Med. **5**(5), 705–712 (2002)

Blane, D., Netuveli, G., Montgomery, S.: Quality of life, health and physiological status and change at older ages. Soc. Sci. Med. **66**(7), 1579–1587 (2008)

Cox, D.R.: Regression models and life-tables. J. Roy. Stat. Soc.: Ser. B (Methodol.) **34**, 187–220 (1972)

Dao, H., Godbout, L., Fortin, P.: On the importance of taking end-of-life expenditures into account when projecting health-care spending. Can. Public Policy **40**(1), 45–56 (2014)

Denuit, M.: Life annuities with stochastic survival probabilities: a review. Method. Comput. Appl. Probab. **11**, 463–489 (2009)

Denuit, M., Frostig, E.: First-order mortality basis for life annuities. Geneva Risk and Insur. Rev. **33**(2), 75–89 (2008)

Hsieh, F., Tseng, Y.-K., Wang, J.-L.: Joint modeling of survival and longitudinal data: likelihood approach revisited. Biometrics **62**, 1037–1043 (2006)

Kalbfleisch, J., Prentice, R.: The statistical analysis of failure time data, 2nd edn, vol. 360. Wiley (2002)

Murphy, M.: Proximity to death and health care costs. Edward Elgar Publishing (2012)

Piulachs, X., Alemany, R., Guillén, M.: A joint longitudinal and survival model with health care usage for insured elderly. UB Riskcenter Working Papers Series 2014-07 (2014)

Proust-Lima, C., Taylor, J.: Development and validation of a dynamic prognostic tool for prostate cancer recurrence using repeated measures of post-treatment PSA: a joint modeling approach. Biostatistics **10**, 535–549 (2009)

Rizopoulos, D.: JM: An R package for the joint modelling of longitudinal and time-to-event data. J. Stat. Softw. **35**(9), 1–33 (2010)

Rizopoulos, D.: Dynamic predictions and prospective accuracy in joint models for longitudinal and time-to-event data. Biometrics **67**(3), 819–829 (2011)

Rizopoulos, D.: Joint Models for Longitudinal and Time-To-Event Data with Applications in R. CRC Press, Boca Ratón (2012a)

Rizopoulos, D.: Fast fitting of joint models for longitudinal and event time data using a pseudo-adaptive Gaussian quadrature rule. Comput. Stat. Data Anal. **56**(3), 491–501 (2012b)

Rizopoulos, D., Lesaffre, E.: Introduction to the special issue on joint modelling techniques. Stat. Methods Med. Res. **23**(1), 3–10 (2014)

Schoen, C., Osborn, R., Squires, D., Doty, M., Pierson, R., Applebaum, S.: How health insurance design affects access to care and costs, by income, in eleven countries. Health Aff. **29**(12), 2323–2334 (2010)

Serrat, C., Rué, M., Armero, C., Piulachs, X., Perpiñán, H., Forte, A., Páez, A., and Gómez, G.: Frequentist and Bayesian approaches for a joint model for prostate cancer risk and longitudinal prostate-specific antigen data. J. Appl. Stat. Appear (2015). doi: 10.1080/02664763.2014. 999032

Tsiatis, A.A., Degruttola, V., Wulfsohn, M.S.: Modeling the relationship of survival to longitudinal data measured with error. Applications to survival and CD4 counts in patients with AIDS. J. Am. Stat. Assoc. **90**, 27–37 (1995)

Verbeke, G., and Molenberghs, G.: Linear Mixed Models for Longitudinal Data. Springer. (2009)

Yue, C.S., Huang, H.C.: A study of incidence experience for Taiwan life insurance. Geneva Pap. Risk Insur. Issues Pract. **36**(4), 718–733 (2011)

# The Commodities Financialization As a New Source of Uncertainty: The Case of the Incidence of the Interest Rate Over the Maize Price During 1990–2014

María-Teresa Casparri, Esteban Otto-Thomasz
and Gonzalo Rondinone

**Abstract** The past decade has witnessed the entry of speculative investors as major participants in commodity markets. This phennomenon arises the question of whether these agents influence price dynamics or not. Therefore, the aim of this paper is to explore evidence in order to determine if commodities have behaved in a similar manner to financial assets. This study will focus specifically in the maize market, analyzing the extent to which financial market variables influence price movements. By means of an autoregressive vector system (VAR) the effect of interest rate changes on maize futures prices will be tested. Implications for countries heavily reliant on commodity exports will be drawn from the results of this study.

**Keywords** Commodities financialization · Maize price risk · Interest rate · Autoregressive vector model

## 1 Introduction

Grain prices have been very volatile in recent years. They reached a peak in 2008 and then decline sharply, and started to raise in 2010. In particular, in 2011, maize price exceeded the level of June 2008 (UN 2011).

Volatility in grain prices is due to several causes, but can be simplify in two mayor correlated factors: market and weather. Climate factors cause traditional

M.-T. Casparri (✉) · E. Otto-Thomasz · G. Rondinone
Center of Quantitative Research Applied to Economics and Management (CMA),
University of Buenos Aires, Av. Córdoba 2122, C1120AAQ Buenos Aires, Argentina
e-mail: mcasparri@econ.uba.ar

E. Otto-Thomasz
e-mail: ethomasz@econ.uba.ar

G. Rondinone
e-mail: gonzalorondinone@gmail.com

© Springer International Publishing Switzerland 2015
J. Gil-Aluja et al. (eds.), *Scientific Methods for the Treatment of Uncertainty in Social Sciences*, Advances in Intelligent Systems and Computing 377,
DOI 10.1007/978-3-319-19704-3_25

uncertainty in grain prices, affecting quantities. Because they are highly unpredictable and have a direct incidence on stocks, they constitute a major source of price uncertainty in the future. On the other hand, market effects are related to the functioning of the market itself: number of producers, productivity, price formation process, speculation, substitution effects, financial effects, etc.

In this particular work, we will focus in one specific financial effect: the commodities financialization process. The purpose of this is to collect evidence to prove that the increase in price volatility is due to actions of new financial investors and not because of traditional market or climate effects.

The major problem of the process is that these participants may add a new kind of uncertainty to the market, making it more complex and volatile. The consequences can be deep, especially on countries dependent on grain exports. Though, studying the financial effects on grain prices can provide more information in order to reduce uncertainty.

In the first section of this work we define the concept of commodities financialization. In the second section, we test one financialization hypothesis for the case of the international maize price: using a vector autoregressive model (VAR) we test the incidence of the interest rate over the maize price during 1990–2014. In the last section, we present some conclusions related to the incidence of this process not only in investment strategies buy mostly in countries dependent on commodities exports.

## 2 Commodities Financialization

During the last years, financial innovations allowed agricultural commodities to become part of portfolio investments. This means that financial investors began to buy these products, having no relation with production or even the commercialization of them. As any financial investment, the objective is to speculate with higher prices in a future horizon.

The process through which these products became part of portfolios of financial investments is called commodities financialization.

According to Tang and Xiong (2010), the presence of large investment funds makes that these markets behave more as a financial market rather than a commodities one. Henderson et al. (2012) defines the process as the relative influence gained by the financial sector in comparison with the real sector in order to determine price levels and daily returns of these markets.

This process was possible through one financial innovation: the Exchange-Traded Funds (ETF's). The ETF's are basically investment funds that replicate the behavior of a certain index, such as a commodity index. The main feature of these funds is that their shares are traded in real time as any other stock, being able to be bought or sold at any time.

Investment in commodity index had an exponential growth during the last years. According to Tang and Xiong (2010), these vehicles try to replicate the return of certain portfolio, including a wide variety of products. However, in all cases the financial engineering is based on long positions on future markets, without possession of the real product. For a deep analysis of the phenomenon see United Nations (2011).

The main question that arises is if the presence of these funds has change commodities price formation process. If they follow traditional investment strategies, the interest rate become a significant variable. Thus, if there is a financialization process, commodities prices might be more sensitive to changes in the interest rate. To test that hypothesis, in the next section we test the influence of the interest rate over the maize price. It is important to mention that Rondinone and Thomasz (2015) have found a strong relation during the last years for the case of soybean.

## 3   The Interest Rate Incidence Over the Maize Price

In order to test the incidence of the interest rate over the maize price we use a vector autoregressive model (VAR).

We select this tool among other options, such as simultaneous equations, because it allows the study of impulse-response functions. These functions analyze the model dynamics, showing the response of the explicative variables. A shock in one variable will affect itself and at the same time will propagate to the rest of the variable through the dynamic structure of the VAR model. To expand about the model see Sims (1980). For information about lags criteria selection, see Ivanov and Kilian (2005).

We use a Bivariate VAR model presented by Urbisaia (2000):

$$Y_{1t} = \delta_1 + \phi_{11}Y_{1,t-1} + \phi_{12}Y_{2,t-2} + \varepsilon_{1t}$$

$$Y_{2t} = \delta_2 + \phi_{21}Y_{1,t-1} + \phi_{22}Y_{2,t-1} + \varepsilon_{2t}$$

The corresponding structural model is:

$$Y_{1t} = b_{10} + b_{12}Y_{2t} + c_{11}Y_{1,t-1} + c_{12}Y_{2,t-1} + \mu_{1t}$$

$$Y_{2t} = b_{20} + b_{21}Y_{2t} + c_{21}Y_{1,t-1} + c_{22}Y_{2,t-1} + \mu_{2t}$$

The vectorial notation of the model is:

$$Y_t = \delta + \phi_1 Y_{t-1} \mid c_t \tag{1}$$

Considering the stationarity of the VAR (1) it can be written as moving average vector ($\infty$):

$$Y_t = \delta + \phi_1 Y_{t-1} + \varepsilon_t = \bar{Y} + \sum_{j=0}^{\infty} \phi_1^j \varepsilon_{t-j}$$

$$\bar{Y} = \begin{bmatrix} \bar{Y}_1 \\ \bar{Y}_2 \end{bmatrix}$$

Combining the previous representation with the structural model,

$$\begin{bmatrix} Y_{1t} \\ Y_{2t} \end{bmatrix} = \begin{bmatrix} \bar{Y}_1 \\ \bar{Y}_2 \end{bmatrix} + \sum_{j=0}^{\infty} \begin{bmatrix} \phi_{11} & \phi_{12} \\ \phi_{21} & \phi_{22} \end{bmatrix}^j \begin{bmatrix} 1 & -b_{12} \\ -b_{21} & 1 \end{bmatrix}^{-1} \begin{bmatrix} \mu_{1t} \\ \mu_{2t} \end{bmatrix}$$

This representation allows an interaction analysis between series. The coefficients $\phi_{ik}$ can be used to generate the effects of $\mu_{1t}$ y $\mu_{2t}$ over the trajectories of $Y_{1t}$ and $Y_{2t}$. These coefficients are the multipliers of the system and their graphic representation constitute impulse-response functions. This functions will be used to test the incidence of the interest rate over the maize price.

## 3.1 VAR Model

The chosen variables are the American treasury bills interest rate, the dollar exchange rate and the stock/consumption relation of the maize:

$$\Delta Pmaize_t = \varnothing_{11}\Delta Pmaize_{t-1} + \varnothing_{12}\Delta Pmaize_{t-2} + \varnothing_{13}Stockcons_{t-1}$$
$$+ \varnothing_{14}Stockcons_{t-2} + \varnothing_{15}\Delta Tasa_{t-1} + \varnothing_{16}\Delta Tasa_{t-2}$$
$$+ \varnothing_{17}\Delta Dollar\ Index_{t-1} + \varnothing_{18}\Delta Dollar\ Index_{t-2} + \partial_{19}$$

$$\Delta Stockcons_t = \varnothing_{21}\Delta Pmaize_{t-1} + \varnothing_{22}\Delta Pmaize_{t-2} + \varnothing_{23}Stockcons_{t-1}$$
$$+ \varnothing_{24}Stockcons_{t-2} + \varnothing_{25}\Delta Tasa_{t-1} + \varnothing_{26}\Delta Tasa_{t-2}$$
$$+ \varnothing_{27}\Delta Dollar\ Index_{t-1} + \varnothing_{28}\Delta Dollar\ Index_{t-2} + \partial_{29}$$

$$\Delta Tasa_t = \varnothing_{31}\Delta Pmaize_{t-1} + \varnothing_{32}\Delta Pmaize_{t-2} + \varnothing_{33}Stockcons_{t-1}$$
$$+ \varnothing_{34}Stockcons_{t-2} + \varnothing_{35}\Delta Tasa_{t-1} + \varnothing_{36}\Delta Tasa_{t-2}$$
$$+ \varnothing_{37}\Delta Dollar\ Index_{t-1} + \varnothing_{38}\Delta Dollar\ Index_{t-2} + \partial_{39}$$

$$\Delta Dollar\ Index_t = \varnothing_{41}\Delta Pmaize_{t-1} + \varnothing_{42}\Delta Pmaize_{t-2} + \varnothing_{43}Stockcons_{t-1}$$
$$+ \varnothing_{44}Stockcons_{t-2} + \varnothing_{45}\Delta Tasa_{t-1} + \varnothing_{46}\Delta Tasa_{t-2}$$
$$+ \varnothing_{47}\Delta Dollar\ Index_{t-1} + \varnothing_{48}\Delta Dollar\ Index_{t-2} + \partial_{49}$$

*Pmaize* represents the maize international price, taken from the future contracts of Chicago Market. These are nominated in dollars per ton (U$/tn) and the information source is the International Monetary Fund.

*Tasa* is the American interest rate and is represented by the 10 year´s treasury bills of constant maturity. The information source is the *Fed of St. Louis.*

*Dollarindex* represents an index that measures the relative valuation of the dollar against a basket of foreign currencies. Such basket is composed as follows: 57.6 % Euro, 13.6 % Yen, 11.9 % Sterling, 11.9 % Canadian Dollar, 9.1 % Swedish Krona and 3.6 % Swiss franc. The information source is the electronic trading platform *DTN PropheteX.*

*Stockcons* represents the relation stock/consumption of maize in USA. The information source is United States Department of Agriculture (USDA).

The variables are monthly from 1990 to 2014. The tests were performed using *Eviews.*

## 3.2 Testing and Results

The methodology is the following: first, the time horizon is divided into two periods: 1990–2003 and 2004–2014. Within each period we perform the following test: first the unit root test on the variables, as to select to use level or first difference in order to gain stability. Then, co-integration test among variables, as to determine if it is possible to use a truly independent VAR model. Then, we perform a stability test to the VAR model itself. If stable, the impulse-response functions are constructed generating a shock of inters rate and a shock of stocks. The tests and figures one to six were performed using *Eviews* software.

## 3.3 Period 1990–2003

The unit root test on all the variables on level does not reject the null hypothesis. Meanwhile, the first difference of every series, does not present unit root (Table 1).

There is no relation among variables according to the co-integration test, so it is possible to apply a VAR model.

**Table 1** Unit root test on variables

| Test Raíz Unitaria Maiz 1990–2003 | |
|---|---|
| Variable | Prob* |
| Maiz | 0.0457 |
| ΔMaiz | 0 |
| Tasa 10y | 0.3626 |
| Δtasa 10y | 0 |
| Dólar Index | 0.5945 |
| Δdólar Index | 0 |
| Stock/cons | 0.0514 |
| Δstock/cons | 0 |

Stability test results stays that all roots are less than one, so the VAR is stable (Fig. 1).

**Fig. 1** Root values of VAR

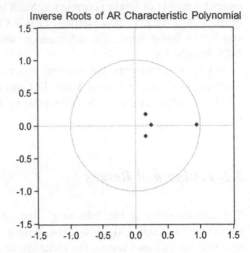

Inverse Roots of AR Characteristic Polynomial

All independence and stability conditions are assure, so we proceed to test the response of the maize price given a shock of interest rate of one standard deviation (Fig. 2).

**Fig. 2** Impulse-response function: interest rate shock

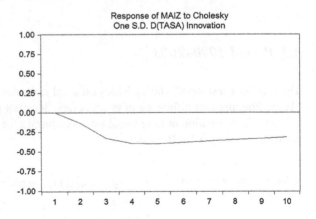

Response of MAIZ to Cholesky
One S.D. D(TASA) Innovation

The impact during the period 1990–2003 matches the theory prediction: it generates a price declination.

If we apply a shock of stocks we obtain the same result (as expected): an increase of stocks generates a price declination (Fig. 3).

**Fig. 3** Impulse-response
function: stock shock

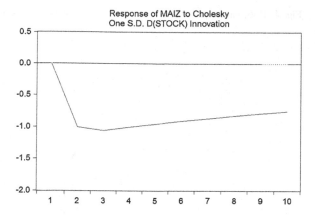

Response of MAIZ to Cholesky
One S.D. D(STOCK) Innovation

## 3.4 Period 2004–2014

The unit root test on all the variables on level does not reject the null hypothesis.
Meanwhile, the first difference of every series, does not present unit root (Table 2).

**Table 2** Unit root test on
variables

| Test Raíz Unitaria Maiz 2004–2014 | |
| --- | --- |
| Variable | Prob* |
| Maiz | 0.6029 |
| ΔMaiz | 0 |
| Tasa 10y | 0.5484 |
| Δtasa 10 y | 0 |
| Dólar Index | 0.116 |
| Δdólar Index | 0 |
| Stock/cons | 0.2827 |
| Δstock/cons | 0 |

There are not relations among variables according to the co-integration test, so it
is possible to apply a VAR model (Fig. 4).

Stability test results stays that all roots are less than one, so the VAR is stable.
All independence and stability conditions are assure, so we proceed to test the
response of the maize price given a shock of interest rate of one standard deviation.

Again, a positive shock of interest rate generates a price declination. Moreover,
during this period the effect is deeper than the previous one. Nevertheless, the effect
disappears after the fifth period (Fig. 5)

In the case of a shock of stocks, the result again shoes a declination of price
(Fig. 6).

**Fig. 4** Root values of VAR

Inverse Roots of AR Characteristic Polynomial

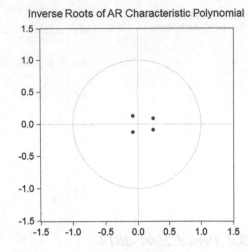

**Fig. 5** Root values of VAR

Response of D(MAIZ) to Cholesky
One S.D. D(TASA) Innovation

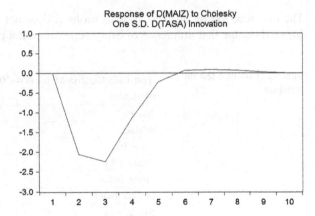

**Fig. 6** Root values of VAR

Response of D(MAIZ) to Cholesky
One S.D. D(STOCK) Innovation

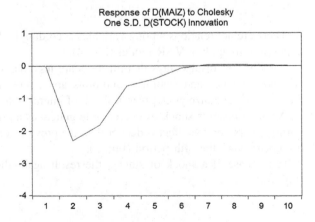

# 4 Conclusions

Testing results are showing that in both periods the interest rate has an effect in the maize price. Nevertheless, the effect is much deeper during 2004–2014. During this period the Commodities Index Funds (in the form of Exchangeable Trade Funds) exhibited a fast growing pace. Thus, the fact that during the period in which many commodities were part of portfolio investment the relation between price and interest rate was stronger, may add more evidence to the commodities financialization hypothesis, in this case particularly to maize. Rondinone and Thomasz (2015) achieved similar results in the case of soybean, although with stronger effects.

We can resume the following consequences. First, the financialization process implies that the incorporation of commodities into a financial portfolio will not diversify risk as it did in the past. Second, at a macro level, changes in the international interest rate will affect not only the capital account of the balance of payments but also the trade balance. This will specially affect countries heavily reliant on commodity exports, and can reinforce the amplification effect of external shocks, as described by Thomasz and Caspari (2010).

# References

Aulerich, N.M., Irwin, S.H., Garcia, P.: The price impact of index funds in commodity futures markets: evidence from the CFTC's daily large trader reporting system. Working Paper (2010)

Basu, P., Gavin, W.T.: What explains the growth in commodity derivatives? Fed. Bank of St. Louis Rev. 93(1), 37–48 (2010)

Gilbert, C.L.: The impact of exchange rates and developing country debt on commodity prices. Econ. J. 99, 773–784 (1989)

Gilbert, C.L.: How to understand high food prices. J. Agric. Econ. 61(2), 398–425 (2010)

Henderson, B.J., Pearson, N.D., Wang, L.: New evidence on the financialization of commodity markets. Working Paper, George Washington University (2012)

Ivanov, V., Kilian, L.: A practitioner's guide to lag order selection for VAR impulse response analysis. Stud. Nonlinear Dyn. Econometr. 9(1) (2005)

Sims, C.A.: Macroeconomics and reality. Econometrica 48, 1–48 (1980)

Sorrentino, A., Thomasz, E.: Incidencia del complejo sojero: implicancias en el riesgo macroeconómico. Revista de Investigación en Modelos Financieros 3(1)1, 9–34 (2014)

Tang, K., Xiong, W.: Index investment and financialization of commodities. Fin. Anal. J. 68(2), 54–74 (2010)

Thomasz, E., Casparri, M.: Chaotic dynamics and macroeconomics shock amplification. Comput. Intell. Bus. Econ. 507–515 (2010)

United Nations: Price Formation in Financialized Commodity Markets: The Role of Information. United Nations, New York (2011)

Urbisaia, H., Drufman, J.: Análisis de series de tiempo: univariadas y multivariadas. Cooperativas, Buenos Aires (2001)

Yang, J., Bessler, D.A., Leatham, D.J.: Asset storability and price discovery in commodity futures markets: a new look. J. Futur. Markets 21(3), 279–300 (2001)

# The Fairness/Efficiency Issue Explored Through El Farol Bar Model

Cristina Ponsiglione, Valentina Roma, Fabiola Zampella
and Giuseppe Zollo

**Abstract** The relationship between fairness and efficiency is a central issue for policy makers. To date there is no agreement among economists whether a public policy that pursues fairness entails a loss of efficiency of an economic system, or an oriented towards efficiency policy is able to facilitate the achievement of a higher fairness. In this paper we present the fairness/efficiency issue through the analysis of four agent-based simulations that implement the El Farol Bar Model proposed by Brian Arthur in (1994). As El Farol Bar models are not interested in exploring the equity issue, we modified them by introducing a set of measures of efficiency and fairness. Particularly, we assume that fairness is a short run issue and we develop time-based measurements of fairness. The computational analysis shows that selfish agents are able to achieve an efficient use of available resources, but they are incapable of generating a fair use of them. The analysis shows that random choices of agents generate an apparent fairness, which occurs only in the long run. Thus, a suitable relationship between efficiency and fairness could be reached by a public policy which defines an appropriate pay-off matrix able to drive agents' choices.

**Keywords** Fairness · Efficiency · Agent-based simulation · El Farol bar model

## 1 Introduction

Efficiency is the extent to which available resources are well-used for intended task. It often refers to the capability of a specific agent to produce an outcome with a minimum amount of waste, expense or unnecessary effort. The notion of fairness, instead, is attributed to a situation that ensures equal distribution of opportunities or goods to members of a community. There is a central question in modern societies:

C. Ponsiglione · V. Roma · F. Zampella · G. Zollo (✉)
Department of Industrial Engineering, University of Naples Federico II, Italy Piazzale
Tecchio 80, Naples, Italy
e-mail: giuseppe.zollo@unina.it

© Springer International Publishing Switzerland 2015
J. Gil-Aluja et al. (eds.), *Scientific Methods for the Treatment of Uncertainty
in Social Sciences*, Advances in Intelligent Systems and Computing 377,
DOI 10.1007/978-3-319-19704-3_26

public policies oriented to fairness are compatible with an efficient use of available resources?

The prevalent answer in literature is that there is a trade-off between fairness and efficiency (Okun 1975). That is, the orientation toward fairness doesn't produce incentives to use resources efficiently, because it doesn't stimulate people to give the best of themselves (Accad 2014). Of course, searching for efficiency inevitably produces social inequality, because it rewards most able or most fortunate people. The conclusion of this viewpoint is that we must resign ourselves to a society where there is a certain degree of inequality. Inequality is the price that society must pay if it wants to stimulate efficiency, economic growth and employment (Bertsimas et al. 2012).

On the other hand, there is a growing number of macroeconomic studies (Andreas 2014; Gruen 2012; Hsieh 2012; Piketty 2013) claiming that a public policy strictly oriented to efficiency is simplistic, and that a fairness-oriented policy generates efficiency in long run. Indeed, a simplicistic efficient viewpoint does not take into account differences generated by economic and social history. Social groups and individuals inherit from the past different access conditions to available resources, regardless of capabilities of each. The diversity of starting points determines negative effects for society as a whole, as it gives rise to a waste of human capital and to an increasing polarization of social inequality. These elements in long run undermine cohesion of social system and its competitiveness.

Furthermore, there are some goods, such as health (Archad et al. 2001; Lay and Leung 2010), education (Brown 1981), security, justice, access to information, sociality that any national constitution considers primary resources to which every citizen should have equal access. Any distortion in access to these resources undermines fundamental rights enshrined in constitutional card. For these services the trade-off between equity and efficiency doesn't exist. Indeed the efficiency of public services is a prerequisite for fairness.

Our approach to the analysis of fairness/efficiency issue is based on the hypothesis that social system is a Complex Adaptive System (CAS), which develops multiple endogenous self-regulating mechanisms continually evolving over time (Sawyer 2005). Government is just one component of those endogenous self-regulating mechanisms. It tries to guide the social system towards desirable ends, by creating an ecosystem of interrelated behaviors (Colander and Kupers 2014). A policy should be designed to play a supporting role in the evolving ecosystem, not to control the system. The main claim of CAS approach is that if people, through their own wishes, do what is socially desirable, control is not needed, as people exhibit self-control. The policy maker should limit to design a norm policy (that is a payoff matrix) to influence behaviors of social actors that encourages self-reliance and concerns about others.

According to CAS viewpoint, a social system shares following characteristics: (i) systems consist of large networks of individuals; (ii) each individual follows relatively simple rules with no central control or leader; (iii) their collective action gives rise to an hard-to-predict and changing pattern of behavior; (iv) systems produce and use information and signals from both internal and external environments; (v) individuals change their behavior through learning or evolutionary processes. Summing up all the above characteristics Mitchell (2009, p. 13) proposes the following definition of a complex adaptive system: "a system in which large networks of components with no central control and simple rules of operation give rise to complex collective behavior, sophisticated information processing, and adaptation via learning or evolution".

To investigate the relationship between fairness and efficiency from the point of view of CAS, we used as reference the El Farol Bar problem created by Arthur (1994) as a device to investigate bounded rationality in economics.

## 2 Efficiency of Social Choices According to the El Farol Bar Model

In El Farol Bar model, every Thursday one hundred agents decide independently whether to go to the bar or not. Going is enjoyable only if the bar is not crowded, otherwise agents would prefer to stay at home. The bar is crowded if more than sixty people show up, whereas it is enjoyable if attendees are sixty or fewer. Every agent knows attendance figures in the past few periods and each of them has a different set of predictors, in form of functions that map the past periods' attendance figures into next week's attendance. The agent decides whether to go (GO) or stay home (STAY) following the direction of the predictor that has proven itself over time more accurate (which is called 'active predictor'). If the active predictor forecasts an attendance below the capacity he/she decides to go, otherwise he/she stays at home. Once decision has been made and real attendance is known, agents calculate the error of their predictors and update predictors' performance indexes according to the accuracy of their forecast. For example, if attendance data in previous weeks are:

$$44\ 78\ 56\ 15\ 23\ 67\ 84\ 34\ 45\ 76\ 40\ 56\ 22\ 35$$

forecasts can be formulated as in Table 1.

**Table 1** Examples of predictors, forecasts and decisions in El Farol Bar Problem

| Predictors | Forecast | Decision |
|---|---|---|
| The same value of last week | 35 | GO |
| 100 minus last week attendance | 65 | STAY |
| last four weeks average | 49 | GO |
| The same value of 2 weeks ago | 22 | GO |
| Etc... | ... | ... |

Arthur creates a pool of predictors, each of which is repeated several times with some variations. Then q of them are randomly distributed to each of 100 agents. Given initial conditions and a set of fixed predictors available to each agent, the

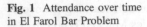 **Fig. 1** Attendance over time in El Farol Bar Problem

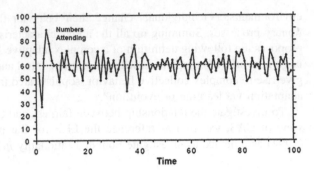

accuracy of all predictors is predetermined. Dynamics in this case is deterministic. The resulting attendance over time is illustrated in Fig. 1.

Results are interesting. Bar attendance always converges to 60: predictors self-organize in a balance, in which on average 40 % of the time active predictors provide a value of attendance greater than 60, and 60 % of the time under 60. Experiments show that these results appear robust to variations both in predictors' types and predictors' quantity assigned to each agent. On the other hand, the model shows that it is impossible to reach a perfectly coordinated state with no fluctuations. Indeed, the variation depends on the limited number of agents. If the number of agents becomes very high (about one million), the fluctuation is canceled.

The El Farol Bar model is important because it shows how agents with bounded rationality - with limited memory and computing capacity - are able to use efficiently resources available to them. Anyway, the model does not tell us if efficiency is obtained at expense of fairness.

We don't know whether agents making wrong predictions eventually give up going to the bar. If this happens then the population of agents will split into two groups: Winners, which often go to the bar, and Losers, which never go. Consequently, the social system would evolve towards an efficient but unfair use of available resources. In order to verify if the efficiency is achieved at the expense of fairness this paper analyzes four computational implementations of El Farol Bar Model, developing and testing appropriate measurements of fairness and efficiency.

## 3  NetLogo Implementations of El Farol Bar Model

The model made by Brian Arthur was subject to numerous revisions and advances (Challet et al. 2008; Farago et al. 2002). We refer exclusively to four implementations made in NetLogo, a multi-agent programmable simulation environment (Wilensky 1999), presented in Table 2.

**Table 2** Comparison between the El Farol Problem models in NetLogo

| | The original El Farol Bar problem (AG9404-Original) (a) | El Farol with weighted predictors (FW9907-Weighted) (b) | El Farol with a cyclic strategy (BSW9903-Cyclic) (c) | El Farol Attack of the coin flippers (FS9914-Random) (d) |
|---|---|---|---|---|
| Conceptualization | Arthur (1994) | Fogel (1999) | Bell and Sethares (1999) | Fogel (1999) |
| Netlogo model | Garofalo (2004) | Rand and Wilensky (2007) | Wilensky (2003) | Stenholm (2014) |
| Predictors | Ten predefined predictors, with at most 20 variants for each predictor, as in Table 1 | Predictors are vectors of past attendances with random weights | Predictors are STAY/GO cycles that vary randomly from 1 to 20 | Predictors are vectors of random weights as (b) plus a random choice |
| Learning system | Predictors are evaluated according to their predictions. The active predictor is that with best score in past evaluations | Predictors are evaluated according to their predictions. The active predictor is that with best score in past evaluations | Duration of STAY is incremented or decremented according to past bar attendances | Predictors are evaluated according to their predictions. The active predictor is that with best score in past evaluations |

## 3.1 The Original El Farol Bar Problem (AG9404-Original)

The original El Farol Bar Problem of Brian Arthur (1994) has been implemented in Netlogo by Garofalo in 2004. The agents (i.e. the customers) are set to 100 and the bar capacity is set to 60. Each agent has a memory ($m$), namely he/she knows attendance values of $m$ previous weeks; each of them has a fixed number of predictors. According to the *active predictor*, each agent predicts the number of customers at bar for next week, and if this value is less or equal of bar capacity, he/she goes to the bar.

### 3.1.1 Predictors

Netlogo implementation of El Farol Bar Problem made by Garofalo (2004) assumes that there are $P_j$ families of predictors, with j = 10. Each family has a number of variants equal to 20. Consequently, 200 predictors are created. Each agent randomly selects the fixed number of predictors among the *20* variants of the 10 families. The number 20 is also the maximum value of memory (m).

The ten predictors' families are:

- TitForTat: This family of predictors predicts next week's attendance by using the same as $i$ (from 1 to 20) weeks ago.
- Mirror: This is a family of predictors using a mirror image around 50 % of $i$ (from 1 to 20) weeks ago as prediction.
- Fixed: The fixed predictor always chooses the same attendance (5 %, 10 %, 15 %, 20 %, 25 %... or 100 % of the number of agents).
- Trend: A $i$ (from 1 to 20) dated 2 day trend applied to the last attendance.
- OppositeTrend: A $i$ (from 1 to 20) dated 2 day opposite trend applied to the last attendance.
- Trend2: A $i$ (from 1 to 20) dated 2 day (3 day spaced) trend applied to the last attendance.
- MovingAverage: A $i$ (from 1 to 20) 5 day Moving average.
- OppositeMovingAverage: A $i$ (from 1 to 20) opposite 5 day Moving average.
- Trend3: A $i$ (from 1 to 20) dated 2 day relative trend applied to the last attendance.
- OppositeTrend3: A $i$ (from 1 to 20) dated 2 day relative trend applied to the last attendance.

### 3.1.2 Learning System

To select agent's *active predictor* at each step, performance error of each predictors is considered. The performance error value is the cumulative sum of absolute differences between actual and predicted attendance over time. The error formula is the following:

$$U_t\left(P_j^i\right) = \lambda \, U_{t-1}\left(P_j^i\right) + (1-\lambda)|P_j^i(A(\eta_t)) - A(t) \tag{1}$$

where:

- $P_j^i$ is the i-th variant of the j-th predictor's family.
- $U_t(P_j^i)$ is the error at the period t of the $P_j^i$ predictor.
- $\eta_t$ is the history of past attendances until the period $t-1$ of the $m$ previous weeks, that is $\eta_t = \{A(t-m), \ldots A(t-2), A(t-1)\}$;
- $P_j^i(A(\eta_t))$ is the attendance prediction made by predictor $P_j^i$ for the period t given the history $\eta_t$;
- $|P_j^i(A(\eta_t)) - A(t)|$ is the current evaluation of predictor $P_j^i$;
- $U_{t-1}(P_j^i)$ is the error until the period $t-1$ of the $P_j^i$ predictor;
- $\lambda$ is a weight between 0 and 1: a low $\lambda$ gives more importance to the present, while a high $\lambda$ to the past.

## 3.2 El Farol with Weighted Predictors (FW9907-Weighted)

The El Farol Bar model revisited by Fogel et al. (1999) was implemented in NetLogo by Wilensky in 2007. The conceptual framework of model is same of the previous one, except for predictors and error evaluation formula.

### 3.2.1 Predictors

Each agent has a fixed number of predictors assigned randomly. Each predictor is a list of k weights (between −1 and 1) assigned to historical attendance data. Each predictor expresses his/her belief on how past attendances influence current attendance. The set of predictors is different from person to person. The history of participation is initialized at random, with values between 0 and 99. Given:

- $m$, the number of weeks for which is known the attendance;
- t, the current week (the previous weeks are $t-1$, $t-2$, etc.);
- A(t), the attendance at period t;
- $c_{ij}$, the initial inclination in the choice to go or not to the bar for the i-th agent according to his/her j-th predictor;

the j-th predictor of the i-th agent has the following form:

$$P_{ij}(t) = \left[w_{ij}^1 * A\,(t-1) + w_{ij}^2 * A\,(t-2) + \ldots + w_{ij}^m * A(t-m) + c_{ij} * 100\right] \tag{2}$$

### 3.2.2 Learning System

The i-th agent chooses the *active predictor* by calculating the cumulative error $U_t(P_{ij})$ at the period t of the j-th predictor as the sum of absolute differences between past predictions and past attendances $|P_{ij}(.) - A(.)|$, as follows:

$$U_t(P_{ij}) = |P_{ij}(t) - A(t)| + |P_{ij}(t-1) - A(t-1)| + \ldots + |P_{ij}(t-m) - A(t-m)|$$

(3)

## 3.3 El Farol with a Cyclic Strategy (BSW9903-Cyclic)

The El Farol Bar Model made by Bell and Sethares (1990) and simulated in NetLogo by Wilensky in 2003 modifies the conceptual framework used by previous models, as it introduces predictors as cycles of STAY and GO. Moreover this model considers different condition of information availability: limited information, when agents know the bar attendance only when they go; perfect information, when agents know always the bar attendance.

### 3.3.1 Predictors

In this model, agents do not make any prediction about future attendance. They don't decide whether or not to go to the bar, but they set how many days should pass before going to the bar, and so doing they set the cycle STAY/GO. Agents have initial inclinations in visiting the bar, expressed as *attendance-frequency*.

### 3.3.2 Learning System

The *attendance-frequency* remains unchanged if the attendance falls within a confidence interval called *dead-zone*. If the value of the bar attendance is greater than its capacity, defined by an *equilibrium* value, agent goes to the bar less frequently, i.e. agent decreases *attendance-frequency* by a *frequency-update* value. If the value of the bar attendance is below the critical value *equilibrium* minus *dead-zone*, i.e. bar is not crowded, agent increases *attendance-frequency* by the *frequency-update* value. The updating of *attendance-frequency* occurs at each step only when the information is perfect. Otherwise, in case of limited information, agents upgrade their cycle only when they go to the bar.

## 3.4 El Farol with a Random Choice (FS9914-Random)

Fogel et al. (1999) realized an extension of El Farol with weighted predictors (FW9907-Weighted) by including a random strategy. The model was implemented in NetLogo by Stenholm (2014).

### 3.4.1 Predictors

Predictors structure is the same of FW9907-Weighted (b). In addition there is a random strategy available for each agent that predicts the bar attendance according to a random number extracted by a uniform distribution INT with settable extremes. The agent choices the random predictor if the attendance in the previous period falls within the interval INT, otherwise the agent choices the best predictor among others. Of course, the width of the interval INT is very important, because wider the interval greater will be the probability to choice the random predictor. For example, if the interval varies from 0 to 100, agents choose always the random predictor.

### 3.4.2 Learning System

In case agent doesn't use the random predictor, he/she choices the best predictor among the others, according to their cumulate prediction errors. The error $U_t(P_{ij})$ of not random predictor $P_{ij}$ at time t is calculated with the following formula:

$$U_t(P_{ij}) = (A(t) - P_{ij}(t))^2 + U_{t-1}(P_{ij}) \tag{4}$$

where:

$U_t(P_{ij})$: prediction error of the j-th predictor of the i-th agent at time t;
A(t): value of attendance at time t;
$P_{ij}(t)$: prediction of the j-th predictor of the i-th agent at time t
$U_{t-1}(P_{ij})$: cumulative error of the j-th predictor of the i-th agent at time t − 1.

## 3.5 Discussion

For every model the attendance oscillates around the satisfaction threshold (Fig. 2). In original El Farol Bar Problem implemented by Garofalo (AG9404-Original) the attendance average (red line in Fig. 2a) stabilize around the threshold level. Even in the El Farol with weighted predictors (FW9907-Weighted) (Fig. 2b) the value of attendance reaches a balance, although the average participation at the bar is near 56 instead of 60. In El Farol with a cyclic strategy and perfect information

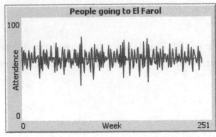

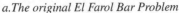

*a.The original El Farol Bar Problem*

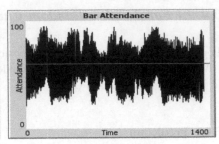

*b.El Farol with weighted predictors*

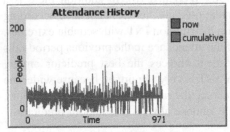

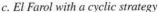

*c. El Farol with a cyclic strategy*

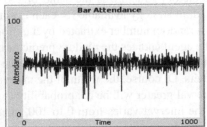

*d. El Farol with a random choice*

**Fig. 2** Attendance history of El Farol Bar models according to a simulation in NetLogo (captions and format are originals)

(BSW9903-Cyclic) (Fig. 2c) attendance oscillates with a much lower variance and the convergence to the satisfaction threshold is relatively rapid and robust. In El Farol with a random choice (FS9914-Random) (Fig. 2d), finally, equilibrium is reached when the random strategy, which happens to be the optimal one, is used by any agent.

# 4 The Fairness Issue

Above models show a lack of interest on the achievement of fairness, as equal access opportunities to the bar. To test the fairness issue we analyzed the behavior of El Farol Bar models by using a well-known measurement of iniquity, the Gini coefficient.

## 4.1 A Measurement of Iniquity: The Gini Coefficient

The Gini coefficient is an index linked to a well-known graph, the Lorenz curve (Fig. 3). In the case of El Farol Bar model, the Lorentz curve (blue line) plots the

**Fig. 3** Gini coefficient and Lorenz curve for the bar presence distribution

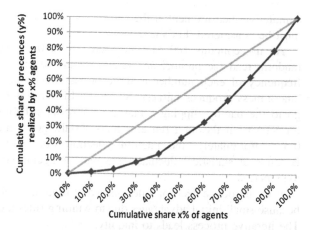

proportion of total presence of agents at bar (y axis) that is cumulatively realized by the bottom x% of agents. The line at 45 degrees (red line) represents the theoretical case of perfect fairness. The Gini coefficient is as the ratio between the area that lies between the fairness line and the Lorenz curve over the triangle area under the fairness line. This coefficient assumes a value between 0 (perfect fairness) and 1 (maximum iniquity).

The variable used to plot the Lorentz curve is the ratio $R_i$ of decision GO of i-th agent in the simulation time. The agents are sorted in crescent order according to $R_i$. Then the cumulative presence is calculated of bottom x % of agents. Finally the Gini coefficient is calculated.

Results are significantly different among models (see Table 3). The model BSW9903-Cyclic (c), in which strategies are cycles STAY/GO, is the most unfair as the value of Gini coefficient settles to the highest value among all. In El Farol with a random choice, FS9914-Random (d), instead, the trend over time of Gini coefficient suggests a fair distribution of attendance. These results are confirmed by one hundred simulations, from which the Gini coefficient is sorted at the thousandth tick.

To verify if there is a significant iniquity, a test on perfect fairness hypothesis is performed. For any model the hypothesis are rejected with a significance level of 0.05.

The results are amazing. No deliberate strategy beats purely random choices. However, this result seems unconvincing. We calculate Gini coefficient as a

**Table 3** Mean and Standard Deviation of Gini coefficient for a sample of 100 simulations for each model

| Models | Mean | Standard deviation |
|---|---|---|
| AG9404-Original (a) | 0.0852* | 0.0107 |
| FW9907-Weighted (b) | 0.2021* | 0.0301 |
| BSW9903-Cyclic (c) | 0.2840* | 0.0235 |
| FS9914-Random (d) | 0.0325* | 0.0041 |

*significant at level 0.05

cumulative measurement at the end of simulation. This fact poses a problem that must be considered. Let's take an example.

An urn contains 100 balls numbered from 0 to 100. You take out a ball at random, read the number and re-enter it in the urn. After many extractions the frequency of each number is more or less the same. That is, in long run, the randomness produces fairness. So, in the El Farol Bar model, if agent decides to go to the bar by flipping a biased coin with probability 0.60 for GO and 0.40 for STAY, in the long run all agents should go to the bar with a frequency of 0.6. Unfortunately, the random selection does not work for short periods. Because of statistical fluctuations a part of agents could remain at home for a long period.

If we introduce decision rules instead of random choice the situation gets worse, because small initial advantages due to winning rules are amplified by the iteration. The iterative process leads to iniquity.

The question is: can we consider fair a situation wherein an agent for many weeks does not access to resources? Consider how dramatic would be this situation if the resource has a vital importance, such as work, health, food. In other words, we believe that fairness must be assessed for short periods. To consider this aspect, we develop new accurate set of measurements.

## 4.2  More Accurate Measurements of Efficiency and Fairness

### 4.2.1  Fair Quantity (FQ) and Fair Period (FP)

Consider the case of a soup kitchen that can distribute only K meals a day for a population of N poor, where K < N. It is clear that every day N - K people do not eat. To achieve fairness in meals distribution several days are necessary, hoping that there will be a poor rotation. What is the minimum number of days necessary to verify the fairness? If K = 60 and N = 100, every poor can expect to receive three meals in five days. The fraction 3/5 is the irreducible fraction of 60/100, obtained by simplifying numerator and denominator by 20. The fraction 3/5 gives us all the information we need. The numerator 3 tells us the Fair Quantity (FQ) of meals per person, and the denominator 5 tells us the Fair Period (FP) needed to get the Fair Quantity (FQ).

It is interesting to note that if K/N is equal to 50/100 the fraction FQ/FP becomes 1/2. That is, it takes only two days to reach a fair distribution of resources. While with 80/100 we have FQ/FP equal to 4/5, and 5 days are necessary to get a fair situation. In short, the duration of the Fair Period does not depend on the amount of resources, but only on the ratio between resources and customers.

The first important implication of this approach is that the fairness requires a minimum period of time to be evaluated. The second important difference is related to what is measured. It is necessary to distinguish between the decision to go

(GO) and the possibility to use the resource (GET). If an agent decides to go to the bar and he/she finds that the bar has already saturated the available seats, he/she does not use the resource. Exceeding customers are chosen at random among the agents who decided to go.

Considering those concepts of Fair Quantity and Fair Period is possible to construct two measures of Efficiency and Fairness.

### 4.2.2 Measurement of Efficiency in the Fair Period

The system is efficient when all resources are used in the Fair Period FP. Given K available resources in each time unit, and a Fair Period FP, the model will be efficient when the number of resources used in one Fair Period is K * FP. For example, if K/N = 60/100 then FP = 5. Consequently, the system is efficient in 5 time units if 60 * 5 = 300 resources are used.

If Y are resources used the Fair Period, we can define an inefficiency index as:

$$\text{Inefficiency Index} = (K * FP) - Y \tag{5}$$

The maximum value of inefficiency for a Fair Period is when resources used are equal to zero, that is:

$$\text{Maximum inefficiency} = (K * FP) \tag{6}$$

Finally, an Efficiency Index for a Fair Period varying between 0 and 1 is defined as follows:

$$\text{Efficiency Index} = 1 - \frac{((K * FP) - Y)}{(K * FP)} = \frac{Y}{(K * FP)} \tag{7}$$

### 4.2.3 Measurement of Fairness in the Fair Period

According to fairness definition, a system is fair when all agents use the same amount of resources.

Considering $X_i$ the amount of resources used by the i-th agent in the Fair Period FP, the index of iniquity is equal to the standard deviation from FQ:

$$\text{Iniquity index} = \sqrt{\frac{\sum_i^N (x_i - FQ)^2}{N}} \tag{8}$$

The maximum value of the iniquity index occurs when K agents use all available resources in the Fair Period FP, while (N−K) agents use nothing. Thus, the index of maximum iniquity (MII) is equal to:

**Table 4** Results of the statistical tests on Efficiency and Fairness

|  | Efficiency index | | Fairness index | |
|---|---|---|---|---|
|  | Mean | Standard dev. | Mean | Standard dev. |
| AG9404-Original (a) | 0.9352* | 0.001142 | 0.5280* | 0.001451 |
| FW9907-Weighted (b) | 0.8422* | 0.003410 | 0.4483* | 0.002125 |
| BSW9903-Cyclic (c) | 0.7834* | 0.013502 | 0.5515* | 0.006822 |
| FS9914-Random (d) | 0.9835* | 0.000691 | 0.2942* | 0.001162 |

*significant at level 0.05

$$\text{MII} = \sqrt{\frac{K * (FP - FQ)^2 + (N - K) * (0 - FQ)^2}{N}} = \sqrt{((FQ * FP) - FQ^2)} \qquad (9)$$

Finally, a Fairness Index for a Fair Period varying between 0 and 1 is defined as follows:

$$\text{Fairness Index} = 1 - \frac{Iniquity\ index}{Maximum\ Iniquity\ Index} \qquad (10)$$

#### 4.2.4 Application of New Measurements of Efficiency and Fairness

To evaluate *Efficiency and Fairness indices* new simulations are performed. The Efficiency Index is high for every model. Instead, the Fairness Index shows a low value for every model (Table 4). It is important to highlight that the Fairness Index is low also for the El Farol Model with a random choice (d).

### 4.3 A Time-Box View of Fairness

Above considerations tell us how to define appropriate measurements for the Fair Period. But a suitable measurement for periods of different length could be defined? An appropriate fairness situation is when the fairness value is high for time windows of variable width. To evaluate this property, the above indices are modified as follows, where the Fair Period is the minimum window and a coefficient 'a' is a positive integer that widens the time box:

$$\text{Time-box Iniquity index (TII)} = \sqrt{\frac{\sum_i^N (x_i - (a * FQ))^2}{N}} \qquad (11)$$

$$\text{Maximum Time-box Iniquity (MTI)} = \sqrt{a^2 * ((FQ * FP) - FQ^2)} \qquad (12)$$

$$\text{Time-box Fairness Index (TFI)} = 1 - \frac{Fractal\ Iniquity\ index}{Maximum\ Fractal\ Iniquity} \qquad (13)$$

For example, suppose that we want to calculate the Time-box Fairness Index (TFI) for three different time windows, with five agents ($N = 5$) and three resources ($K = 3$). In this simple case $FQ = 3$ and $FP = 5$ (Table 5).

By using the above formulas for different time windows, we obtain:

- For $a = 1$ and for the first window ($t_1$ to $t_5$):
  Amount of resources for the Agent $ag_i = [5,2,2,3,3]$

$$TII = \sqrt{\frac{((5-3)^2 + (2-3)^2 + (2-3)^2 + (3-3)^2 + (3-3)^2}{5}} = 1.09$$

$$MTI = \sqrt{((3*5) - 3^2)} = 2.45$$

$$TFI = 1 - \frac{1.09}{2.45} = 1 - 0.44 = 0.56$$

For the second and third window TFI is equal to 0.40
- For $a = 2$ and for the first window ($t_1$ to $t_{10}$):
  Amount of resources for the Agent $ag_i = [7,3,4,4,7]$

$$TII = \sqrt{\frac{((7-6)^2 + (3-6)^2 + (4-6)^2 + (4-6)^2 + (7-6)^2}{5}} = 1.84$$

$$MTI = \sqrt{((6*10) - 6^2)} = 4.90$$

$$TFI = 1 - \frac{1.84}{4.90} = 1 - 0.37 = 0.63$$

For the second windows FFI is equal to 0.52
- For $a = 3$ ($t_1$ to $t_{15}$) it easy to calculate that TFI is equal to 0.68:

It is easy to verify that the average value of TFIs for each period increases as the time window grows, passing from 0.48 to 0.58 and to 0.68.

In order to verify this behavior for any El Farol Bar model a new set of generative experiments was developed. Figure 4 shows TFI values for time windows of different widths. It clearly shows that small windows present lower values of TFI.

It's important to note that El Farol Bar model with a random choice (model d), which presented a low value of Gini coefficient, for small windows shows an TFI lower than other models, while in long run is almost perfectly fair.

Those results are confirmed by one hundred simulations, from which values of the TFI are calculated (Table 6).

**Table 5** The Time-box Fairness Index (TFI) for time windows of different width

| Time | Agents | | | | | Time-box fairness index | | |
|---|---|---|---|---|---|---|---|---|
| | ag1 | ag2 | ag3 | ag4 | ag5 | a = 1 | a = 2 | a = 3 |
| 1 | 1 | 1 | 1 | 0 | 0 | 0.56 | 0.63 | 0.68 |
| 2 | 1 | 1 | 1 | 0 | 0 | | | |
| 3 | 1 | 0 | 0 | 1 | 1 | | | |
| 4 | 1 | 0 | 0 | 1 | 1 | | | |
| 5 | 1 | 0 | 0 | 1 | 1 | | | |
| 6 | 1 | 0 | 1 | 0 | 1 | 0.4 | | |
| 7 | 1 | 0 | 0 | 1 | 1 | | | |
| 8 | 0 | 0 | 1 | 0 | 0 | | 0.52 | |
| 9 | 0 | 0 | 0 | 0 | 1 | | | |
| 10 | 0 | 1 | 0 | 0 | 1 | | | |
| 11 | 0 | 1 | 0 | 1 | 0 | 0.4 | | |
| 12 | 1 | 0 | 1 | 0 | 0 | | | |
| 13 | 0 | 0 | 1 | 0 | 0 | | | |
| 14 | 0 | 0 | 1 | 1 | 1 | | | |
| 15 | 0 | 1 | 1 | 1 | 0 | | | |
| Average fractal fairness index | | | | | | 0.45 | 0.58 | 0.68 |

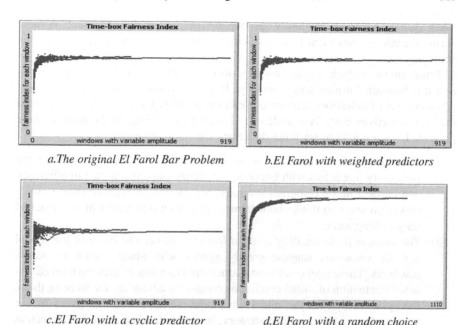

a.The original El Farol Bar Problem        b.El Farol with weighted predictors

c.El Farol with a cyclic predictor        d.El Farol with a random choice

**Fig. 4** The time-box fairness index

**Table 6** Results of the statistical tests on the Time-box Fairness Index

|  | TFI1** | | TFI2*** | |
|---|---|---|---|---|
|  | Mean | Standard deviation | Mean | Standard deviation |
| AG9404-Original | 0.5339* | 0.0001904 | 0.7877* | 0.0006350 |
| FW9907-Weighted | 0.4485* | 0.0011696 | 0.5669* | 0.002116 |
| BSW9903-Cyclic | 0.5427* | 0.0003841 | 0.6343* | 0.0006916 |
| FS9914-Random | 0.2917* | 0.0000108 | 0.9468* | 0.0012673 |

*significant at level 0.05
**TFI1: values calculated on a time window equal to FP
***TFI2: values calculated on a time window of maximum width

## 5 Conclusions and Guidelines for Future Research

The analysis carried out on the four El Farol Bar models shows that agents with bounded rationality, who decide on the basis of experience, are able to reach an acceptable level of efficiency in the use of available resources. Indeed, agents are able to select a set of strategies that are complementary. In so doing they develop an ecology of efficient behaviors. However the attainment of efficiency does not guarantee a fair use of resources. In the long run there seems to be some fairness when choices are made randomly. In fact, the fairness is only apparent. A more accurate analysis shows that models of efficient behavior fail to achieve fairness in short periods.

From above analysis we are not able to draw any conclusion about the relationship between fairness and efficiency. In other words we are not able to say whether a set of behaviors that pursues efficiency can reach fairness, or if between the two objectives there is a trade-off. However, according to the analysis here presented, we are able to set three directions for future research:

(i) *The fairness is a short run issue.* Any model that seeks to explore the possibility that actors with bounded rationality can reach a state of efficiency and fairness in the resources use must address the problem of short term. It makes no sense to think about fairness as a promised land which is reached only in long term.

(ii) *The fairness is the result of a social learning game.* The El Farol Bar model and its variations suggest greedy agents who always want to access resources. These agents have no disincentive to mitigate their own greed. As small fluctuation of initial conditions creates an advantage for some of them, the system amplifies the initial advantage, which encourages them to use more and more their winning strategy. In contrast, when an initial fluctuation creates a disadvantage for an agent, and this has no chance to learn because the limitation of owned strategies, then the system triggers them to surrender. This dynamics creates two groups: the winners and losers. It is clear that a social system that pursues an objective of fairness must avoid the formation of these two groups. To do so, it is necessary to develop a system of incentives that avoids both the formation of greedy agents and defeatist agents. To this end attention should be given to the mechanism of agent's learning.

(iii) *The fairness is the result of a public policy.* In light of described results it seems clear that agents left to themselves have no incentive to the pursuit of fairness. It is therefore necessary a public policy that defines the system of incentives and disincentives to guide actors' behavior. In terms of modeling this fact means to build an adequate pay-off matrix for the decisions GO and STAY in the case of a undercrowded bar and overcrowded one. It is our belief that an adequate pay-off matrix would push agents to achieve behaviors suitable to obtain both efficiency and fairness.

This will be the aim of future research.

# 6  NetLogo Models

NetLogo models are available at the following web address: https://sites.google.com/site/equitaedefficienza/home.

# References

Accad, M.: L. Equity-Efficiency 1: evolution and robinhood economics. http://malayangdiskusyon. blogspot.it/2014/02/equity-efficiency-1-evolution-and.html. Accessed 10 Feb 2014

Andreas, J.: When is equity efficient? Usually. http://medianism.org/2014/03/05/when-is-equity-efficient-usually/. Accessed 5 March 2014

Archad, L., Le Grand, J., Sassi, F.: Equity versus efficiency: a dilemma for the NHS. BMJ, 323 (2001)

Arthur, B.W.: Inductive reasoning and bounded rationality. Am. Econ. Rev. **84** (1994)

Bell A.M., Sethares W.A.: The El Farol problem and the internet: Congestion and Coordination Failure. Dissertation, Boston (1999)

Bertsimas, D., Farias, V.F., Trichakis, N.: On the efficiency-fairness trade-off. Manag. Sci. **58** (2012)

Brown, W.J.: Education Finance in Canada. Canadian Teachers' Federation, Ottawa (1981)

Challet, D., Marsili, M., Ottino, G.: Shedding light on El Farol. Phys. A: Stat. Mech. Appl. **332** (2008)

Colander, D., Kupers, R.: Complexity and the Art of Public Policy. Solving Society's Problems from Bottom Up. Princeton University Press, Princeton (2014)

Farago, J., Greenald, A., Hall, K.: Fair and efficient solutions to the Santa Fe bar problem. In: Proceeding of the Grace Hopper Celebration of Women in Computing ACM (2002)

Fogel, D.B., Chellapilla, K., Angeline P.J.: Inductive reasoning and bounded rationality reconsidered. IEEE Trans. Evol. Comput. **3** (1999)

Garofalo, M.: Modeling the El Farol bar problem in NetLogo. Preliminary Draft (2004)

Gruen, N.: Where equity and efficiency thrive together: can you propose some more exmples?. http://clubtroppo.com.au/2012/11/24/where-equity-and-efficiency-thrive-together-can-you-propose-some-more-examples/. Accessed 24 Nov 2012

Hsieh, C., Hurst E., Jones, C.I., Klenow P.J.: The Allocation of Talent and U.S. Economic Growth. unpublished paper (2012)

Lay, T.Y.Y., Leung, G.M.: Equity and efficiency in healthcare: are they mutually exclusive? Hong Kong Med. J. **16** (2010)

Mitchell, M.: Complexity A Guided Tour. Oxford University Press, New York (2011)

Okun, A.: The Big Trade-off, Underlined edn. Brookings Inst Pr, Washington, DC (1975)

Piketty, T.: Le capital au XXIe siècle. SEUIL, Paris (2013)

Rand, W., Wilensky, U.: NetLogo El Farol model. http://ccl.northwestern.edu/netlogo/models/ElFarol (2007)

Sawyer, R.K.: Social Emergence: Societies as Complex Systems. Cambridge University Press, Cambridge (2005)

Stenholm, R.: NetLogo El Farol Attack of the coin flippers. http://ccl.northwestern.edu/netlogo/models/community/El%20Farol%20Attack%20of%20the%20coin%20flippers. Accessed 24 Jan 2014

Wilensky, U.: NetLogo. http://ccl.northwestern.edu/netlogo/ (1999)

Wilensky, U.: NetLogo El Farol Network Congestion model. http://ccl.northwestern.edu/netlogo/models/ElFarolNetworkCongestion (2003)

# Part VI
# Neural Networks and Genetic Algorithms

# Part VI
## Neural Networks and Genetic Algorithms

# Comparative Analysis Between Sustainable Index and Non-sustainable Index with Genetic Algorithms: Application to OECD Countries

Martha-del-Pilar Rodríguez-García, Klender Cortez-Alejandro and Alma-Berenice Méndez-Sáenz

**Abstract** This study analyses the differences in financial portfolio metrics between sustainable index and non-sustainable firms in the market index through the use of the portfolio theory and genetic algorithms from 2007 to 2013. The sample consists in 926 firms of four regions (1) Europe: Germany, Austria, Denmark, Spain, Finland, Italy, Norway, Sweden and United Kingdom, (2) Asia: Japan, (3) America: Canada, United States of America and Mexico and (4) Oceania: Australia. To measure the performance of the portfolio two classical metrics: Jensen's alpha and Sharpe ratio were considered. We also calculate a conditional metric that measures the number of times the return of a given portfolio exceeds the average market return. The goal is to find a portfolio that maximizes these three metrics using a weighted ratio and compare the results between the sustainable and non-sustainable portfolios. Due to a nonlinear programming problem, we use genetic algorithms to obtain the optimal portfolio. The results show a better performance in sustainable portfolios in eight countries, although the amount of countries increases if only the conditional metric is considered.

**Keywords** Genetic algorithm · Jensen's alpha · Sustainable index · Portfolio theory

## 1 Introduction

Corporate social responsibility (CSR) has shown strong growth in the last two decades. Also, has witnessed an astounding ascendancy and global resonance in recent years. This has created pressures on companies to adopt voluntary CSR

M.-d.-P. Rodríguez García (✉) · K. Cortez-Alejandro · A.-B. Méndez-Sáenz
Facultad de Contaduría Pública y Administración, Universidad Autónoma
de Nuevo León, Manuel L. Barragan y Pedro de Alba, Ciudad Universitaria,
San Nicolás de los Garza, NL, Mexico
e-mail: marthadelpilar2000@yahoo.com

© Springer International Publishing Switzerland 2015                         331
J. Gil-Aluja et al. (eds.), *Scientific Methods for the Treatment of Uncertainty
in Social Sciences*, Advances in Intelligent Systems and Computing 377,
DOI 10.1007/978-3-319-19704-3_27

policies, and consequently, the notion of "doing well by doing good" is moving incrementally into mainstream business practice (Long 2008). The differences of CSR implementation within the strategy of the companies are mainly from cultural aspects. In countries with poor socioeconomic conditions, they are likely to focus only on making short-term profits to survive rather than being socially responsible and taking morality into account like developed countries (Azmat and Ha 2013).

There exist ethical or sustainable indexes that serve as a benchmark for the investors who want to consider the sustainable practices. There have been studies focused on fund performance, the usage of ethical criteria, relationship between corporate disclosures, financial performance of ethical indexes. This last topic is the one that is going to be studied in this paper by making a comparison between the Sustainable Index (SI) and the firms in the Market Index not included in the Sustainable Index (NSI) of OECD countries. The complexity of this research is that the countries' sample is wide and therefore to find a pattern of sustainable indexes against non-sustainable indexes could be different due to issues such as culture, socio demography, legal, etc.

This study contributes to existing research in two ways. First, we study CSR by different regions and countries: (1) Europe: Germany, Austria, Denmark, Spain, Finland, Italy, Norway, Sweden and United Kingdom (UK), (2) Asia: Japan, (3) America: Canada, United States of America (USA) and Mexico and (4) Oceania: Australia. For this, we explored the legal, cultural, and socioeconomic divergences and convergences. Second, we analyzed the relations between sustainable index portfolios and financial performance.

## 2 RSC by Regions

In a particular context, in US and Europe, Matten and Moon (2008) suggest that convergence of CSR has largely been driven by global institutional pressures, while divergence is catalysed by differences in national business systems. For instance, employee health plans represent an explicit form of US firms' CSR policies, programs, and strategies (Jamali and Neville 2011).

European countries have a longstanding tradition of integrating economic and social policies. Matten and Moon (2008) mentioned that pertinent social obligations are seen in Europe as the purview of government. Taking employee relations as an example, stringent European regulations and protections "implicitly" determine business organizations level of CSR.

Based on Steurer et al. (2012), the high levels of activity in UK and Ireland can be explained historically with the origins of CSR as a liberal concept under right-wing governments, and the similarly high levels of activity in Sweden, Denmark, Finland and Norway. In Addition, CSR in UK is dominated by market logic; to illustrate this, the State encourages the private provision of welfare, social benefits are modest, often means-tested, and stigmatising (Steurer et al. 2012).

Other examples come from Austria and Germany where CSR has been characterized by granting social rights considers existing class, and status differentials (with a focus on work-related, insurance-based benefits), redistributive effects are limited, social policies aim to preserve traditional family structures (Steurer et al. 2012). Finally, Italy and Spain are fragmented and 'clientelistic' support focusing on income maintenance (pensions), still under development, making older systems of social support (family, church) indispensable (Steurer et al. 2012). Similarly, France has a mandatory law according to which companies with more than 300 employees must draft social responsibility reports.

Codes of conduct, associated inspections and audits in Asia, has made CSR a common practice but in most cases had failed. In these mentioned companies, labour issues and rights of workers are seen as the most important aspects and benefits of CSR, including risk reduction, staff recruitment/retention, cost savings, and building good relationship within stakeholders (Gorostidi and Xiao 2013).

The findings in Australia argued that CSR is becoming increasingly popular, which is reflected in CRS-related studies conducted by commercial firms and professional bodies (Truscott et al. 2009). In addition, a variety of indices have been developed to evaluate companies' performance in social and environmental issues.

Welford (2004) mentioned that high degree of philanthropy, pay special attention to child labour, and local community in USA and Canada. In USA, universal health care financed through taxation represents implicit CSR that is contained within the institutional environment of European institutions (Jamali and Neville 2011). For example, employee relations and business organizations level of CSR is less regulated in US context, it allows firms the freedom to "explicitly" design CSR policies and standards in this domain. Finally, Matten and Moon (2008) argue that US' style of CSR is characterized by less regulation and more incentive and opportunity for business organizations to fill social niches.

CSR' history in Latin America has identified several social issues which are important to address, like inequality and poverty alleviation, corruption, working health and safe conditions, inadequate public services, environmental degradation, etc. (Casanova and Dumas 2010). Other general trends like relatively weak private sector activity and a weak involvement and CSR promotion by the governments were also identified in this entire region. In other hand, Casanova and Dumas (2010) mentioned that business leaders associations and foundations have facilitated the adoption of CSR policies in Mexico, focusing mainly in philanthropy.

# 3 Literature Review

Since the early 1990s, the SRI industry has experienced strong growth in the US, Europe, and the rest of the world. An important factor behind this growth was ethical consumerism, where consumers pay a premium for products that are consistent with their personal values. Remembering Markowitz (1952), the process of selecting a portfolio may be starts with the relevant beliefs about future performances and ends with the choice of portfolio, this is, it should contain the "expected returns-variance of returns" rule. The preferences of SRI can be represented by a conditional multi-attribute utility function, in the sense that they appear to derive utility from being exposed to the SR attribute (Bollen 2007).

Derwall et al. (2005) found that socially responsible investments produce a significant increase in stock performance. Bollen (2007) compared the relation between annual fund flows and lagged performance in SR funds with a matched sample of conventional funds. For the 1980 through 2002 period, SR investors exhibit a significantly larger response to positive returns than investors in conventional funds, but a smaller response to negative returns than investors in conventional funds. Otherwise, Statman and Glushkov (2009) found that the portfolios generated with the best qualified firms in the dimensions of CSR outperform portfolios with lower-rated shares in the dimension of CSR. In Herzel et al. (2011) used as sample the S&P500 and the Domini 400 Social Index (DSI) from 1992–2008 and found that sustainable portfolios generate better measures performance than non-sustainable portfolios.

While other analysis concluded that portfolios with responsible stocks have lower financial performance than traditional (Hamilton et al. 1993). Hamilton et al. (1993) used estimations of Jensen's Alpha to examine the risk-adjusted return of all funds of socially responsible investment contained in the database Lipper from December 1990 and found that socially responsible funds tend to show a similar or lower performance than unrestricted funds on a risk-adjusted basis.

Moreover, several studies as Schroder (2007) and Becchetti and Ciciretti (2009) have concluded that there is no significant statistical difference between socially responsible companies that the ones that do not perform this type of investment. For example, Schroder (2007) analyzed 29 SRI stock indexes and found that these rates do not lead to a significant improvement or underperform compared to their reference's indices. In the research of Becchetti and Ciciretti (2009) estimate the daily returns from 1990 to December 2003 for two portfolios one that includes a socially responsible companies from Domini 400 Social Index (DSI400) developed by Kinder, Lydenberg and Domini (KLD) have on average lower returns and unconditional variance than and another with non-responsible companies.

## 4 Theoretical Framework

The Portfolio Selection Theory established by Markowitz in 1952 is the foundation of the Modern Portfolio Theory (MPT) which was later expanded by Sharpe in 1964 with the Capital Asset Pricing Model (Mangram 2013). The MPT refers to an investment framework for the construction and selection of portfolios that will give the highest level of portfolio return minimizing its risk. The mean-variance fundamental theorem of a portfolio proposed by Markowitz (1952) maximizes an expected return maintaining a given variance or minimizes a variance maintaining a given return. If every possible combination of the investments is presented in a graph of risks and returns, the points define a region limited by an Efficient Frontier (EF) (Markowitz 1952).

In this research we will use the MPT as a method to find the combination of shares that maximize the risk- return tradeoff of a portfolio and determine whether the risk- return ratio is different between the responsible and non-responsible firms. The equation to evaluate a portfolio performance is expressed as follows:

$$\overline{R}_p = \sum_{i=1}^{n} x_i \overline{R}_i / \sum_{i=1}^{n} x_i \tag{1}$$

where:

$x_i$ = Weight of each share in the Portfolio.
$\overline{R}_i$ = Expected return of each share.
$\overline{R}_p$ = Portfolio's expect return.

While the variance of the portfolio is given by the following equation:

$$\sigma_p^2 = \sum_{j=1}^{N} \sum_{j=1}^{N} x_j x_k \sigma_{j,k} \tag{2}$$

where:

$x_j, x_k$ = Represent the weights within the portfolio of shares $j$, $k$.
$N$ = Number of firms in the portfolio.
$\sigma_{j,k}$ = Covariances $j,k$.
$\sigma_p^2$ = Variance of the portfolio.
$\sigma_{j,k} = \sigma_j^2 = \sigma_k^2$, if $j = k$.

Early studies used a variety of techniques to evaluate the portfolio such as Sharpe ratio (Sharpe 1966) which is a risk- adjusted portfolio performance and Jensen's alpha, which uses beta as the proper measure of risk (Jensen 1968). These two metrics are described below.

Jensen's alpha is a measure of an abnormal, the main purpose is to compare the performance of a certain portfolio (or index) to a benchmark portfolio in this case the degree to which the expected return of the responsible portfolio exceeds the performance of the market portfolio, which is expressed in the following equation:

$$\alpha = R_p - [R_F + \beta(R_m - R_f)] \tag{3}$$

where:

$\alpha$ = Jensen's alpha
$R_p$ = Portfolio's expect return
$R_f$ = Return on risk-free rate
$R_m$ = Market portfolio's return

Sharpe ratio considers the return gained relative to the risk assumed. This ratio can be used to express how well the return of an asset compensates the assumed risk by an investor and is define as:

$$S = \frac{R_P - R_f}{\sigma_P} \tag{4}$$

where:

$R_p$ = Portfolio's expect return
$R_f$ = Return on risk-free rate
$\sigma_P$ = Expected risk measured as portfolio's standard deviation

If $R_f$ is risk free then the variance of the returns is zero, this is the excess of standard deviation is the same as the share returns. Each share will have different risk-return profiles or different Sharpe Ratios.

Another metric considered to evaluate a portfolio's performance is the number of times the return of a given portfolio exceeds the average market return. We denote this metric as Total Profit (TP) and it is given as follows:

$$TP = \left(\frac{1}{N}\right) \sum_{t=1}^{N} v_t \tag{5}$$

where:
N = is the number of observations.

$$V_t = \begin{cases} 1 & \text{si} \ (R_P - R_M)_t > 0 \\ 0 & \text{si} \ (R_P - R_M)_t \leq 0 \end{cases}$$

# 5 Genetic Algorithm

The Genetic Algorithms (GA) is a stochastic optimization algorithms based in the biological concepts of genetic re-combination, genetic mutation, and natural selection (Lai and Li 2008). They were developed by Holland (1975) and Michalewicz (1996). Genetic algorithms are considered a valid approach to solve many complex problems in finance such as optimization of trading rules, forecasting returns, portfolio optimization, among others. There are several researches that have used GA as a method to solve a portfolio optimization such as Lai and Li (2008) among others.

The GA start with an initial population randomly generated with a constant number of chromosomes and in a portfolio's optimization problem each chromosome represents the weight of each stock. Then the selection, crossover and mutation operation occurred until the population converges to containing only chromosomes with good fitness (Krishnasamy 2011).The basic steps to create a GA are:

Step 1: Initialize with a randomly generated population.
Step 2: Evaluate each chromosome.
Step 3: Apply elitist selection: carry on the best individuals to the next generation.
Step 4: Replace the current population by the new population.
Step 5: If the termination condition is satisfied then stops, if not go to Step 2.

There may be some problems in using the GA for optimization portfolios, while in theory this optimization process is true; in practice there are some problems due to the assumption made by Holland that the population is infinite (El-Mihoub et al. 2006). In practice a finite population is used and the interaction of the "genes" is very small which affect the sampling ability and therefore its performance (El-Mihoub et al. 2006). The use of a local search method, which introduces new "genes", can help to combat the problem cause by the accumulation of stochastic errors derived from a finite population and also can improve the results of the optimization.

## 6  Sample

Two portfolios were designed for each of the fourteen countries country, one for the sustainable firms (SF) and another for the non-sustainable firms (NSF). To do this, we use the Stock Exchange (SE) database for each country. The sample is composed of those firms that have been considered within SI and the NSI during 2007 to 2013 (see Table 1). Of the total sample, the 35 % correspond to sustainable firms and 65 % to the non-sustainable. In total there are 926 firms. The countries with a higher number of companies are Canada, Australia and Italy and with the lowest Germany, Denmark and Mexico. The country with the higher number of SF is UK and the lowest is Denmark. In Japan, the number of SF is equal that the NSF.

For the purposes of this study, we considered the monthly prices for each firm and we obtained our data from yahoo finance and google finance. For the risk-free rate, we used interest rates from each month that were reported by the Central Banks of each country. To determinate the sustainable index portfolio we use the firms within the maket index that are consider in the sustainable index provided mainly by NASDAQ, FTSE, EIRIS. This stock selection considers the firms with the leadership role in sustainability performance. To select the non-sustainable index we used the firms within the index market that are not in the sustainable index.

To build the Efficient Frontier we used monthly price data to calculate prices yields and standard deviation of each portfolio. Once we constructed the EF of investment portfolios, two per country, one for SI and one for the NSI, we estimate the three metrics to evaluate the portfolios' performance indicated in Eqs. (3), (4)

**Table 1** Sample of OECD
countries by Regions

| Country | NSF | SF | Total |
|---|---|---|---|
| **Europe** | | | |
| Austria | 8 | 12 | 20 |
| Denmark | 3 | 15 | 18 |
| Finland | 4 | 18 | 22 |
| Germany | 7 | 10 | 17 |
| Italy | 117 | 38 | 155 |
| Norway | 17 | 4 | 21 |
| Spain | 12 | 16 | 28 |
| Sweden | 7 | 18 | 25 |
| UK | 17 | 68 | 85 |
| **Asia** | | | |
| Japan | 70 | 71 | 141 |
| **Oceania** | | | |
| Australia | 136 | 27 | 163 |
| **America** | | | |
| Canada | 178 | 6 | 184 |
| Mexico | 5 | 13 | 18 |
| USA | 14 | 15 | 29 |
| Overall sample | *595* | *331* | *926* |

and (5). The idea is to find the shares' weight combination within the portfolio that maximizes these three metrics. As we have three goals, one for each metric, creating a function that combines the three objectives into one is required. For this, a weighted ratio (WR) which represents the weighted average of the three metrics was used.

This leads to a nonlinear programming problem, since the function we want to maximize includes metric calculated from a linear regression (Jensen's alpha) as well as the metric determinant of Eq. (5). In addition it was considered as the main constraint to take into the extent possible results with statistically significant metric of Jensen's alpha values. The use of linear programming is why we opted for genetic algorithms proposed by Holland (1975) which has been used to evaluate investment portfolios. The optimization problem is defined as follows:

$$Max: \frac{R_p - [R_F + \beta(R_m - R_f)]}{3} + \frac{R_P - R_f}{3\sigma_P} + \frac{\sum_{t=1}^{N} v_t}{3N} \tag{6}$$

$$s.t$$
$$p - value(\alpha) \le 0.1$$
$$\sum_{i=1}^{N} x_i = 1$$
$$0 \le x_i \le 1$$

# 7 Software

To determinate the weights of each firm we use Evolver 6 of Palisade Decision Tools which has the advantage of applying genetic algorithms together with the mathematical optimization generator *OptQuest*[1] to find optimal solutions to a specific problem. However due to the complexity of the sample and the large amount of time needed we decided to apply just the genetic algorithm method of Evolver that did not have a lot of difference with the *OptQuest* in the first results. Taking into account the programming problem, the same weights of each firm were considered as initial values. Also we set the stopping parameter in a threshold of 0.00001 % of change in the weights; i.e., if the target cell is not modified into a 0.00001 % then the algorithm stops, the above order reduce the optimization time.

The Evolver default parameters were considered. The population size indicates how many organisms (or complete groups of variables) are stored in memory at all times. Palisade (2013) notes that there is much debate and research regarding the optimal population size for different problems, but is recommended between 30 and 100; in our case we use the value of 80. The crossover rate reflects the likelihood that future scenarios or "organisms" contain a mixture of information from the previous generation of original organisms; this parameter sets to a value of 0.50, i.e., a descendant organism contains about 50 % of its equity of a "parent" and the rest of the values of the other "parent." Rate mutation reflects the likelihood that future scenarios contain some values random; for this research a value of 0.1 was considered.

# 8 Results

The EF of the countries show a polygon of feasible combinations, i.e. outside this area, it is not possible to achieve a risk-return trade-off. To trace the upper boundary of the polygon we solved the following linear programming problem as follows:

$$Min \quad \sigma_p^2$$
$$s.t : \overline{R}_p = r; \quad 0 \le x \le 1; \quad \sum_{i=1}^{T} x_i \tag{7}$$

where: $r = r_j^{min}, \ldots, r_k^{max}, r_j^{min}$ = stock price value with minimum return and $r_k^{max}$ = stock price value with maximum return.

---

[1]*OptQuest* is a registered product of *OptTek Systems*. *OptQuest* integrate metaheuristics methods such as tabu search procedures, artificial neural networks and scatter search on a single composite method. Since this generator is based on scatter search for combinations and apply local search methods, Martí and Laguna (2002) point that can be considered included in the so-called Lamarckian algorithms.

On the other hand, the inner border of the polygon is obtained by joining the points where the entire investment is placed in a particular stock portfolio. Only three countries: Sweden, UK and Denmark for any given level of expected risk have a higher expected return on sustainable index compared with the non-sustainable index considering the same risk (Fig. 1).

In contrast, four countries have better performance in non-sustainable index than the sustainable index: Norway, Finland, Spain and Japan (see Fig. 2).

However there are cases where the EF curves intersect such as Australia, Austria, Canada, Germany and US where the SI is better than MI to a level of risk, then changes inversely (see Fig. 3). The opposite situation occurs in Mexico and Italy where at first the SI has lower return than MI given a certain level of risk and then this situation reverses (see Fig. 4).

As a result of the GA optimization, Table 2 shows the average time needed for each country optimization portfolio.

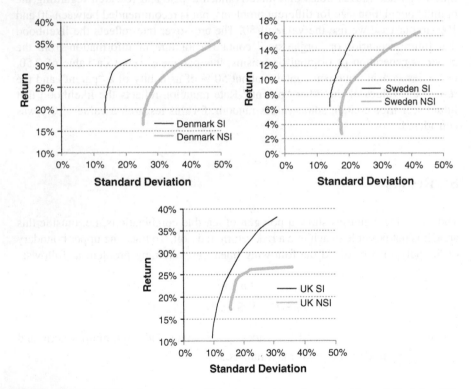

**Fig. 1** Efficient frontiers for countries with best performance in sustainable portfolio index (SI)

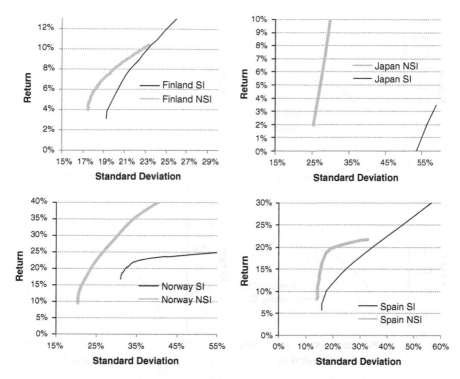

**Fig. 2** Efficient frontiers for countries with best performance in non sustainable portfolio in the market index (MI)

In Table 3 we present the risk (standard deviation) and return of the portfolios considering the multi objective maximization of Eq. (6). These results suggest no apparent difference between sustainable (SI) and non sustainable index portfolios (NSI) since 50 % of the countries have higher expected returns in sustainable portfolios and 50 % have less risk.

Nevertheless to determine the risk-return trade-off of the portfolios we analyzed the performance metrics discussed in the theoretical framework. Table 4 shows each of the metrics used to measure portfolios' performance by region and country as a result of the GA optimization. First, considering the percentage that the portfolio exceeds the market index, we find that the SI beats the NSI in 71 % of the countries. In contrast, focusing in this metric only in Austria, Finland, Norway and Japan the SI performance is lower than the NSI.

Secondly, taking into account the inclusion of the risk in the performance metrics the results change. SI performance is better than NSI in only 43 % of the countries due to the systematic risk (Beta) in Jensen's Alpha and total risk (standard deviation) in Sharpe Ratio.

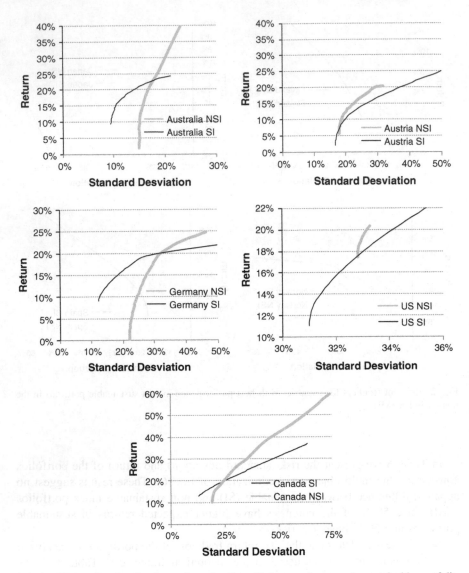

**Fig. 3** Efficient frontiers intersects for countries with best performance in sustainable portfolio index (SI) given a low risk

Finally, in order to determine the multi objective optimization of the three metrics the weighted ratio of Eq. (6) was considered. The weighted ratio shows definitely that the best performing index through the GA optimization is the sustainable firms in the market index (SI) in 57 % of the countries sample, i.e. Austria Denmark, Italy, Sweden, Spain, Australia, Canada and Mexico. The countries with no evidence of higher performance in SI than NSI are Finland, Germany, Norway, UK, Japan and USA.

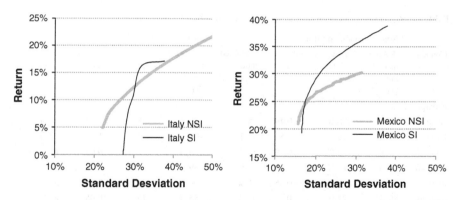

**Fig. 4** Efficient frontiers intersects for countries with best performance in non sustainable portfolio in the market index (MI) given a low risk

**Table 2** Time average for optimization

|                        | Non sustainable portfolio | Sustainable portfolio |
|------------------------|---------------------------|-----------------------|
| Average number of test | 2,177                     | 3,628                 |
| Valid test             | 2,046                     | 3,450                 |
| Average time           | 0:03:58                   | 0:08:00               |

**Table 3** Risk-return in non sustainable vs sustainable portfolios

| Country   | Non sustainable portfolio | | Sustainable portfolio | |
|-----------|-------|--------------------|-------|--------------------|
|           | Media | Standard deviation | Media | Standard deviation |
| **Europe** |       |                    |       |                    |
| Austria   | 22.51 % | 41.73 % | 38.01 %  | 77.11 %  |
| Denmark   | 22.45 % | 37.76 % | 30.67 %  | 26.45 %  |
| Finland   | 9.84 %  | 27.53 % | 3.11 %   | 36.93 %  |
| Germany   | 18.80 % | 29.01 % | 13.41 %  | 24.70 %  |
| Italy     | −8.24 % | 29.32 % | 11.68 %  | 32.58 %  |
| Norway    | 25.38 % | 26.20 % | 23.45 %  | 47.39 %  |
| Spain     | 16.88 % | 18.15 % | 112.62 % | 226.02 % |
| Sweden    | 6.15 %  | 26.20 % | 12.46 %  | 25.17 %  |
| UK        | 22.07 % | 13.35 % | 17.71 %  | 25.56 %  |
| **Asia**  |       |                    |       |                    |
| Japan     | 6.89 %  | 34.28 % | 4.98 %   | 30.40 %  |
| **Oceania** |     |                    |       |                    |
| Australia | 28.34 % | 59.85 % | 13.32 %  | 14.46 %  |
| **America** |     |                    |       |                    |
| Canada    | 14.97 % | 30.73 % | 14.56 %  | 15.44 %  |
| Mexico    | 23.11 % | 18.69 % | 31.45 %  | 17.81 %  |
| USA       | 15.59 % | 12.71 % | 15.73 %  | 19.73 %  |

**Table 4** Risk-return ratios in non sustainable vs sustainable portfolios

| Country | Non sustainable portfolio | | | | Sustainable portfolio | | | |
|---|---|---|---|---|---|---|---|---|
| | % Above market index (TP) | Jensen´s alpha | sharpe ratio | Weighted ratio | % Above market index (TP) | Jensen´s alpha | Sharpe ratio | Weighted ratio |
| **Europe** | | | | | | | | |
| Austria | 89 % | 28 % | 0.49 | 56 % | 76 % | 48 % | 0.47 | 57 % |
| Denmark | 81 % | 16 % | 0.53 | 50 % | 100 % | 25 % | 1.06 | 77 % |
| Finland | 72 % | 9 % | 0.29 | 37 % | 46 % | 3 % | 0.03 | 17 % |
| Germany | 81 % | 15 % | 0.58 | 51 % | 89 % | 9 % | 0.47 | 48 % |
| Italy | 53 % | 5 % | −0.35 | 8 % | 100 % | 26 % | 0.30 | 52 % |
| Norway | 85 % | 21 % | 0.90 | 65 % | 85 % | 15 % | 0.45 | 48 % |
| Spain | 85 % | 18 % | 0.82 | 62 % | 100 % | 105 % | 0.49 | 85 % |
| Sweden | 61 % | 3 % | 0.14 | 26 % | 100 % | 9 % | 0.42 | 50 % |
| UK | 99 % | 20 % | 1.52 | 90 % | 100 % | 16 % | 0.62 | 59 % |
| **Asia** | | | | | | | | |
| Japan | 99 % | 7 % | 0.20 | 42 % | 61 % | 5 % | 0.16 | 27 % |
| **Oceania** | | | | | | | | |
| Australia | 86 % | 37 % | 0.41 | 55 % | 97 % | 13 % | 0.65 | 58 % |
| **America** | | | | | | | | |
| Canada | 79 % | 27 % | 0.45 | 50 % | 90 % | 21 % | 0.86 | 66 % |
| Mexico | 85 % | 16 % | 0.95 | 65 % | 97 % | 24 % | 1.47 | 90 % |
| USA | 86 % | 13 % | 1.19 | 73 % | 92 % | 12 % | 0.77 | 60 % |

# 9 Conclusions

The findings of this study are varied and can be used to observe international trends. Investors make their decisions fundamentally based on the risk-return tradeoff. Although for Bollen (2007) the investor faces a utility function where in addition to pursuing a good relationship of the risk-return performance also has an incentive to invest in firms with responsible, ethical and environmental principles. These ideas are reflected in the experienced growth as responsible investment US increased 76 % in 2014 the SRI and Europe 91 % between 2011 and 2013.

On the other hand, the firms are increasingly active in creating and marketing targeted products for sustainable investors. In Long (2008) mentioned that the firms search a CSR management to enter a global market. Every day cultural, social, legal differences etc. not differentiate tends of financial markets on liability because the rules of the stock exchange are similar.

Regarding the portfolio performance situation generally we find that if we only consider the performance, the SI financial performance is better in 71 % than the NSI, but when we include the risk measured by beta or standard deviation for Jensen's Alpha and Sharpe Ratio the results reversed that is the NSI beat the SI in 67 %, this is due to the visibility of companies within a sustainability index.

Companies within the SI should be careful in their risk's policies. If a firm left the SI the press prejudices their image. Therefore, we consider important to use a weighted metric of all measures, and with almost 60 % SI is better than the NSI. This is a reflection of the importance that investors have given to firms that keep attention in practices from the perspective of Corporate Governance, Environmental and Social.

Perform schemes for SMEs in favor to entering the Stock Exchange markets as a way to be more visible for consumers and investors in each country. Their strategy should be careful in risk variable, for example control the fiscal weaknesses, environment and management via the corporate government. The growth of social responsible investments will continue in the large companies sector. These efforts help to create a national framework in which environmental, social and governance considerations in investing are able to become the norm.

# References

Azmat, F., Ha, H.: Corporate social responsibility, customer trust, and loyalty-perspectives from a developing country. Thunderbird Int. Bus. Rev. **55**(3), 253–270 (2013)

Becchetti, L., Ciciretti, R.: Corporate social responsibility and stock market performance. Appl. Finan. Econ. Taylor and Francis J. **19**(16), 1283–1293 (2009)

Bollen, N.: Mutual fund attributes and investor behavior. J. Finan. Quan. Anal. **42**, 683–708 (2007)

Casanova, L., Dumas, A.: Corporate social responsibility and Latin American multinationals: Is poverty a business Issue? Universia Bus. Rev. **25**, 132–145 (2010)

Derwall, J., Guenster, N., Bauer, R., Koedijk, K.: The Eco-Efficiency Premium Puzzle. Finan. Anal. J. **61**(2), 51–63 (2005)

El-Mihoub, T., Hopgood, A., Nolle, L., Battersby, A.: Hybrid genetic algorithms: a review. Eng. Lett. **13**(2), 124–137 (2006)

Gorostidi, H., Xiao, Z.: Review on contextual corporate social responsibility: towards a balanced financial and socially responsible performance. Int. J. Hum. Soc. Sci. **3**(18), 102–114 (2013)

Hamilton, S., Jo, H., Statman, M.: Doing well while doing good? The investment performance of socially responsible mutual funds. Finan. Anal. J. **49**, 62–66 (1993)

Herzel, S., Nicolosi, M.: A socially responsible portfolio selection strategy. Sustainable Investment Research Platform, Working Paper 11-09 (2011)

Holland, J.: Adaptation in Natural and Artificial Systems: An Introductory Analysis with Applications to Biology, Control and Artificial Intelligence. University of Michigan Press, Michigan (1975)

Jamali, D., Neville, B.: Convergence versus divergence of CSR in developing countries: an embedded multi-layered institutional lens. J. Bus. Ethics **102**, 599–621 (2011)

Jensen, M.: The performance of mutual funds in the period 1945–1964. J. Finan. **23**(2), 389–415 (1968)

Krishnasamy, V.: Genetic algorithm for solving optimal power flow problem with UPFC. Int. J, Softw. Eng Appl. **5**(1), 39 49 (2011)

Lai, S., Li, H.: The performance evaluation for fund of funds by comparing asset allocation management of mean- variance model or genetic algorithms to that of fund managers. Appl. Finan. Econ. **18**, 483–499 (2008)

Long, J.: From Cocoa to CSR: finding sustainability in a cup of hot chocolate. Thunderbird Int. Bus. Rev. **50**(5), 315–320 (2008)

Mangram, M.: A simplified perspective of the Markowitz Portfolio theory. Global J. Bus. Res. **7** (1), 59–70 (2013)

Markowitz, H.: Portfolio selection. J. Finan. **7**(1), 77–91 (1952)

Martí, R., Laguna, M.: Búsqueda dispersa. 1er Congreso español sobre algoritmos evolutivos y bioinspirados. Mérida (España), 302–307 (2002)

Matten, D., Moon, J.: Implicit and explicit CSR: a conceptual framework for a comparative understanding of corporate social responsibility. Acad. Manag. Rev. **33**(2), 404–424 (2008)

Michalewicz, Z.: Genetic Algorithms + Data Structures = Evolution Programs. Springer, Berlin (1996)

Palisade: User's guide Evolver, the genetic algorithm solver for Microsoft Excel, Version 6. Palisade Corporation, Ithaca (2013)

Schroder, M.: Is there a difference? The performance characteristics of SRI Equity Indices. J. Bus. Finan. Acc. **34**(1), 331–348 (2007)

Statman, M., Glushkov, D.: The wages of social responsibility. Finan. Anal. J. **65**(4), 33–46 (2009)

Steurer, R., Martinuzzi, A., Margula, S.: Public policies on CSR in Europe: themes, instruments, and regional differences. Corp. Soc. Responsib. Environ. Manag. **19**, 206–227 (2012)

Sharpe, W.: Mutual fund performance. J. Bus. **39**, 119–138 (1966)

Truscott, R., Barrtlett, J., Tywoniak, S.: The reputation of corporate social responsibility industry in Australia. Australas. Mark. J. **17**(2), 84–91 (2009)

Welford, R.: Corporate social responsibility in Europe, North America and Asia. J. Corp. Citizensh. **17**, 33–52 (2004)

# Sovereign Bond Spreads and Economic Variables of European Countries Under the Analysis of Self-organizing Maps

Antonio Terceño-Gómez, Lisana B. Martinez,
M. Teresa Sorrosal-Forradellas and M. Belén Guercio

**Abstract** This paper presents an empirical analysis related to sovereign bond spreads and a set of economic variables since 1999 until 2013 for a sample of European countries. The analysis is carried out using an original tool in the financial literature: Self-Organizing Maps. This representation is able to cluster countries-years according to the similarities between their main macroeconomic fundamentals. We find interesting groups of countries and we relate them with their level of sovereign bond spreads. The results reflect the incidence of the last financial crisis over the economies and the effect over the eurozone.

**Keywords** European countries · Sovereign bond spreads · Self-organizing maps · Financial crisis

A. Terceño-Gómez (✉) · M.T. Sorrosal-Forradellas
Department of Business Management Faculty of Business and Economics, Universitat Rovira
i Virgili, Av. Universitat, 1, 43204 Reus, Spain
e-mail: antonio.terceno@urv.cat

M.T. Sorrosal-Forradellas
e-mail: mariateresa.sorrosal@urv.cat

L.B. Martinez · M. Belén Guercio
Universidad Nacional del Sur, National Scientific and Technical Research Council
(CONICET), Institute of Social and Economic Researchs of South (IIESS), Universidad
Provincial del Sudoeste, 12 de Octubre 1198 – 3° Piso (B8000CPB), Bahía Blanca, Argentina
e-mail: lbmartinez@iiess-conicet.gob.ar

M. Belén Guercio
e-mail: mbguercio@iiess-conicet.gob.ar

© Springer International Publishing Switzerland 2015
J. Gil-Aluja et al. (eds.), *Scientific Methods for the Treatment of Uncertainty
in Social Sciences*, Advances in Intelligent Systems and Computing 377,
DOI 10.1007/978-3-319-19704-3_28

347

# 1 Introduction

During the last years, some European countries showed specific changes over their economies and financial markets. Since the inception of the 2008 financial crisis, sovereign bond spreads in European countries increased and some economic variables changed their tendencies. The global financial crisis modifies investors' perception of risk and the diversification of investment portfolios propagates market risks; more than enough reasons to study financial crisis consequences, contagion effects and bond spreads determinants.

According to the literature related to financial crisis there are many theories that attempt to explain its meaning, the main channels of propagation and the importance of several factors interacting in the market (investors, political, commercial and financial issues).

Eichengreen et al. (1996) consider the contagion effect as a situation in which it is known that there is a crisis in one place and it increases the chances that occur in another country. Other authors (Kaminsky and Schmukler 2002; Basu 2002) specify the concept and the phenomenon that for reasons not so obvious in time, such as currency devaluation or the announcement of default of sovereign debt obligations, triggering a series of immediate and subsequent events among countries in the same region and in some cases, beyond a certain region.

Therefore, European countries suffered severe consequences in their economies and financial system as a result of the contagion effect. So, the aim of this paper is to analyze the evolution and coincidences of a set of sovereign bond spreads and economic variables of European countries, considering the incidence of the euro currency and the impact of the 2008 financial crisis. The group of explanatory variables included was chosen bearing in mind the literature related to the determinants of sovereign bond spreads.

This paper contributes to the literature in different aspects. First, we expand empirical studies by analyzing the evolution of economics variables and sovereign bond spreads during a period of 15 years and for 17 European countries, which allows to identify key variables sensitive to the financial crisis. Second, it is possible to identify the incidence of the euro currency over the sample of countries. Third, the methodology applied is novel and endeavours to the goal of the research. And fourthly, we analyze the European market, where the most visible effects have taken place and which allow us to identify the main variables affected by the financial crisis. The paper is organized as follows. Section 2 presents a brief review of the literature on sovereign bond spreads and the incidence of the last financial crisis over European economies. Section 3 describes the data used in the paper. Section 4 sets out the empirical results. Finally, Sect. 4 contains the main conclusions.

## 2 Literature Review

Since the last financial crisis and the followed contagion effect over the whole of the economies, European countries become more unstable and the single currency becomes more vulnerable. The extent of the crisis forced European policy makers to undertake rigorous fiscal measures, to inject large amounts of money into financial institutions (Grammatikos and Vermeulen 2011) and to set rescue packages for economies such as Greece, Ireland, Portugal and to some extent to Spain.

De la Dehesa (2009) identifies some of the effects of the crisis on the real economy caused by various channels. One is the derivative of the wealth effect, since increasing the prices of financial assets, owners are considered richer and tend to increase consumption. But the crisis has declined considerably the financial wealth, which was a straight drop in the level of consumption. The second channel is derivative from the called financial accelerator, which shows how the wealth effect can be amplified by other financial channels. If families have less wealth, their solvency is lower, which leads to increase the cost of external financing, further reducing consumption capacity and amplifying the economic downturn. The third channel mentioned is the derivative of the solvency of major financial institutions, through a reduction in lending activity and the consequent improvement in capital ratios.

In this context, numerous articles focus their attention over the main determinants of sovereign bond spreads in the euro area. The first of these determinants is credit risk, which includes default risk, downgrade risk and credit spread risk. During the crisis, debt and deficit indicators increased. Governments had greater difficulties in coping with larger debt and deficit. As a consequence, the market's perception of default changed. This situation led to a decrease in rating qualifications of these economies and to an increase in the credit risk spread.

The second determinant is liquidity risk. Liquid markets allow investors to make decisions at any time. Consequently, the number of financial operations should be considered to determine the size and depth of the market, which in turn determines the liquidity premium level. Liquidity risk and credit risk are interconnected (Barrios et al. 2009; Arghyrou and Kontonikas 2011). On the one hand, if a government increases its bonds supply, the pressure on the liquidity premium decreases. On the other hand, a high supply is associated with an increase in public debt and deficit, which increases credit risk premium.

Finally, the third determinant of sovereign bond spreads is risk aversion. Bond spreads are affected by the amount of risk that investors are willing to bear when they invest in financial markets. Hence, an increase in the risk perception of the economy will increase the bond spread. Furthermore, according to Barrios et al. (2009), the combination of high risk aversion and large current account deficits tend to magnify the incidence of deteriorated public finances on government bond yield spreads.

Moreover, a common and prevailing view of the literature of euro government bond indices is that spreads are driven by a common global factor (Codogno et al. 2003; Geyer et al. 2004; Barrios et al. 2009; Manganelli and Wolswijk 2009;

Sgherri and Zoli 2009) represented by international factors, such as risk perception. Sgherri and Zoli (2009) suggest that the euro area sovereign risk premium tends to commove over time in relation to the global risk and that sovereign spreads in the euro area have been increasing in the same direction influenced by debt sensitivity. They point out that liquidity of sovereign bond markets play a significant, although limited, role in explaining spreads.

## 3  Methodology and Data

Our database comprises seventeen European countries: European Monetary Union (EMU) countries, except Luxemburg (Austria, Belgium, Finland, France, Germany, Greece, Ireland, Italy, Netherlands, Portugal and Spain), and non-EMU countries (Czech Republic, Denmark, Hungary, Poland, Sweden and United Kingdom).

Data is obtained from DataStream, for the period 1999 to 2013 measured at the end of each year. We use the Government Bond Index (GBI) calculated by JPMorgan to calculate sovereign bond spreads. We select the GBI with annual frequency that represents government bonds with a maturity between seven and ten years. Germany is our benchmark to estimate bond spreads.

There are some exceptions with respect to the period included of three countries, given the unavailability of data. The time series of Poland and Czech Republic begin in 2000, Hungary time series starts in 2003 and Belgium in the 2001 year.

As has been commented previously, the group of economic variables included was chosen bearing in mind the literature related to the determinants of sovereign bond spreads (Martinez et al. 2013; Min et al. 2003; Gande and Parsley 2003; Hilscher and Nosbusch 2010; Baldacci et al. 2011; González Rozada and Levy-Yeyati 2006; Arora and Cerisola, 2000; Eichengreen and Mody 1998). These sets of variables are: Unemployment rate, Stock Market Index, Inflation, Gross Domestic Product and main aggregates and International Reserves.

With the available data, we work with a total of 247 patterns. Each pattern represents a country in a particular year. They are defined as vectors of 6 components or variables. The first variable is the unemployment rate. The second one is the annual profitability of the most representative equity index in each country, calculated as the difference between the index price in two consecutive years (t and t − 1), and divided by the price in year t. The third component is the percentage of the general government gross debt over the Gross Domestic Product. Variable number four is the level of inflation. The fifth variable is the relative increment of Gross Domestic Product (at current prices). Finally, the last component is the percentage of change of international reserves.

SOM network is implemented in Matlab using the toolbox developed by the Laboratory of Information and Computer Science in the Helsinki University of Technology.

The input layer of the SOM, in all cases considered, consists of 6 units. Each of them contains the value of one of the variables that define the patterns.

The implementation of the SOM network for the whole sample generates as output a map of 12 rows and 7 columns. Patterns are distributed in the map according to its similarity. That is, if two patterns are near in the map, the country-year that they represent share a common macroeconomic situation. As the distance between two patterns increases more different is the economical situation between them. In order to make groups of patterns with similar characteristics, we apply in a first step a dual criteria objective function, which minimizes the number of groups and maximizes the homogeneity of the patterns within each group. We obtain 10 different groups. However, we prioritize the interpretative value of the groups and in a second step we have reduced the number of groups. We add one new criterion in the SOM that allow us to group the patterns in only 5 groups. We analyze the results in the next section.

# 4  Results

We consider all the economic variables of each country during the whole period of analysis and we compute the results applying SOM. Figure 1 shows the location of the patterns in the Kohonen map and the grouping related to this analysis. Each

| | | | | | | |
|---|---|---|---|---|---|---|
| PO00 PO05<br>PO01 PO06<br>PO04 PO07 | FN00<br>ES06 | CZ03 HN05<br>CZ04 HN06<br>CZ05 SD99 | DK00 PT99<br>DK05 NL99 | IR03 IR05<br>IR04 | NL00 IR02<br>NL01 IR99 | CZ01 PT00<br>HN03 IR00<br>HN04 IR01 |
| | PO08<br>ES05 | | DK06<br>IR06 | CZ06<br>CZ07 | UK04 | CZ00 DK99 UK00<br>CZ02 UK99 |
| PO02 ES99<br>PO03 ES00<br>GR00 ES01 ES02 | FN99 ES04<br>FR00<br>ES03 | (1) | FN04 | SD06 FN07<br>SD07 NL06<br>SD10 ES07 | DK04 UK07<br>UK05<br>UK06 | DK03 UK01<br>HN07<br>SD00 |
| | FR99<br>BD99 | FN05<br>FN06 | SD05 | SD04<br>NL07 | DK07 | DK01 UK03<br>DK02 NL02<br>UK02 PT01 PT02 |
| GR99 GR04<br>GR01 GR05<br>GR03 IT99 | | FR03 BD04<br>FR05 BD05<br>BD03 BD06 | FN10 PT07<br>PT05<br>PT06 | PT04 | SD03 NL05<br>OE00<br>NL04 | SD02<br>NL03 |
| BG03 IT04<br>BG04 GR06 | BG05<br>BG06 | BD07 | OE03<br>OE04  (3) | OE99 OE02<br>OE01 | (2) | SD01 FN03 BD00<br>FN01 FR01 BD01<br>FN02 FR02 BD02<br>PT03 |
| BG01 GR07 IT02<br>BG02 IT00<br>GR02 IT01 | IT05 IT06 | FR04 FR07 | HN10 | DK10 OE06<br>OE05 NL10 | UK08<br>NL08 | FR08<br>BD08 |
| IT03 | BG07<br>IT07 | HN08 FR06<br>UK10 BD10 | PO10 FN11<br>PO11 | (4) | FN08 | CZ08 PO09 SD11<br>CZ11 PO12 OE08<br>DK08 SD08 ES08<br>IR07 |
| BG08 GR08<br>FR11 | BG11<br>IT08 | HN11 FR10<br>BG10 BD11 | | DK12 PO13<br>DK13 | SD12 | DK11 FN12<br>OE07 OE11 |
| HN12 IT11<br>IT09 IT12 | UK12<br>IT10 | HN13 BG12<br>UK11 BG13<br>UK13 | BD12 PT08<br>BD13 | CZ12 OE12 NL13<br>OE09 OE13<br>OE10 FN13 | CZ10 NL11 | CZ13 NL12 |
| GR10 PT12<br>GR11 IR10<br>PT11 IR11 | IT13 | IT13  (5) | FR12 | | CZ09<br>SD13 | |
| GR12 ES13<br>GR13 IR12<br>ES12 IR13 | PT13 ES10 | ES11 | BG09 GR09<br>FR09 PT10 | BD09 IR09<br>ES09 | DK09 SD09 | HN09 NL09<br>UK09 PT09<br>FN09 IR08 |

**Fig. 1** Kohonen Map

pattern has named with two letters that identify the country and two numbers corresponding to the year (Austria: OE, Belgium: BG, Finland: FN, France: FR, Germany: BD, Greece: GR, Ireland: IR, Italy: IT, Netherlands: NL, Portugal: PT, Spain: ES, Czech Republic: CZ, Denmark: DK, Hungary: HN, Poland: PO, Sweden: SD and United Kingdom: UK). There are five groups to be interpreted.

In order to study the behavior of each country-year pattern related with its position in the map, it is necessary to evaluate the value of each variable in each area. This information is showed in Fig. 2, through the scale values represented by colors in the numeric scale on the right of each map. In all cases, the dark blue color indicates the minimum values of each variable, while the red indicates the highest ones.

Nonetheless, it is important to note that the best expected value for some variables is clearly the opposite for others. For example, the best situation for a country is represented by lower level of unemployment, public debt over GDP and inflation, so considering the components maps scale colors the best situation is represented by the color blue. On the other side, for stock market indices, GDP growth and International Reserves is desirable a high level; therefore the best color in the scale is the red one.

Briefly, we can summarize the characteristics of each group as follows. Group 1 is representative of an economic growth situation but with some imbalance that is shown in a high level of unemployment. Group 2 is the most favorable one in terms of growth and economic balance although the stock markets behave in a moderate

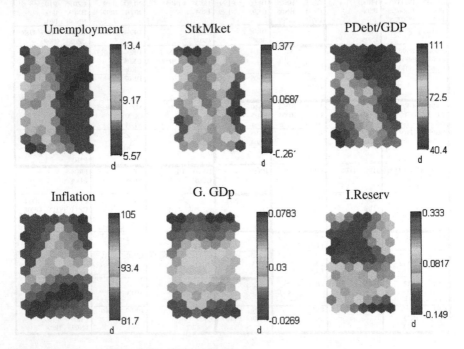

**Fig. 2** Components Maps (color figure online)

way. In group number 3, the economy starts to go down (variation of GDP decays, the percentage of public debt increases), but inflation and unemployment is still medium. Group 4 is characterized by low values of GDP, high levels of inflation and negative profitability in the stock markets. Nevertheless, some positive signals (low levels of unemployment and a low government debt) imply that in this group we find countries, as we will see, just before the crisis or during the crisis if the country was strong enough. Finally, group 5 corresponds to the weakest economic situation and it can be associated with the worst period of financial crisis.

Results from the map show the financial crisis effect on all European countries. Most of the patterns belonging to groups number 4 and 5 represent countries from year 2008 to 2013 considering the exception of the year 2010, when some countries improved their economic situation and are situated in the group 2 and 3 (Sweden, UK, Finland, Hungary and Germany).

In general, countries not belonging to the Monetary Union are relatively less affected by the financial crisis. Inside this set of countries it is possible to distinguish two subgroups: Sweden, Denmark and UK, remained jointly in the same groups until 2009. During the year 2010 these countries recovered their economy situation and moved to different groups. Since that year, Sweden and Denmark move to group 4 and UK goes to group 5, which indicates a deeper effect of the contagion over this last country.

Moreover, Poland, Czech Republic and Hungary present similar behaviors related to their economic variables. Even so, Czech Republic presents best improvements before the financial crisis and it suffered faint contagion effect. Poland shows smooth affectation before and after the turmoil. Hungary is the most affected economy of this set of countries, moving to the group 5 in 2011.

Related to the UME countries, they follow similar patterns associated with the inception of the financial crisis. Previous years were in general of high growth (group 1 and 2) or moderated (group 3). From the contagion effect of the 2008 financial crisis, almost all of them moved to group 4 and 5, although suffering clearly different intensified effects.

Belgium, Greece and Italy are countries that followed a sequence of moderate growth and deep fall since the beginning of the crisis. Another group of countries that present similar characteristics between them is formed by Spain, Ireland and Portugal. The evolution of these countries since 2008 is homogenous, move since group number 2 to number 4 and 5 progressively.

Austria, Finland, Germany and Netherlands are the countries less affected by the financial contagion. Patterns from these countries never move to the group 5, with the exception of Germany in 2011.

Related to the start of the euro currency in the European Union, and considering the previous analysis, nearly all the countries maintained their economic situation and only three of them, Austria, France and Germany, moved from group 2 to group 3; converging with Belgium, Greece and Italy in group 3 (great increase in sovereign debt). This situation goes on to the start of the financial crisis in 2008.

Although Netherlands belongs to the Monetary Union presents similar evolution to the group of Denmark, Sweden and UK.

**Table 1** Bonds spreads levels in each country according to the SOM groups

| C/Y | 1999 | 2000 | 2001 | 2002 | 2003 | 2004 | 2005 | 2006 | 2007 | 2008 | 2009 | 2010 | 2011 | 2012 | 2013 |
|---|---|---|---|---|---|---|---|---|---|---|---|---|---|---|---|
| Czech Rep | | | 2,06/2 | -0,04/2 | -0,49/1 | -0,03/1 | 0,50/1 | 0,23/2 | -0,22/2 | 0,28/4 | 1,15/4 | 0,60/4 | 0,92/4 | 1,73/4 | 0,67/4 |
| Denmark | 0,12/2 | 0,13/2 | 0,18/2 | 0,04/2 | 0,02/2 | -0,09/2 | 0,10/2 | -0,01/2 | -0,03/2 | 0,10/4 | 0,39/4 | 0,30/4 | 0,19/4 | -0,20/4 | -0,06/4 |
| Hungary | | | | | 2,09/2 | 3,69/2 | 3,73/1 | 3,66/1 | 2,95/2 | 2,85/3 | 5,91/4 | 4,73/3 | 5,21/5 | 8,07/5 | 4,95/5 |
| Poland | | 3,95/1 | 5,87/1 | 3,65/1 | 0,99/1 | 1,91/1 | 2,33/1 | 1,82/1 | 1,25/1 | 1,67/1 | 2,46/4 | 2,91/4 | 3,19/4 | 4,13/4 | 2,45/4 |
| Sweden | -0,22/1 | 0,12/2 | -0,37/2 | 0,11/2 | 0,13/2 | -0,01/2 | 0,26/2 | 0,04/2 | -0,14/2 | 0,02/4 | -0,55/4 | -0,02/2 | 0,41/4 | -0,20/4 | 0,38/4 |
| UK | 0,10/2 | 0,03/2 | -0,20/2 | -0,13/2 | -0,13/2 | 0,20/2 | 1,06/2 | 0,86/2 | 0,89/2 | 0,30/4 | 0,13/4 | 0,68/3 | 0,45/5 | 0,08/5 | 0,45/5 |
| Austria | -0,30/2 | -0,15/2 | -0,01/2 | -0,10/2 | -0,40/3 | -0,33/3 | 0,03 | 0,00 | 0,02 | 0,08/4 | 0,72/4 | 0,32/4 | 0,37/4 | 1,18/4 | 0,44/4 |
| Belgium | -0,33 | -0,22 | 0,04/3 | -0,10/3 | -0,31/3 | -0,36/3 | 0,02/3 | 0,03 | 0,04 | 0,14/5 | 0,81/5 | 0,23/5 | 1,02/5 | 2,21/5 | 0,74/5 |
| Finland | -0,40/1 | -0,16/1 | -0,07/2 | -0,10/2 | -0,43/2 | -0,44/2 | -0,02/1 | -0,04/1 | -0,01/2 | 0,08/4 | 0,56/4 | 0,18/3 | 0,27/4 | 0,59/4 | 0,18/4 |
| France | -0,52/1 | -0,28/1 | -0,18/2 | -0,22/2 | -0,40/3 | -0,43/3 | -0,01/3 | 0,01/3 | 0,03/3 | 0,07/4 | 0,35/5 | 0,15/5 | 0,31/5 | 1,25/5 | 0,57/5 |
| Germany | /1 | /2 | /2 | /2 | /3 | /3 | /3 | /3 | /3 | /4 | /4 | /3 | /5 | /4 | /4 |
| Greece | 2,58/3 | 0,74/1 | 0,31/3 | 0,03/3 | -0,22/3 | -0,28/3 | 0,14/3 | 0,18/3 | 0,21/3 | 0,27/5 | 2,22/5 | 2,41/5 | 10,09/5 | 36,55/5 | 39,69/5 |
| Italy | -0,27/3 | -0,06/3 | 0,11/3 | -0,07/3 | -0,28/3 | -0,30/3 | 0,07/3 | 0,17/3 | 0,21/3 | 0,23/5 | 1,32/5 | 0,49/5 | 1,75/5 | 4,88/5 | 3,05/5 |
| Netherlands | -0,02/2 | 0,06/2 | 0,08/2 | 0,05/2 | -0,04/2 | -0,09/2 | 0,02/2 | -0,01/2 | 0,02/2 | 0,08/4 | 0,56/4 | 0,16/4 | 0,18/4 | 0,43/4 | 0,22/4 |
| Portugal | -0,27 | -0,21/2 | 0,06/2 | -0,09/2 | -0,32/2 | -0,36/2 | 0,02/3 | 0,08/3 | 0,13/3 | 0,18/4 | 1,00/4 | 0,68/5 | 3,68/5 | 12,48/5 | 5,53/5 |
| Spain | -0,39/1 | -0,12/1 | -0,02/1 | -0,10/1 | -0,36/1 | -0,41/1 | -0,02/1 | -0,03/1 | 0,01/2 | 0,09/4 | 0,78/4 | 0,51/5 | 2,47/5 | 3,22/5 | 3,94/5 |
| Irland | -0,39/2 | -1,69/2 | 0,25/2 | -0,10/2 | 0,74/2 | -0,33/2 | -0,01/2 | -0,08/2 | 0,01/4 | 0,08/4 | 1,20/4 | 1,47/5 | 5,96/5 | 6,52/5 | 3,38/5 |

Note: Bond spreads value/SOM group

In order to compare earlier results with bonds spreads of each country, we present yearly spreads levels in Table 1, with their respective references to identify the belonging group of the Kohonen Map. Additionally, in Table 2 we summarize the mean of spreads in each group, distinguishing between EMU countries and non-EMU countries.

**Table 2** Mean group spreads

| | | |
|---|---|---|
| **Group 1**: 0,127 | *Non EMU*: 2,087 | *EMU*: −0,142 |
| **Group 2**: 0,973 | *Non EMU*: 0,389 | *EMU*: −0,135 |
| **Group 3**: 0,237 | *Non EMU*: 2,754 | *EMU*: 0,021 |
| **Group 4**: 0,707 | *Non EMU*: 1,104 | *EMU*: 0,352 |
| **Group 5**: 4,384 | *Non EMU*: 3,202 | *EMU*: 4,587 |

Note: Mean of spreads considering GBI index

As could be interpreted, it is possible to assert that groups 1 and 2 assemble negative spreads means, related with economic growth. In group 1 all countries present negative spreads. Greece is an exception in the year 2000. This country present a continuous instability since that period, worsen during the financial crisis.

The group 2 is extensive and gathers many countries and years, so there are great spreads dispersion, between 1.06 and −1.685. Moreover, there is an interesting aspect to highlight. The 71 % of the total of positive spreads are data from non-EMU countries, while EMU countries only represent the 29 %. The main reason that justified negative spreads for EMU members, differently than non-EMU countries, is the convergence related with the beginning of the common currency, since data in this group coincide with years after 2002.

Group number 3, is formed by countries-years with positive spreads. Nonetheless, the mean of spreads in this group is very low (0,237) and correspond to moderate growth situations. In this context, appear some financial problems. All values are nearly cero, positives or negatives, except Greece in 1999 that shows a spread level of 2.58.

Respect to the group 4, the mean of spreads is moderate (0.707) but in this case hardly all the variables correspond to the pre-crisis period and positive spreads represent the 90 % of the group. Non EMU countries are the only which present negative spreads. This result is a consequence of the homogenization of the euro countries, even in this case when effects are negatives.

Moreover, related to group 5, the spreads mean is elevated, according to the wide variety of countries in this subsample. This group includes countries with high sovereign bonds spreads, such as: Greece, Portugal and Ireland and others countries as Spain and Italy with lower levels, but all of them have in common a deep impact of the financial crisis.

# 5 Conclusions

We apply a novel technique in finance, Self Organizing Maps, in order to analyze the relationship between economic variables and sovereign bond spreads, considering the possible incidence of the 2008 financial crisis and the introduction of the common currency. Specifically, we consider a set of 17 European countries (11 belonging to the EMU and 6 non-EMU countries) during a period of 15 years, from 1999 to 2013, taking into account the following variables: unemployment rate, stock market index variation, public debt over GDP ratio, inflation, GDP growth and international reserves. These variables are among the most used in literature about the determination of sovereign bond spreads.

We obtain different groups of countries-years according to the similarity between their economic variables. Considering the components maps it is possible to examine apiece economic variable level during the whole period of analysis. The 2008 financial crisis clearly affected all countries of the sample, although the effect was different between the eurozone countries and the rest of European economies.

Group 1 represents an improvement in the economic context, despite some exceptions specifically related with high unemployment rates. This group is mainly formed by Poland, Czech Republic and Spain, until the year 2006. Although the behavior in macroeconomic terms is similar, it is possible to appreciate a clear divergence in the spreads levels among EMU countries (negative mean spread) compared with non-EMU members (high positive means).

The second group represents better economic situation than the first one, related with GDP growth, lower unemployment rates and reduced public debt over GDP ratios. In this group are located Denmark, Sweden and UK since the beginning date of the sample until 2007 (stronger countries which do not belong to the EMU) with Netherlands and, to a lesser extent, Portugal and Ireland during the years previous to the crisis, when they experimented a great economic development.

The mean of the spreads of this group is the lowest compared with all the groups obtained. EMU countries present negative mean spreads and non-EMU countries present an important decrease in the mean spreads of these countries.

In the group 3 we mainly find EMU countries during the years before the crisis (there are almost entirely Belgium, Greece and Italy since the beginning of the review period until 2007) and also some economies that had an economic rebound in 2010. This group is characterized by economies that began to notice a slowdown in GDP growth and an increasing in public debt ratios. Nonetheless, unemployment and inflation rates maintain stable. This translates into increased bond spreads levels, for EMU countries the mean of the spreads become positive and the mean is even higher for countries outside EMU. However, the mean of the overall spreads for this group maintains a moderate level.

Group 4 is already related with crisis features (most of the analyzed countries are in this group during 2008 and 2009). However, it still maintains low level of unemployment and public debt, so to this group also belong the strongest European countries during the recent years of the sample, signaling a gradual exit from the crisis. The difference between the mean of the spreads for countries belonging to the EMU with those that do not belong to the EMU is smaller than in groups 1 and 3. This is related with an increase in the spreads level of EMU countries as well as a decrease of the mean spreads of non-euro countries. In both sets of countries, the differential is higher in this group than in group 2.

Finally, the group 5 is the worst performance in terms of macroeconomic variables. Thus, we find a high level of unemployment rate, public debt/GDP ratio and inflation rate, while negative changes in GDP, lower levels of international reserves and poor stock market behaviour.

In this group we find entirely Belgium, Greece and Italy from 2008 together with Portugal, Spain, Ireland and Hungary (the countries more affected by the crisis). Obviously, the overall level of the spreads average for this group is the highest, and it is the only group in which the spreads average for EMU countries is higher than the set of non-EMU countries (explained by the high levels reached by Greece from 2011).

The analysis highlights the different situations and groupings in which the economies of the European Union have stayed since 1999–2013, with respect to macroeconomic variables and spread levels, especially in two specific events: the financial crisis and the introduction of the euro.

# References

Arghyrou, M.G., Kontonikas, A.: The EMU sovereign-debt crisis: fundamentals, expectations and contagion European economy. J. Int. Finan. Mark. Inst. Money 22(4), 658–677 (2011)

Arora, V., Cerisola, M.: How does U.S. monetary policy influence economics conditions emerging markets? International monetary fund working paper /00/48 (2000)

Baldacci, E., Gupta, S., Mati, A.: Political and fiscal risk determinants of sovereign spreads in emerging markets. Rev. Dev. Econ. 15, 251–263 (2011)

Barrios, S., Iversen, P., Lewandowska, M., Setze, R.: Determinants of intra-euro area government bond spreads during the financial crisis. Eur. Econ.-Econ. Pap. 388, 1–28 (2009)

Basu, R.: Financial contagion and investor "Learning": an empirical investigation. IMF working papers, 02/218, pp. 1–37 (2002)

Codogno, L., Favero, C., Missale, A.: Yield spreads on EMU government bonds. Econ. Policy, 503–532 (2003)

De la Dehesa, G.: La primera gran crisis financiera del siglo XXI: orígenes, detonantes, efectos, respuestas y remedios. Alianza Editorial S.A, Madrid (2009)

Eichengreen, B., Mody, A.: What explains changing spreads on emerging market debt? National Bureau of Economic Research, working paper, 6408, 1–48 (1998)

Eichengreen, B., Rose A., Wyplosz C.: Contagious currency Crises. National Bureau of Economic Research, working paper, 5681, 1–48 (1996)

Gande, A., Parsley, D.: News spillover in the sovereign debt market. J. Finan. Econ. 75, 691–734 (2003)

Geyer, A., Kossmeier, S., Pichler, S.: Measuring systematic risk in EMU spreads. Rev. Finan. 8(2), 171–197 (2004)

González, R.M., Levy Yeyati, E.: Global factors and emerging market spreads. Econ. J. 118, 1917–1936 (2008)

Grammatikos, T., Vermeulen, R.: Transmission of the financial and sovereign debt crises to the EMU: stock prices, CDS spreads and exchange rates. J. Int. Money Finan. **31**(3), 517–533 (2011)

Hilscher, J., Nosbusch, Y.: Determinants of sovereign risk: macroeconomic fundamentals and the pricing of sovereign debt. Rev. Finan. **14**, 235–262 (2010)

Kaminsky, G., Schmukler, S.: Emerging market instability: do sovereign ratings affect country risk and stock returns? World Bank Econ. Rev. **16**(2), 171–195 (2002)

Manganelli, S., Wolswijk, G.: What drives spreads in the euro area government bond market? Econ. Policy, 193–240 (2009)

Martinez, L.B., Terceño, A., Teruel, M.: Sovereign bond spreads determinants in Latin American countries: before and during the XXI financial crisis. Emerg. Mark. Rev. **17**, 60–75 (2013)

Min, H.G., Lee, D.H., Nam, C., Park, M.C., Nam, S.H.: Determinants of emerging market bond spreads: cross-country evidence. Global Finan. J. **14**, 271–286 (2003)

Sgherri, S., Zoli E.: Euro area sovereign risk during the crisis. IMF working papers /09/222, 1–22 (2009)

# Using Genetic Algorithms to Evolve a Type-2 Fuzzy Logic System for Predicting Bankruptcy

Vasile Georgescu

**Abstract** In this paper, we use GAs to design an interval type-2 fuzzy logic system (IT2FLS) for the purpose of predicting bankruptcy. The shape of type-2 membership functions, the parameters giving their spread and location in the fuzzy partitions and the set of fuzzy rules are evolved at the same time, by encoding all together into the chromosome representation. Type-2 FLSs have the potential of outperforming their type-1 FLSs counterparts, because a type-2 fuzzy set has a footprint of uncertainty that gives it more degrees of freedom. The enhanced Karnik-Mendel algorithms are employed for the centroid type-reduction and defuzzification stage. The performance in predicting bankruptcy is evaluated by multiple simulations, in terms of both in-sample learning and out-of sample generalization capability, using a type-1 FLS as a benchmark.

**Keywords** Interval Type-2 fuzzy sets and fuzzy logic systems · Enhanced Karnik-Mendel (KM) algorithms · Genetic algorithms · Bankruptcy prediction

## 1 Introduction

Type-1 Fuzzy Sets (T1FSs), pioneered by Zadeh, use a crisp membership function to capture the variation in measurements of a given feature for different instances, disregarding the possibility of variation in the membership degree itself. Type-2 Fuzzy Sets (T2FSs) have also been introduced by (Zadeh 1975), in order to overcome this problem. The concept has been further developed by (Karnik et al. 1999; Mendel 2001) and others. It consists of considering a secondary membership function that represents the uncertainty in the primary membership assignment at each measurement point. In practice, general T2FSs (GT2FSs) are less useful because of the users' inability to correctly specify the secondary memberships.

V. Georgescu (✉)
Department of Statistics and Informatics, University of Craiova, Craiova, Romania
e-mail: v_geo@yahoo.com

© Springer International Publishing Switzerland 2015
J. Gil-Aluja et al. (eds.), *Scientific Methods for the Treatment of Uncertainty in Social Sciences*, Advances in Intelligent Systems and Computing 377,
DOI 10.1007/978-3-319-19704-3_29

Interval Type-2 Fuzzy Sets (IT2FS) proved to be more tractable in applications since their secondary membership function is simply a crisp interval, assigning a membership degree equal to unity for each membership curve embedded between an Upper and a Lower Membership Function (UMF and LMF). The research interest has been also extended to Type-2 Fuzzy Logic Systems (T2FLSs), characterized by T2FSs. T2FLSs are expected to provide better performance in prediction based on their representational advantage over Type-1 FLSs. Most applications of T2FLSs involve only IT2FSs, because of the computational complexity of GT2FSs. T2FLSs also include a fuzzifier, fuzzy rules and a fuzzy inference engine, but differ from T1FLSs in respect of the output processing part, where the defuzzifier is preceded by a type-reducer. The commonly used type-reduction and defuzzification method in IT2FLSs is based on the centroid of an IT2FS, originally developed by (Karnik and Mendel 2001). The Karnik-Mendel (KM) algorithms proved to play a crucial role in designing IT2FLSs, by providing an efficient way to carrying out the centroid type-reduction step. (Mendel and Liu 2007) proved monotonicity and super-exponential convergence of the algorithms. (Wu and Mendel 2009) proposed Enhanced KM (EKM) algorithms to reduce the computational cost of the standard KM algorithms.

The rest of this paper is organized as follows. Section 2 describes T2FSs. Section 3 presents IT2FLSs and their main components such as the fuzzifier, the fuzzy inference engine, the type-reducer and the defuzzifier, with a special emphasis on the enhanced Karnik-Mendel algorithms. In Sect. 4, a GA-based approach to evolving an IT2FLS for bankruptcy prediction is discussed. The last section is devoted to concluding remarks.

## 2  Type-2 Fuzzy Sets

In this section, we introduce the GT2FSs and IT2FSs.

**Definition 1:** A *General Type-2 Fuzzy Set* (GT2FS) $\tilde{A}$ is characterized by a membership function $\mu_{\tilde{A}}: X \times J_X \to [0, 1]$, also called type-2 membership function, which itself is fuzzy (Mendel and John 2003). Usually, the fuzzy set $\tilde{A}$ is expressed as:

$$\tilde{A} = \left\{ \left( (x, u), \mu_{\tilde{A}}(x, u) \right) | x \in X, \forall u \in J_x \subseteq [0, 1] \right\} \tag{1}$$

where $x \in X$ is the primary variable, $u \in J_x \subseteq [0, 1]$ is the secondary variable, $J_x$ is called the primary membership of $x$ and $\mu_{\tilde{A}}(x, u) \in [0, 1]$. Using the notation $f_x(u) = \mu_{\tilde{A}}(x, u) \in [0, 1]$, an alternative representation of the GT2FS is

$$\tilde{A} = \int_{x \in X} \int_{u \in J_x} \mu_{\tilde{A}}(x, u) / (x, u) = \int_{x \in X} \left( \int_{u \in J_x} f_x(u) / u \right) \Big/ x, \quad J_x \subseteq [0, 1] \quad (2)$$

The $\int \int$ denotes union over all admissible $x$ and $u$ (Mendel and John 2002).

**Definition 2:** At each value of $x$, say $x = x'$, the two-dimensional plane whose axes are $u$ and $\mu_{\tilde{A}}(x', u)$ is called a *vertical slice* of $\mu_{\tilde{A}}(x, u)$. A *secondary membership function* is a vertical slice of $\mu_{\tilde{A}}(x', u)$, i.e.,

$$\mu_{\tilde{A}}(x = x', u) \equiv \mu_{\tilde{A}}(x') = \int_{u \in J_{x'}} f_{x'}(u) / u, \quad for \quad x' \in X \quad and \quad \forall u \in J_{x'} \subseteq [0, 1] \quad (3)$$

where $0 \leq f_{x'}(u) \leq 1$. The *domain* $J_x$ of a secondary membership function is called the *primary membership* of $x$. In (3) $J_{x'}$ is the primary membership of $x'$. The *amplitude* of a secondary membership function is called *secondary grade*. In (3) $f_{x'}(u)$ is a secondary grade.

**Definition 3:** Uncertainty in the primary membership of a type-2 fuzzy set, $\tilde{A}$, consists of a bounded region, called *footprint of uncertainty* (FOU) (Mendel and John 2002), which is conveyed by the union of all primary memberships, i.e.,

$$FOU(\tilde{A}) = \bigcup_{x \in U} J_x \quad (4)$$

The Upper Membership Function (*UMF*) and Lower Membership Function (*LMF*) of $\tilde{A}$ are two type-1 *MFs* that bound the *FOU* and are denoted by $\underline{\mu}_{\tilde{A}}(x)$ and $\bar{\mu}_{\tilde{A}}(x)$ respectively (Figs. 1 and 2).

**Definition 4:** $\tilde{A}$ is called an *interval type-2 fuzzy set* (*IT2FS*) if all the secondary grades of a type-2 fuzzy set $\tilde{A}$ are equal to 1, i.e.,

$$\mu_{\tilde{A}}(x, u) = 1, \quad \forall x \in X, \forall u \in J_x \subseteq [0, 1] \quad (5)$$

# 3 Interval Type-2 Fuzzy Logic Systems

## 3.1 Main Components

An IT2FLS is shown in Fig. 3. It is similar to a type-1 FLS, containing a fuzzifier, rule base, fuzzy inference engine, and output processing. The main difference is that a type-2 FLS has a type-reducer in the output processing. The type-reducer has the

**Fig. 1** Primary and secondary membership functions of a general type-2 fuzzy set

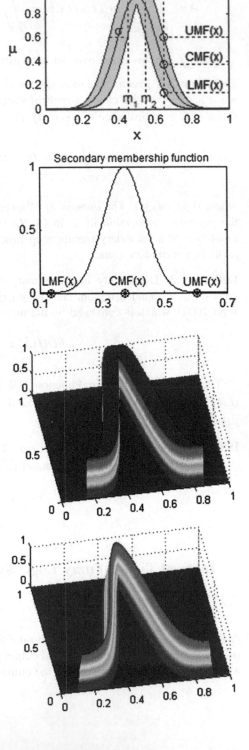

**Fig. 2** 3D representation ($x$, $\mu(x)$, $\mu(x,u)$) of *interval* (left) and *general* (right) type-2 fuzzy sets

ability to generate a type-1 fuzzy set from a type-2 fuzzy set. The defuzzifier then can defuzzify this type-1 fuzzy set to a crisp number. IT2FLSs have demonstrated better ability to handle uncertainties than their type-1 counterparts (Mendel 2007a, b).

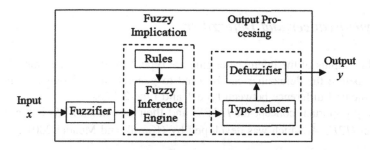

**Fig. 3** Interval type-2 fuzzy logic system (IT2FLS)

## 3.2 Fuzzifier

Consider the rule base of a type-2 FLS consisting of $N$ rules of the form ($\tilde{R}^{\ell}$: IF $x_1$ is $\tilde{A}_1^{\ell}$ and $x_2$ is $\tilde{A}_2^{\ell}$ and ... $x_p$ is $\tilde{A}_p^{\ell}$, THEN $y$ is $\tilde{B}^{\ell}$), where $\tilde{A}_i^{\ell}$ ($i = 1, ..., p$) and $\tilde{B}^{\ell}$ are type-2 fuzzy sets, and $x = (x_1,..., x_p)$ and $y$ are linguistic variables.

The fuzzifier maps a crisp point $x = (x_1' ... x_p')$ into a type-2 fuzzy set $\tilde{A}_x$.

## 3.3 Fuzzy Inference Engine

The inference engine matches the fuzzy singletons with the fuzzy rules in the rule base. To compute unions and intersections of type-2 sets, compositions of type-2 relations are needed. Just as the sup-star composition is the key computation for a type-1 FLS, the extended sup-star composition is defining for a type-2 FLS.

The first step in the extended sup-star operation consists of obtaining the firing set $\prod_{j=1}^{p} \mu_{\tilde{x}_j}(x_j) \equiv F^i(x)$ by performing the input and antecedent operations. As only interval type-2 sets are used and the *meet* operation is implemented by the minimum $t$-norm, the firing set is the type-1 interval set $F^i(x) = \left[ \underline{f}^i(x), \overline{f}^i x) \right] \equiv \left[ \underline{f}^i, \overline{f}^i \right]$, where $\underline{f}^i(x) = \min\left( \mu_{\tilde{x}_1^i}(x_1), \mu_{\tilde{x}_2^i}(x_2) \right)$ and $\overline{f}^i(x) = \min\left( \overline{\mu}_{\tilde{x}_1^i}(x_1), \overline{\mu}_{\tilde{x}_2^i}(x_2) \right)$. The terms $\underline{\mu}_{\tilde{x}_j^i}(x_j)$ and $\mu_{\tilde{x}_j^i}(x_j)$ are the lower and upper membership grades of $\mu_{\tilde{x}_j^i}(x_j)$. Next, the firing set, $F^i(x)$, is combined with the consequent fuzzy set of the

$i$th rule, $\mu_{\tilde{Y}^i}$, using the minimum $t$-norm to derive the fired output consequent sets. The combined output fuzzy set may then be obtained using the maximum $t$-conorm.

## 3.4 Type-Reducer and Defuzzifier

Since the outputs of the inference engine are type-2 fuzzy sets, they must be type-reduced before the defuzzifier can be used to generate a crisp output. This is the main structural difference between type-1 and type-2 *FLSs*.

A very important concept for *IT2FSs* and their associated *FLSs* is the *centroid* $C_{\tilde{A}}$ of an *IT2FS* $\tilde{A}$, which was developed by (Karnik and Mendel 2001).

**Definition 5:** The centroid $C_{\tilde{A}}$ of an *IT2FS* $\tilde{A}$ is the union of the centroids of all its embedded *T1FSs* (denoted by $A_e$), i.e.,

$$C_{\tilde{A}} = [c_l(\tilde{A}), c_r(\tilde{A})], \quad \text{where} \quad c_l(\tilde{A}) = \min_{\forall A_e} c(A_e), \quad c_r(\tilde{A}) = \max_{\forall A_e} c(A_e) \quad (6)$$

Let $x_1 \leq \ldots \leq x_i \leq \ldots \leq x_N$, be a discretization of the primary variable of an *IT2FS* $\tilde{A}$. The centroid of *IT2FS* $\tilde{A}$, $c_{\tilde{A}} = [c_l, c_r]$, can be computed as the optimal solutions of the following interval weighted average problems (Karnik and Mendel 2001):

$$c_l = \min_{\forall \theta_i \in [\underline{\mu}_{\tilde{A}}(x_i), \overline{\mu}_{\tilde{A}}(x_i)]} \frac{\sum_{i=1}^{N} x_i \theta_i}{\sum_{i=1}^{N} \theta_i} = \frac{\sum_{i=1}^{k_l} x_i \overline{\mu}_{\tilde{A}}(x_i) + \sum_{i=k_l+1}^{N} x_i \underline{\mu}_{\tilde{A}}(x_i)}{\sum_{i=1}^{k_l} \overline{\mu}_{\tilde{A}}(x_i) + \sum_{i=k_l+1}^{N} \underline{\mu}_{\tilde{A}}(x_i)} \quad (7)$$

$$c_r = \max_{\forall \theta_i \in [\underline{\mu}_{\tilde{A}}(x_i), \overline{\mu}_{\tilde{A}}(x_i)]} \frac{\sum_{i=1}^{N} x_i \theta_i}{\sum_{i=1}^{N} \theta_i} = \frac{\sum_{i=1}^{k_r} x_i \underline{\mu}_{\tilde{A}}(x_i) + \sum_{i=k_r+1}^{N} x_i \overline{\mu}_{\tilde{A}}(x_i)}{\sum_{i=1}^{k_r} \underline{\mu}_{\tilde{A}}(x_i) + \sum_{i=k_r+1}^{N} \overline{\mu}_{\tilde{A}}(x_i)} \quad (8)$$

where $k_l$ and $k_r$ are called "switch points", with $x_{k_l} \leq c_l \leq x_{k_l+1}$ and $x_{k_r} \leq c_r \leq x_{k_r+1}$. The determination of $k_l$ and $k_r$ can be performed by using either the KM algorithms or the Enhanced KM (EKM) algorithms, proposed by (Wu and Mendel 2009), given in Table 1, which improve the KM algorithms with better initializations, computational cost reduction techniques, and stopping rules.

To obtain a crisp output in a *T2FLS*, the type-reduced set is defuzzified using the same method used in a *T1FLS*. The type-reducer reduces the outputs of the rules to the type-1 output of the system as an interval-valued fuzzy set $[y_l, y_r]$. This type-1 interval-valued fuzzy set can be defuzzified as $y = (y_l + y_r)/2$.

**Table 1** EKM algorithm for computing the centroid end-points of an *IT2FS*

| Step | EKM algorithm for $c_l$ | EKM algorithm for $c_r$ |
|---|---|---|
| 1. | Set $k = [N/2.4]$ (the nearest integer to $N/2.4$) and compute $$\alpha = \sum_{i=1}^{k} x_i \overline{\mu}_{\tilde{A}}(x_i) + \sum_{i=k+1}^{N} x_i \underline{\mu}_{\tilde{A}}(x_i),$$ $$\beta = \sum_{i=1}^{k} \overline{\mu}_{\tilde{A}}(x_i) + \sum_{i=k+1}^{N} \underline{\mu}_{\tilde{A}}(x_i).$$ Compute $c' = \alpha/\beta$ | Set $k = [N/1.7]$ (the nearest integer to $N/1.7$) and compute $$\alpha = \sum_{i=1}^{k} x_i \underline{\mu}_{\tilde{A}}(x_i) + \sum_{i=k+1}^{N} x_i \overline{\mu}_{\tilde{A}}(x_i),$$ $$\beta = \sum_{i=1}^{k} \underline{\mu}_{\tilde{A}}(x_i) + \sum_{i=k+1}^{N} \overline{\mu}_{\tilde{A}}(x_i).$$ Compute $c' = \alpha/\beta$ |
| 2. | Find $k' \in [1, N-1]$ such that $x_{k'} \le c' \le x_{k'+1}$ | |
| 3. | Check if $k' = k$. If yes, **stop** and set $c' = c_l$. If no, go to step 4 | Check if $k' = k$. If yes, **stop** and set $c' = c_r$. If no, go to step 4 |
| 4. | Compute $s = sign(k' - k)$ and $$\alpha' = \alpha + s \cdot \sum_{i=\min(k,k')+1}^{\max(k,k')} x_i \left[ \overline{\mu}_{\tilde{A}}(x_i) - \underline{\mu}_{\tilde{A}}(x_i) \right]$$ $$\beta' = \beta + s \cdot \sum_{i=\min(k,k')+1}^{\max(k,k')} \left[ \overline{\mu}_{\tilde{A}}(x_i) - \underline{\mu}_{\tilde{A}}(x_i) \right]$$ Compute $c''(k') = \alpha'/\beta'$ | Compute $s = sign(k' - k)$ and $$\alpha' = \alpha - s \cdot \sum_{i=\min(k,k')+1}^{\max(k,k')} x_i \left[ \overline{\mu}_{\tilde{A}}(x_i) - \underline{\mu}_{\tilde{A}}(x_i) \right]$$ $$\beta' = \beta - s \cdot \sum_{i=\min(k,k')+1}^{\max(k,k')} \left[ \overline{\mu}_{\tilde{A}}(x_i) - \underline{\mu}_{\tilde{A}}(x_i) \right]$$ Compute $c''(k') = \alpha'/\beta'$ |
| 5. | Set $c' = c''(k')$, $\alpha = \alpha'$, $\beta = \beta'$ and $k = k'$ and go to Step 2 | |

# 4 Evolving an *IT2FLS* by Genetic Algorithms for Predicting Bankruptcy

Genetic algorithms (GAs) are general-purpose optimization methods and have been widely used to evolve *T1FLSs* (Shi et al. 1999). Here, we extend the usage of GAs for evolving *IT2FLSs*, as a solution of a mixed integer optimization problem. The shapes and positions of the *IT2MFs*, along with the rules of an *IT2FLS*, are simultaneously evolved. Then the designed *IT2FLS* is applied to bankruptcy prediction and the classification accuracy is compared with that of a *T1FLS*, which is used as a benchmark.

Four types of *IT2MFs* are used in our implementation: IT2zmf, IT2gaussmf, IT2trimf and IT2smf. We encode the type of *IT2MFs* by an integer, $tmf \in \{1, 2, 3, 4\}$. Let $[x_a, x_b] \subseteq X$ be the interval spanned by an *IT2FS*. If we denote by $m = (x_a + x_b)/2$, $s = (x_b - x_a)/2$ and $e$, respectively, the *middle point*, the *spread* and the *extent of uncertainty*, each of the four *IT2MFs* mentioned above can be uniquely encoded by means of four parameters: $(tmf, m, s, e)$. See Fig. 4 for more details.

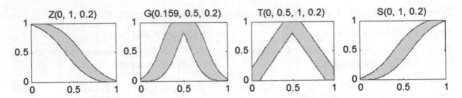

**Fig. 4** IT2MFs, with $x_a = 0, x_b = 1$ and $e = 0.2$, or $m = (x_a + x_b)/2 = 0.5$, $s = (x_b - x_a)/2 = 0.5$. We have: IT2zmf $(m-s, m+s, e)$; IT2gaussmf $(s/\pi, m, e)$; IT2trimf $(m-s, m, m+s, e)$; IT2smf $(m-s, m+s, e)$

In order to evolve an *IT2FLS* by means of a GA, we need to encode each candidate solution of the optimization problem into a chromosome. Let us assume that the *IT2FLS* consists of *NI* input variables, one output variable and *NR* rules. We consider that the universes of discourse of all input and output variables are covered by fuzzy partitions with the same number of *IT2FSs*, say *NF*. The meaning of each gene depends upon its position in the chromosome. The first $(NI + 1) \cdot NF \cdot 4$ genes will encode the fuzzy space defined by all the *IT2FSs* that generate the fuzzy partitions of the input and output variables. Each fuzzy rule in the rule base is also encoded by $(NI + 1)$ genes, with values represented by an integer in the following set: $\{-NF, -NF+1, \ldots, -1, 0, 1, \ldots, NF-1, NF\}$. An integer in the first *NI* positions will indicate an *IT2FS* selected for the corresponding input variable, while an integer in the last position, $NI + 1$, will indicate an *IT2FS* selected for the output variable. The integer 0 points out the absence of a certain variable in that rule and a negative integer shows that the negation of an *IT2FS* is to be selected for the corresponding variable in the rule. A rule without a nonzero antecedent and consequent part is not a feasible rule and will not be included in the rule base.

Finally, each chromosome consists of $(NI + 1) \cdot NF \cdot 4 + (NI + 1) \cdot NR$ genes. It is part of a population of chromosomes represented in the hyperspace of potential solutions. First, an initial chromosome population is randomly generated. Then, the population will undergo genetic operations such as *selection, crossover,* and *mutation* to evolve and optimize chromosomes. Every chromosome is assigned a fitness value, and then a selection operator is applied to choose relatively 'fit' chromosome to be part of the reproduction process. The crossover and mutation operators are closely related to the encoding scheme of *MFs* and rules. Crossover operator blends the genetic information between chromosomes to explore the search space, whereas mutation operator is used to maintain adequate diversity in the population of chromosomes to avoid premature convergence. In our case, we have to solve a mixed integer optimization problem and thus it is important to choose the genetic operators accordingly. *Tournament selection, Extended Laplace crossover* and *Power mutation* are the most appropriate choices for integer variables (Deep et al. 2009). A *truncation procedure* is also used to ensure that, after crossover and mutation operations have been performed, the integer restrictions are satisfied. Each chromosome is presented as a vector argument to the objective function. We split the vector argument into two sub-vectors: a vector $x \in \mathfrak{R}^{n_1}$ of real-valued variables and a vector $y \in Z^{n_2}$ of integer-valued variables. Our mixed integer optimization problem is then defined as follows: min $f(x, y)$, subject to: $x_i^L \leq x_i \leq x_i^U$, with $x_i$ real-valued, $i = 1, \ldots, n_1$; $y_i^L \leq y_i \leq y_i^U$, with $y_i$ integer-valued, $i = 1, \ldots, n_2$. Constraint handling is based on a parameter free, penalty function approach (Deep et al. 2009). The fitness function uses the error rate between desired outputs and estimated outputs. Therefore, the fitness value is evaluated based on simulating the *IT2FLS* on a given learning dataset.

For testing the performance of the designed *IT2FLS*, we consider the problem of financial distress prediction, which is a hard binary classification problem as it is high-dimensional, most data distribution is non-Gaussian and exceptions are common. A sample of 130 Romanian companies has been drawn from those listed on Bucharest Stock Exchange (BSE). The binary variable to be predicted is *entering insolvency or not*. As predictors, a selection of the most relevant 8 financial ratios has been used.

Given the complexity of the designing procedure of a *FLS*, it is difficult to say whether there is a unique optimal configuration of the system or not. Almost surely, there is not, considering the so many degrees of freedom involved in design. Clearly, starting with a random initialization of the population and using stopping criteria that ensure approaching more ore less the "global" solution in reasonable time, a different configuration is evolved at each tentative design, with different performances for both the *IT2FLS* and *TIFLS*. A particular configuration of an *IT2FLS* evolved using GA with our available training dataset is shown in Fig. 5. It consists of 8 rules, 8 antecedent variables and one consequent variable.

An important advantage of *TIFLSs* is the possibility of designing them much faster than *IT2FLSs*, because of the computational cost of type-reducing procedure.

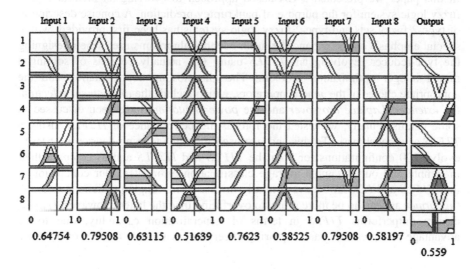

**Fig. 5** Rule viewer of the IT2FLS

However, *IT2FLSs* have more degrees of freedom and this gives them the power of outperforming their *T1FLSs* counterparts, especially in terms of generalization capability (out-of-sample prediction accuracy). Our experimental setup is to use the same training dataset in order to evolve 100 configurations for both the *T1FLS* and *IT2FLS* and to compare their in-sample and out-of-sample average classification error rates.

The in-sample average classification error rates for *T1FLS* and *IT2FLS* are rather similar: the percentage of misclassified companies, which results, in average, using *T1FLS* (5.71 %), is only slightly lesser than that using *IT2FLS* (6.14 %). This little difference in their average forecasting performances is actually an indication of overfitting, due to a specialization in excess on reclassifying the training dataset.

By contrast, the out-of-sample average classification error rates show that the forecasting performance of *T1FLS* deteriorates to a greater extent than that of *IT2FLS*. Thus *IT2FLS* outperforms significantly *T1FLS* in terms of generalization capability, giving rise to more balanced misclassification errors on training and test datasets (Table 2).

**Table 2** In-sample and out-of-sample average classification error rates

|                                                      | T1FLS   | IT2FLS  |
| ---------------------------------------------------- | ------- | ------- |
| In-sample average classification error rate          | 5.71 %  | 6.14 %  |
| Out-of-sample average classification error rate      | 9.79 %  | 7.35    |

## 5 Conclusion

In this paper we presented a GA-based approach to evolving an interval type-2 fuzzy logic system for the purpose of bankruptcy prediction. A concise description of type-2 fuzzy sets and type-2 fuzzy logic systems has been first provided, with a special emphasis on the enhanced Karnik-Mendel algorithms used when designing the type-reducer. Evolving an *IT2FLS* using GA actually consists of solving a mixed integer optimization problem. The chromosome representation and the appropriate choice of the genetic operators, which include the *tournament selection*, the *extended Laplace crossover* and the *power mutation*, have been discussed in some details. Finally, we tested the forecasting performance of *IT2FLS*, using a *T1FLS* as a benchmark. To this end, we used the same training dataset in order to evolve 100 configurations for both the *T1FLS* and *IT2FLS* and to compare their in-sample and out-of-sample average classification error rates. Our conclusion is that *T1FLS* and *IT2FLS* behave similarly in case of in-sample prediction, although *the former* appears to be more prone to overfitting than the latter. However, *IT2FLS* clearly outperforms *T1FLS* in terms of generalization capability, due to its advantage of representing and capturing uncertainty with more degrees of freedom.

# References

Deep, K., Singh, K.P., Kansal, M.L., Mohan, C.: A real coded genetic algorithm for solving integer and mixed integer optimization problems. Appl. Math. Comput. **212**(2), 505–518 (2009)

Karnik, N.N., Mendel, J.M., Qilian, L.: Type-2 fuzzy logic systems. IEEE Trans. Fuzzy Syst. **7**(6), 643–658 (1999)

Liang, Q., Mendel, J.M.: Interval type-2 fuzzy logic systems: theory and design. IEEE Trans. Fuzzy Syst. **8**(5), 535–550 (2000)

Karnik, N.N., Mendel, J.M.: Centroid of a type-2 fuzzy set. Inf. Sci. **132**(1–4), 195–220 (2001)

Mendel, J.M.: Uncertain Rule-Based Fuzzy Logic Systems. Prentice Hall, Upper-Saddle River (2001)

Mendel, J.M.: Advances in type-2 fuzzy sets and systems. Inf. Sci. **177**(1), 84–110 (2007a)

Mendel, J.M.: Type-2 fuzzy sets and systems: an overview. IEEE Comput. Intell. Mag. **2**(1), 20–29 (2007b)

Mendel, J.M., John, R.I.: Type-2 fuzzy sets made simple. IEEE Trans. Fuzzy Syst. **10**(2), 117–127 (2002)

Mendel, J.M., Liu, F.: Super-exponential convergence of the Karnik-Mendel algorithms for computing the centroid of an interval type-2 fuzzy set. IEEE Trans. Fuzzy Syst. **15**(2), 309–320 (2007)

Shi, Y., Eberhart, R., Chen, Y.: Implementation of evolutionary fuzzy systems. IEEE Trans. Fuzzy Syst. **7**(2), 109–119 (1999)

Wu, D., Mendel, J.M.: Enhanced Karnik-Mendel algorithms. IEEE Trans. Fuzzy Syst. **17**(4), 923–934 (2009)

Zadeh, L.A.: The concept of a linguistic variable and its application to approximate reasoning—1. Inf. Sci. **8**, 199–249 (1975)

# References

[1] ...

# Part VII
# Optimization and Control

# Hedge for Automotive SMEs Using An Exotic Option

Javier-Ignacio García-Fronti and Julieta Romina-Sánchez

**Abstract** The automotive firms (usually SMEs) work as suppliers for a big automaker, so the former have financial dependence on the latter's structure. Each of these SMEs, working as supplier for a brand, is likely to find its sales falling or its gross margin shrinking when a depreciation occurs in the automaker's stock price. Therefore, fluctuation in automaker's stock price can impact negatively in its suppliers. This paper uses a stochastic model to calculate the premium that the SME must pay for hedge against these losses. Mathematically, it calculates the probability at time cero of automaker's stock price hitting a specific barrier before the option expires. For these purposes, 2014 intraday quotes have been used.

**Keywords** Automotive industry · Exotic option · Hedge instrument · Stochastic model

## 1 Introduction

In the automotive market, SMEs profits dependent on the automaker's turnover. This generates a liquidity problem when the automaker has period of low production because SMEs receive less orders. In an extreme case, this could culminate in bankruptcy. Using automaker's stock prices as indicator of economic and financial welfare, it is assumed that a decreases of stock price implies a decline in orders to SMES.

This paper proposes a stochastic model to calculate the premium that the SME must pay for hedge against these losses. Mathematically, it calculates the probability

J.-I. García-Fronti (✉) · J. Romina-Sánchez
Facultad de Ciencias Económicas, Universidad de Buenos Aires. Av,
Córdoba 2122, Buenos Aires, Argentina
e-mail: javier.garciafronti@economicas.uba.ar

J. Romina-Sánchez
e-mail: julietarsanchez@gmail.com

© Springer International Publishing Switzerland 2015
J. Gil-Aluja et al. (eds.), *Scientific Methods for the Treatment of Uncertainty in Social Sciences*, Advances in Intelligent Systems and Computing 377,
DOI 10.1007/978-3-319-19704-3_30

at time cero of automaker's stock price hitting a specific barrier before the option expires. For these purposes, 2014 intraday quotes have been used.

This hedge instrument could also be implemented as part of a public policy, helping employment policies. From this point of view, the purchase of this insurance would allow the government to help SMEs companies when its sales drop, avoiding employment problems.

To achieve the goal of this paper, the first section presents the stochastic model which calculates the expected value of an exotic option called "Option barrier". This instrument pays if the stock price of the automaker hit a certain (lower) barrier before maturity (Hull 2009, p. 550). The second section runs the model using 2014 intraday stock prices of Renault, Peugeot and Ford and presents the relationship between certain relevant barriers and the price of fair premium for its coverage. Finally, some conclusions are discussed.

## 2  Stochastic Model

We model the stock price of the automobile company as a geometric Brownian motion with drift (Lin 2006, p. 127):

$$S(t) = S(0)e^{\mu t + \sigma W(t)} \tag{1}$$

where $\mu > 0$ y W(t) is a standard Brownian motion.

We define the Stopping Time $\tau_L$ as:

$$\tau_L = \inf\{t; S(t) \leq L\} \tag{2}$$

$\tau_L$ is a random variable indicating the first time the geometric Brownian motion with drift hit barrier L.

To understand how $\tau_L$ behaves; we calculate its Laplace's transform (Baxter and Rennie 1996, p. 8).

For a given real value $z \geq 0$, we define $\{Z(t)\}$ as

$$Z(t) = e^{-zt}[S(t)]^{\xi} \tag{3}$$

$$Z(t) = [S(0)]^{\xi} e^{(-z + \xi\mu)t + \xi\sigma W(t)} \tag{4}$$

Given that $\{Z(t)\}$ is a geometric Brownian motion with drift, it will be martingale if and only if (Lin 2006, p. 117):

$$\mu^* + \frac{1}{2}[\sigma^*]^2 = 0 \tag{5}$$

So,

$$\mu^* = -z + \xi\mu \tag{6}$$

$$\sigma^* = \xi\sigma \tag{7}$$

Subsequently, $\{Z(t)\}$ is martingale if:

$$-z + \xi\mu + \frac{1}{2}\xi^2\sigma^2 = 0 \tag{8}$$

Solving (8) we found that:

$$\xi = \frac{-\mu \pm \sqrt{\mu^2 + 2\sigma^2 z}}{\sigma^2} \tag{9}$$

$$\xi = \frac{\mu}{\sigma^2}\left[-1 \pm \sqrt{1 + 2\frac{\sigma^2}{\mu^2}z}\right] \tag{10}$$

Since the Brownian motion is bounded from above (L barrier), the negative $\xi$ is choose to impose an upper bound. Thus, $\{Z(t)\}$ is a geometric Brownian motion with drift bounded for all t.

$$\xi = \frac{\mu}{\sigma^2}\left[-1 - \sqrt{1 + 2\frac{\sigma^2}{\mu^2}z}\right] \tag{11}$$

Using the Optional Sampling theorem (Lin 2006, p. 122):

$$E\{Z(\tau_L)\} = Z(0) \tag{12}$$

After some algebra,

$$E\{e^{-z\tau_L}\} = e^{b\frac{\mu}{\sigma^2}\left[-1 - \sqrt{1 + 2\frac{\sigma^2}{\mu^2}z}\right]} \tag{13}$$

Thus, the Laplace transform of the inverse Gaussian distribution is obtained.

$$\tau_L \sim inverse\ gaussian\left(\alpha^*, \beta^*\right)$$

With parameters:

$$\alpha^* = b\frac{\mu}{\sigma^2} \ and\ \beta^* = \frac{\mu^2}{\sigma^2} \tag{14}$$

The theoretical price of the exotic option (barrier option) is the expected present value of the payoff, considering a risk-free rate "r".

$$\text{Present value of expected payoff} = E\{e^{-z\tau_L}\mathbb{I}_{\{\tau_L < T\}}\} \tag{15}$$

Since we know the distribution of $\tau_L$ and t > 0, hence:

$$E\{e^{-z\tau_L}\mathbb{I}_{\{\tau_L < T\}}\} = e^{\frac{b}{\sigma^2}\left[\mu - \sqrt{\mu^2 + 2r\sigma^2}\right]} \int_0^T \frac{b}{\sqrt{2\pi t^3}\sigma} e^{-\left[t\sqrt{\mu^2 + 2r\sigma^2} - b\right]^2 \frac{1}{2\sigma^2 t}} dt \tag{16}$$

Gaussian inverse distribution density function with

$$\alpha^* = \frac{b}{\sigma^2}\sqrt{\mu^2 + 2r\sigma^2} \text{ and } \beta^* = \frac{\mu^2 + 2r\sigma^2}{\sigma^2} \tag{17}$$

The Inverse Gaussian distribution can be expressed in terms of the standard normal distribution (Panjer and Willmot 1992, p. 114). Thus, where N(x) is the function of the Standard Normal distribution, we obtain:

$$\text{Present Value of Expected Payoff} = E\{e^{-z\tau_L}\mathbb{I}_{\{\tau_L < T\}}\}$$

$$= e^{\frac{b}{\sigma^2}\left[\mu - \sqrt{\mu^2 + 2r\sigma^2}\right]} N\left(\frac{\sqrt{\mu^2 + 2r\sigma^2}T - b}{\sigma\sqrt{T}}\right)$$

$$+ e^{\frac{b}{\sigma^2}\left[\mu + \sqrt{\mu^2 + 2r\sigma^2}\right]} N\left(-\frac{\sqrt{\mu^2 + 2r\sigma^2}T + b}{\sigma\sqrt{T}}\right) \tag{18}$$

Practitioners use risk neutral probability. After a change of measure ($\mu = r - \frac{1}{2}\sigma^2$), the formula becomes:

$$\text{Present Value of Expected Payoff} = E\{e^{-z\tau_L}\mathbb{I}_{\{\tau_L < T\}}\}$$

$$= \frac{L}{S(0)} N\left(\frac{\left(r + \frac{1}{2}\sigma^2\right)T - ln\left(\frac{S(0)}{L}\right)}{\sigma\sqrt{T}}\right)$$

$$+ \left(\frac{S(0)}{L}\right)^{\frac{2r}{\sigma^2}} N\left(-\frac{\left(r + \frac{1}{2}\sigma^2\right)T + ln\left(\frac{S(0)}{L}\right)}{\sigma\sqrt{T}}\right) \tag{19}$$

Noting the equation to which the proposed model (Eq. 19) arrives, it can be seen that if we assume the fact that barrier L approaches zero, the value of the risk premium would be very low. This is so because it implies that the barrier option will pay when the share price is close to zero, which is unlikely.

## 3 A Hedging Tool

The aim of the model is to obtain the fair premium (transactional cost excluded) that an automotive firm should pay for hedge against a reduction of demand due to a decrease in the automaker's stock price.

For calibration purposes, we use intraday stock price of three automakers: Ford Motor Company (F-NYSE), Renault (RNO.PA-Paris) and Peugeot (UG.PA-Paris) from the first of January 2014 until the first of January 2015. Moreover, the risk-free rate is, one-year LIBOR for FORD and one-year EURIBOR for Renault and Peugeot.

Consider a given collection of $k + 1$ stock prices: $S(0), S(\Delta t) = S_1, S(2\Delta t) = S_2, \ldots, S(k\Delta t) = S_k$ at the moments $0, \Delta t, 2\Delta t, \ldots, k\Delta t = T$ in the time period $[0, T]$. In each subperiod $[(j-1)\Delta t, j\Delta t]$ with $1 \leq j \leq k$, we calculate each return on stock price $U_j$ as:

$$S^2 = \frac{1}{k-1} \sum_{j=1}^{k} \left(U_j - \overline{U}\right)^2 \tag{21}$$

where $\overline{U} = \sum_{j=1}^{k} U_j$.

The stock price of the automobile company S(t) is modeled as a geometric Brownian motion with drift (see Eq. 1), we calculate the estimated standard deviation of the intraday stock prices as:

$$\hat{\sigma} = \frac{S}{\sqrt{\Delta t}} \tag{22}$$

In particular, we use $\Delta t = \frac{1}{252}$.

### 3.1 Ford Motor Company

Calibration for the Ford Motor Company gives these parameters (Table 1):

| Table 1 Ford Motor Company's parameters | | |
|---|---|---|
| T= | | 1 |
| σ= | | 21 % |
| r= | | 0.6 % |
| S(0)= | | 15.34 |

Replacing these values into Eq. (19), the equation that determines the value of the premium for Ford Motor Company is as follows:

Expected present value of the payoff $= E\{e^{-z\tau_L}I_{\{\tau_L < T\}}\}$

$$= \frac{L}{15.34}N\left(\frac{\left(0.006 + \frac{1}{2}0.21^2\right) - \ln\left(\frac{15.34}{L}\right)}{0,21}\right)$$

$$+ \left(\frac{15.34}{L}\right)^{\frac{2 \cdot 0.006}{0.21^2}} N\left(-\frac{\left(0.006 + \frac{1}{2}0.21^2\right)T + \ln\left(\frac{15.34}{L}\right)}{0.21}\right) \quad (23)$$

For different values of L barrier, Fig. 1 shows that the premium grows exponentially as the barrier increases. Moreover, for barrier less than 8, the premium to
be paid is close to zero. Despite the fact that the SME is the one who chooses the
barrier at its own risk, Fig. 1 can be used as a tool for making comparisons between
the fair premium and the market price.

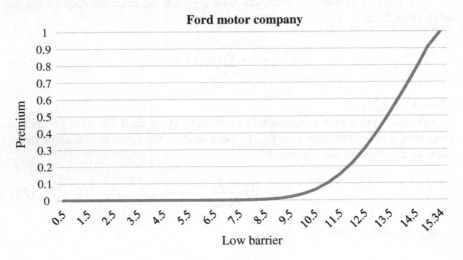

**Fig. 1** Ford Motor Company's premium vs low barrier

## 3.2 Renault

Calibration for the Renault gives these parameters (Table 2):

**Table 2** Renault's parameters

| T= | 1 |
|---|---|
| σ= | 29 % |
| r= | 0.3 % |
| S(0)= | 60.53 |

Replacing these values into Eq. (19), the equation that determines the value of
the premium for Renault is as follows:

Expected present value of the payoff $= E\{e^{-z\tau_L}I_{\{\tau_L < T\}}\}$

$$= \frac{L}{60.53}N\left(\frac{(0.003 + \frac{1}{2}0.29^2) - ln\left(\frac{60.53}{L}\right)}{0.29}\right)$$

$$+ \left(\frac{60.53}{L}\right)^{\frac{2*0.003}{0.29^2}}N\left(-\frac{(0.003 + \frac{1}{2}0.29^2) + ln\left(\frac{60.53}{L}\right)}{0.29}\right) \quad (24)$$

For different values of L barrier, Fig. 2 shows that the premium grows exponentially as the barrier increases. Moreover, for barrier less than 27, the premium to be paid is close to zero. Despite the fact that the SME is the one who chooses the barrier at its own risk, Fig. 2 can be used as a tool for making comparisons between the fair premium and the market price.

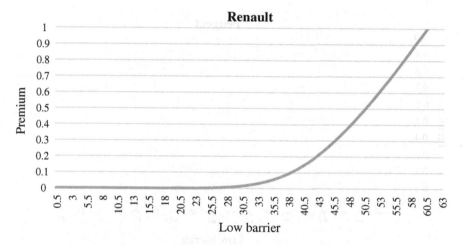

**Fig. 2** Renault's premium vs low barrier

## 3.3 Peugeot

Calibration for the Peugeot gives these parameters (Table 3):

| Table 3 Peugeot's parameters | | |
|---|---|---|
| T= | | 1 |
| σ= | | 38 % |
| r= | | 0.3 % |
| S(0)= | | 10.22 |

Replacing these values into Eq. (19), the equation that determines the value of the premium for Peugeot is as follows:

Expected present value of the payoff $= E\left\{e^{-z\tau_L}\mathrm{I}_{\{\tau_L < T\}}\right\}$

$$= \frac{L}{10,22} N\left(\frac{\left(0.003 + \frac{1}{2}0.38^2\right) - ln\left(\frac{10.22}{L}\right)}{0.38}\right)$$

$$+ \left(\frac{10.22}{L}\right)^{\frac{2*0.003}{0.38^2}} N\left(-\frac{\left(0.003 + \frac{1}{2}0.38^2\right)T + ln\left(\frac{10.22}{L}\right)}{0.38}\right) \quad (25)$$

For different values of L barrier, Fig. 3 shows that the premium grows exponentially as the barrier increases. Moreover, for barrier less than 4, the premium to
be paid is close to zero. Despite the fact that the SME is the one who chooses the
barrier at its own risk, Fig. 3 can be used as a tool for making comparisons between
the fair premium and the market price.

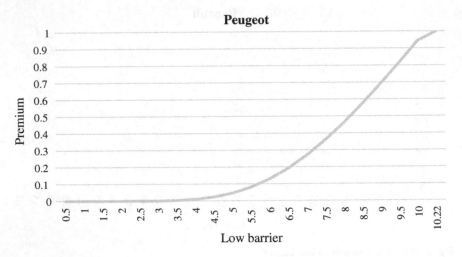

**Fig. 3** Peugeot's premium vs low barrier

## 4 Conclusion

In the economy, one of the biggest global issues is unemployment. In particular, in
a scenario of decrease in cars demand, small SMEs react immediately reducing the
amount of employees. Profits and stability of the companies that produce auto parts
depend heavily on what happens in the automaker's business and, therefore,
profitability of the small one can be explained (in part) by big one's stock price.

In this paper, we uses a stochastic valuation model for a barrier option for
hedging against SMEs' liquidity problems generated at periods of low production.
The proposed coverage will pay a dollar if the stock price of the automaker hit
certain barrier before maturity. It is important to note that this coverage could be

available in market as a part of a governmental policy to prevent employment problems in the sector.

We calibrate the model for three companies and we calculate the fair premium values for different barrier levels. The lower the value of the barrier, the lower the premium payable, since the probability that the value of a stock reaches very low values is negligible.

# References

Hull, J.: Options, Futures, and Other Derivative Securities, 7th edn. Prentice-Hall, Englewood Cliffs (2009)

Lin, X.S.: Introductory Stochastic Analysis for Finance and Insurance. Wiley, Hoboken (2006)

Baxter, M.B., Rennie, A.J.O.: Financial Calculus, An Introduction to Derivative Pricing. Cambridge University Press, Cambridge (1996)

Murray, R.: Spiegel: Laplace Transforms. Schaum's Outline Series. McGraw-Hill, New York (1965)

Panjer, H.H., Willmot, G.E.: Insurance Risk Models. Society of Actuaries, Schaumburg (1992)

# Obtaining Classification Rules Using LVQ +PSO: An Application to Credit Risk

Laura Lanzarini, Augusto Villa-Monte, Aurelio Fernández-Bariviera
and Patricia Jimbo-Santana

**Abstract** Credit risk management is a key element of financial corporations. One of the main problems that face credit risk officials is to approve or deny a credit petition. The usual decision making process consists in gathering personal and financial information about the borrower. This paper present a new method that is able to generate classifying rules that work no only on numerical attributes, but also on nominal attributes. This method, called LVQ+PSO, combines a competitive neural network with an optimization technique in order to find a reduced set of classifying rules. These rules constitute a predictive model for credit risk approval. Given the reduced quantity of rules, our method is very useful for credit officers aiming to make decisions about granting a credit. Our method was applied to two credit databases that were extensively analyzed by other competing classification methods. We obtain very satisfactory results. Future research lines are exposed.

**Keywords** Credit risk · Classification rules · Learning vector quantization (LVQ) · Particle swarm optimization (PSO)

L. Lanzarini (✉)
Instituto de Investigación en Informática LIDI, Universidad Nacional de la Plata,
50 y 120, La Plata, Buenos Aires, Argentina
e-mail: flaural@lidi.info.unlp.edu.ar

A. Villa-Monte · A. Fernández-Bariviera
Departament of Business, Universitat Rovira i Virgili, Avenida de la Universitat, 1,
Reus, Spain
e-mail: avillamonte@lidi.info.unlp.edu.ar

A. Fernández-Bariviera
e-mail: aurelio.fernandez@urv.net

P. Jimbo-Santana
Dpto. Ciencias de la Computación, ESPE Universidad de las Fuerzas Armadas,
Campus Politécnico, Av. Gral. Rumiñahui s/n, Sangolquí, Ecuador
e-mail: pjimbo@pcpcsolutions.com

© Springer International Publishing Switzerland 2015
J. Gil-Aluja et al. (eds.), *Scientific Methods for the Treatment of Uncertainty in Social Sciences*, Advances in Intelligent Systems and Computing 377,
DOI 10.1007/978-3-319-19704-3_31

# 1 Introduction

Data mining comprises a set of techniques that are able to model available information. One of the most important stages in the process is knowledge discovery. It is characterized by obtaining new and useful information without assuming prior hypothesis. One of the preferred techniques by decision makers is the association rule.

An association rule is an expression: IF *condition1* THEN *condition2*, where both conditions are conjunctions of propositions of the form (attribute = value), whose solely restriction is that attributes in the antecedent must no be present in the consequent. When a set of association rules present in the consequent the same attribute it is called a set of classification rules (Witten et al. 2011; Hernández et al. 2004).

This paper presents a new method for obtaining classification rules that combine a neural network with an optimization technique. We focus on reaching good results using a reduced number of rules.

Section 2 describes briefly previous literature on credit risk. Section 3 and 4 describe the neural network and the metaheuristics, respectively. Section 5 describes the proposed method. Section 6 gives the results of the empirical application and Sect. 7 concludes.

# 2 Related Work

There are several methods of rule construction. If the goal is to obtain association rules, the a priori method (Agrawal and Srikant 1994) or some of its variants could be used. This method identifies the most common sets of attributes and then combines them to get the rules. There are variants of the a priori method, are usually oriented reduce computation time.

Under the topic classification rules, the literature contains various construction methods based on trees such as C4.5 (Quinlan 1993) or pruned trees as the PART method (Frank and Witten 1998). In both cases, the key is to get a set of rules that covers the examples fulfilling a preset error bound. The methods of construction rules from trees are partitives and are based on different attributes' metrics to assess its ability to cover the error bound.

This paper presents a different approach based on Particle Swarm Optimization (PSO) to determine the rules. There are methods of obtaining rules using PSO (Wang et al. 2007). However, when operating with nominal attributes, a sufficient number of examples to cover all areas of the search space is required. If this situation is not feasible, its consequence is a poor initialization of the population, leading to premature convergence. As a way to bypass this problem, while reducing the turnaround time, is to obtain the initial state from a competitive LVQ neural network (Learning Vector Quantization). There is some literature that uses PSO as a means to determine the optimal quantity of competitive neurons to be used in the network, such as Hung and Huang (2010). The aim of this paper is the inverse process, since the neural network is the starting point for the optimization technique.

It is also important to mention previous studies used other techniques to predict credit risk. Yu et al. (2010) concludes that Support Vector Machine (SVM) is a technique that achieves the most accurate prediction. However, their solution is not comparable to our proposal. SVM is a nonlinear model that does not offer as a result a set of rules. As a consequence, it prevents the justification of the decision. A similar case is the use of neural networks in order to make decisions, mentioned also in Yu et al. (2010).

## 3 Learning Vector Quantization (LVQ)

Learning Vector Quantization (LVQ) is a supervised classification algorithm based on centroids or prototypes (Kohonen 1990). It can be interpreted as a three layer competitive neural network. The first layer is only an input layer. The second layer is where the competition takes place. The third layer performs the classification. Each neuron in the competitive layer has an associated numerical vector of the same dimension as the input examples and a label indicating the class they will represent. These vectors are the ones that, at the end of the adaptive process, will contain information about the classification prototypes or centroids. There are different versions of the training algorithm. We will describe the one used in this article.

When starting the algorithm, a number of $K$ centroids should be indicated. This allows defining the network architecture, given that number of inputs and outputs are defined by the problem.

Centroids are initialized taking $K$ random examples. Then, examples are entered one at a time in order to adapt the position of the centroids. In order to do this, the closest centroid to the example is determined, using a preset distance measure. Since this is a supervised process, it is possible to determine whether the example and the centroid correspond to the same class. If the centroid and the example belong to the same class, the centroid to moved closer to the example with the aim of strengthening the representation. Conversely, if the classes are different, the centroid is moved away from the example. These movements are performed using a factor or adaptation rate.

This process is repeated either until changes are less than a preset threshold or until the examples are identified with the same centroids in two consecutive iterations, whichever comes first

In the implementation used in this article, we also examine the second nearest centroid and, in case it belongs to a different class of the example and be at a distance of less than 1.2 times the distance to the first centroid, it is moved away. Several variants of LVQ can be consulted in Kohonen et al. (2001).

## 4  Obtaining Classification Rules with Particle Swarm Optimization (PSO)

Particle Swarm Optimization (PSO) is a population-based metaheuristic proposed by Kennedy and Eberhart (1995). In it, each particle in the population represents a possible solution to the problem and adapts following three factors: knowledge on the environment (fitness value), historical knowledge or past experience (memory) and historical knowledge or previous experiences of particles located in its neighborhood (social knowledge).

PSO was originally defined to work on continuous spaces. In order to operate with it on a discrete space it is necessary to take into account some precautions. Kennedy and Eberhart (1997) defined a binary version of PSO method. One of the central problems of the latter method is its difficulty changing from 0 to 1 and from 1 to 0 once it has stabilized. This has led to different versions of binary PSO, looking to improve the exploratory capacity. In particular, this work will use a variant defined by Lanzarini et al. (2011).

Obtaining classification rules using PSO, when operating on nominal and numeric attributes, requires a combination of the methods mentioned above. This is so, because it is necessary to say which attributes will be part of antecedent and what value or range of values it may take (a combination of discrete and continuous spaces).

Since it is a population technique, it should be analyzed the required information in each particle of the population. A decision between representing a single rule or the full rules set per particle should be made. At the same time, the representation scheme of each rule should be chosen. Tanking into account the aim of this work, we follow the Iterative Rule Learning (IRL) approach developed by Venturini (1993), in which each individual represents a single rule and the solution is constructed from the best individuals obtained in a sequence of executions. Consequently, using this approach implies that the population technique be applied iteratively until the desired coverage, obtaining a single rule for each iteration: the best individual of the population. It has also been decided to use a fixed length representation where only the antecedent of the rule is coded and given this approach, an iterative process will associate all individuals in the population with a default class, which does not require coding the consequent. A detailed representation of the particles could be seen in Lanzarini et al. (2015).

The fitness value of each particle is computed as follows:

$$fitness = \alpha * Balance * support * confidence - \beta * lengthAntecedent$$

where:

- support: is the support of the rule. It is ration of the quantity of examples that fulfils the rule to the total number of examples under examination.

- confidence: it is the confidence of the rule. It is the ratio of the quantity of examples that fulfils the rule to the quantity of examples that fulfil the antecedent.
- lengthAntecedent: it is the ratio of the quantity of conditions used in the antecedent to the total quantity of available attributes. It is noteworthy that each attribute can only appear once in the antecedent of a rule.
- Balance: assumes values in the interval (0, 1] and aims to offset the imbalance between the two classes. If the difference between the quantities of objects in each class is less than 20 %, it is equal to 1 and does not modify the fitness function. In other cases, its value is (H − E)/H, being H the quantity of examples in the opposite class that appears in the consequent and E is the quantity of misclassified examples by the rule.
- $\alpha$, $\beta$: are two coefficients that reflect the importance of each term.

# 5 LVQ+PSO. Proposed Method for Obtaining Rules

Rules are obtained through an iterative process that analyzes examples not covered in each class, beginning by the more populated classes. Whenever a rule is obtained, examples covered by such rule are removed from the set of input data. The process continues until covering all examples, or until the amount of uncovered examples in each class examples is either below the respective minimum established support or until they the maximum number of attempts to obtain a rule have been reached. It is important to note that, since examples are removed from the set of input data once they are covered by the rules, they constitute a classification. This is to say that, in order to classify a new example, rules must be applied in the order in which they were obtained and the example will be classified according to the corresponding class of the consequent of the first rule whose antecedent verifies for the example under examination.

Before starting the iterative process of obtaining rules, the method starts with the supervised training of a LVQ neural network, using the full set of examples and the algorithm described in Sect. 2. The goal of this step is to identify the most promising areas of space search.

Since neural networks operate only with numeric data, nominal attributes are represented by a dummy coding using as many binary digits as the different options of the nominal attribute. In addition, before starting the training, each numeric attribute is linearly scaled in the interval [0, 1]. The similarity measure used is the Euclidean distance. Once training is complete, each centroid will contain approximately the average of the examples it represents.

In order to obtain each of the rules, it is determined firstly, which is corresponding class of the consequent. With the aim of obtaining rules with high support, the proposed method analyzes the classes having a greater number of uncovered examples. The minimum support that a rule must meet is proportional to the amount of non covered examples of the class by the time that was obtained. In other words, the minimum support required for each class decreases along iterations, as examples of the corresponding class are covered. Thus, it is expected that the first rules have more support than the last rules.

Once the class is selected, the consequent is determined by the rule. In order to obtain the antecedent, a swarm population will be optimized, using the algorithm described in Sect. 3, initialized with the information of all centroids able to represent a minimum number of examples from the selected class and its immediate neighbors. The information of the centroid is used to determine the velocity of the particle. In both cases, it is intended to operate with a value between 0 and 1 that measures the degree of participation of the attribute (if numeric) or attribute value (if nominal) in building the antecedent of the rule. In the case of nominal attributes, it is clear that the average indicates the ratio of elements represented by the centroid that match the same value. However, when it is numeric, this ratio is not present in the centroid but the deviation of the examples (considering a specific dimension). If the deviation in a certain dimension is zero, all examples coincide in the value of the centroid, but if it is too large, it should be understood that it is not representative of the group. Therefore, it would not be appropriate to include it in the antecedent of the rule. If deviation is large, using ($1-1.5 * deviation_j$), the speed value (argument of the sigmoid function) will be lower and the probability that the attribute be used is reduced. In all cases the speed is initialized randomly in a range previously defined. Figure 1 shows the pseudocode of the proposed method.

```
Train LVQ network using all training examples
Compute the minimum support for each class
While (the end criteria is not reached)
      Choose the class with largest number of
      non covered examples
      Construct a reduced population of the
      particles, based on centroids
      Evolve the population using PSO according
      section 4
      Obtain the best population rule
      If (the rule fulfils with support and
      confidence required) then
        Add the rule to the set of rules
      Consider as covered the examples cor-
      rectly classified by the previous rule
      Recalculate the minimum support for this
      class
    End if
    End while
```

Fig. 1 Pseudocode of the proposed method

# 6 Results

In this section we compare the performance of the proposed method, LVQ+PO, *vis-à-vis* C4.5 methods defined by Quinlan (1993) and PART defined by Frank and Witten (2011). Both alternative methods allow classification rules. C4.5 is a pruned tree whose branches are mutually exclusive and allow classifying examples. PART gives as a result a list of rules equivalent to those generated by the proposed classification method, but in a deterministic way. PART operation is based on the construction of partial trees. Each tree is created in a similar manner to that proposed for C4.5 but during the process construction errors of each branch are calculated. These errors allow the selection of the most suitable combinations of attributes. For a detailed description of the method see PART [1].

We worked on two credit risk data bases belonging to UCI repository (Lichman 2013): German commercial credit staff and personnel data set Australian credit data set. The German credit data set is totally of 1000 instances, treats including 300 "bad" and 700 "good" cases. Each sample has 20 attributes (7 numerical and categorical 13), treats including account status, loan purposes, loan amount, age, property status, etc. The Australian credit data set Consist of 690 instances, 307 "bad" and 383 "good" cases, and each sample has 14 attributes (8 June numerical and categorical).

We performed 30 independent runs of each method. For LVQ+PSO, we use a LVQ network of 30 neurons distributed between classes in proportion to the examples used.

PART method was executed with a confidence factor of 0.3 for the pruned tree. For other parameters default values were used.

Tables 1 and 2 summarize the results obtained by applying the three methods. In each case was considered not only the accuracy of coverage of the rule set, but also the "transparency" of the obtained model. This "transparency" is reflected in the average number of rules obtained and the average number of terms used to form the antecedent.

In addition, given the nature of the problem under study, i.e. credit risk predition, we computed Type I error for each of the methods. This error measures the ratio of examples that the model predicts as belonging to the class GOOD (clients that will not have problem in paying back the credit), when in fact does not belong to such class, to the total number of examples.

**Table 1** Experiment results on German credit data set

|  | Accuracy | Average length of antecedent | Size of rule set | Type I Error |
|---|---|---|---|---|
| C4.5 | 0.706 ± 0.04 | 4.95 ± 0.242 | 64.211 ± 7.287 | 0.17 |
| PART | 0.704 ± 0.013 | 2.81 ± 0.106 | 53.533 ± 3.14 | 0.14 |
| LVQ+PSO | 0.709 ± 0.015 | 3.657 ± 0.152 | 7.833 ± 0.400 | 0.14 |

**Table 2** Experiment results on the Australian crédito data set

|        | Accuracy        | Average length of antecedent | Size of rule set | Type I Error |
|--------|-----------------|------------------------------|------------------|--------------|
| C4.5   | 0.852 ± 0.012   | 4.085 ± 0.456                | 18.277 ± 4.821   | 0.07         |
| PART   | 0.794 ± 0.03    | 2.149 ± 0.195                | 25.955 ± 3.041   | 0.11         |
| LVQ+PSO| 0.859 ± 0.010   | 2.723 ± 0.233                | 5.80 ± 0.38      | 0.07         |

Our results confirm that in both cases the proposed method achieves higher accuracy higher than PART but equivalent to that achieved by C4.5 method. However, if we observe the size of the rule set to achieve such accuracy, the proposed method uses a much smaller number of rules. As a consequence, our model is more simple and straightforward to interpret by the decision maker.

We would like to highlight that the important simplification in the set of rules does not only achieves an accuracy comparable to previous methods, but also keep Type I error in line with the selected benchmark models.

# 7 Conclusions

We introduce a new method of obtaining classification rules using a variant of binary PSO, whose population is initialized with information from the centroids of a network previously trained LVQ neural network. This combination allows operating with numerical and nominal attributes.

Experimental results on two public databases show that the LVQ+PSO method obtains a simpler model . It uses, on average, on average about 17.15 % of the number of rules generated by PART and 16.53 % of the rules generated by C4.5, with a history formed by few conditions and equivalent accuracy.

Although based on the conducted tests, no evidence of dependence between results and the initial size of the LVQ network has arisen, it is considered desirable to repeat the measurements using an LVQ network of minimum size and a version of variable population PSO to adequately explore the solution space.

**Acknowledgments** We thank UCI Machine Learning Repository for the generous provision of data for this paper.

# References

Agrawal, R., Srikant, R.: Fast algorithms for mining association rules in large databases. In: Proceedings of the 20th International Conference on Very Large Data Bases, VLDB '94, pp. 487–499. Morgan Kaufmann Publishers Inc., San Francisco (1994)

Frank, E., Witten, I.H.: Generating accurate rule sets without global optimization. In: Proceedings of the Fifteenth International Conference on Machine Learning, ICML '98, pp. 144–151. Morgan Kaufmann Publishers Inc., San Francisco (1998)

Hernández Orallo, J., Ramírez Quintana, M.J., Ferri Ramírez, C.: Introducción a la Minería de Datos. 1ra Edición, Pearson (2004)

Hung, C., Huang, L.: Extracting rules from optimal clusters of self-organizing maps. ICCMS '10. Second International Conference on Computer Modeling and Simulation, vol. 1, 2010, pp. 382–386 (2010)

Kennedy, J., Eberhart, R.C.: Particle swarm optimization. In: Proceedings of the IEEE International Conference on Neural Networks, pp. 1942–1948 (1995)

Kennedy, J., Eberhart, R.C.: A discrete binary version of the particle swarm algorithm. In: Proceedings of the IEEE International Conference on Systems, Man, and Cybernetics, vol. 5, pp. 4104–4108 (1997)

Kohonen, T.: The self-organizing map. Proc. IEEE **78**(9), 1464–1480 (1990)

Kohonen, T., Schroeder, M.R., Huang, T.S. (eds.): Self-organizing Maps, 3rd edn. Springer, New York (2001)

Lanzarini, L., Lopez, J., Maulini, J.A., Giusti, A.: A new binary pso with velocity control. In: Tan, Y., Shi, Y., Chai, Y., Wang, G. (eds.) Advances in Swarm Intelligence, Lecture Notes in Computer Science, pp. 111–119. Springer, Heidelberg (2011)

Lanzarini, L., Villa Monte, A., Ronchetti, F.: SOM+PSO. A Novel Method to Obtain Classification Rules. J. Comput. Sci. Technol. (JCS&T) **15**(1), (2015) (forthcoming)

Lichman, M.: UCI Machine Learning Repository. University of California, School of Information and Computer Science, Irvine, CA. http://archive.ics.uci.edu/ml. Accessed 5 Jan 2015 (2013)

Quinlan, J.R.: C4.5: Programs for Machine Learning. Morgan Kaufmann Publishers Inc., San Francisco (1993)

Venturini, G.S: A supervised inductive algorithm with genetic search for learning attributes based concepts. In: Brazdil, P. (ed.) Machine Learning: ECML-93, Lecture Notes in Computer Science, pp. 280–296. Springer, Heidelberg (1993)

Wang, Z., Sun, X., Zhang, D.: A pso-based classification rule mining algorithm. In: Proceedings of the 3rd International Conference on Intelligent Computing: Advanced Intelligent Computing Theories and Applications. With Aspects of Artificial Intelligence, ICIC '07, pp. 377–384. Springer, Heidelberg (2007)

Witten, I.H., Eibe, F., Hall, M.A.: Data Mining: Practical Machine Learning Tools and Techniques, 3rd edn. Elsevier (2011)

Yu, H., Huang, X., Hu, X., Cai, H.: A comparative study on data mining algorithms for individual credit risk evaluation. In: Fourth International Conference on Management of e-Commerce and e-Government, ICMeCG 2010, pp. 35–38 (2010)

# Optimization of Securitized Cash Flows for Toll Roads

Susana Sardá and M. Carmen Molina

**Abstract** In this paper we propose a methodology that can assist in the design of the quantitative aspects of a process of securitization to finance toll roads. Our goal is to provide the process with a series of mechanisms to streamline it, and to optimize the overall result of securitization for both the originator and for the bondholders. While we choose to apply it to toll roads, we believe that with simple adjustments, the proposed methodology can be extended to many securitization processes, and especially those related to infrastructure funding. We define the objective functions to be optimized to obtain the optimum volume of cash flows to securitize for each investment period. After defining the goals consistent with the demands of the highway investor (maximizing results and minimizing the investment recovery time), we can conceptualize the problem of determining the optimal level of funding through securitization. It is a bi-objective programme, in which the company aims to maximize the surplus generated by the project and simultaneously to maximize liquidity. We define, develop and justify the constraints related to the problem and once the bi-objective optimization problem has been set, we approach the resolution of the optimization problem. We opt for multiobjective fuzzy programming to resolve this programme.

**Keywords** Securitization · Infrastructures · Fuzzy programming · Toll roads · Financing

S. Sardá (✉) · M.C. Molina
Department of Business Administration, Universidad Rovira i Virgili Avda,
Universitat, 1 43204 Reus (Tarragona), Spain
e-mail: susana.sarda@urv.cat

M.C. Molina
e-mail: mcarmen.molina@urv.cat

© Springer International Publishing Switzerland 2015
J. Gil-Aluja et al. (eds.), *Scientific Methods for the Treatment of Uncertainty in Social Sciences*, Advances in Intelligent Systems and Computing 377,
DOI 10.1007/978-3-319-19704-3_32

393

# 1   Introduction

Asset securitization allows financial resources to be obtained by issuing assets backed by future cash flows. Despite its involvement in the financial crisis (Gorton 2009; Nadauld and Sherlund 2013), which led to a consequent reduction in the volume of issues, securitization is still a financing instrument meriting consideration, especially in Europe, where defaults remained low despite the recession. Even the European Central Bank, with the ABS Purchase Programme, is encouraging the issuance of this kind of assets.

In this paper, we propose a methodology that can assist in the design of the quantitative aspects of a process of securitization to finance toll roads. Our goal is to provide the process with a series of mechanisms to streamline it, and to optimize the overall result of securitization for both the originator and for the bondholders. While we choose to apply it to toll roads, we believe that with simple adjustments the proposed methodology can be extended so many securitization processes, and especially those related to infrastructure funding.

This methodology can contribute to the selection of the portfolio of securitized future cash flows. The originator must decide not only on the amount of funding that is being sought through securitization, but must also consider the temporal distribution of the securitized cash flows that this funding is going to provide. Since the high construction costs and long payback periods characterizing large public works projects tend to be a powerful disincentive for this type of investment, we will start from objectives consistent with these assumptions, i.e., maximizing the results and minimizing the payback time of the investment. Our approach is applicable to infrastructure projects where the private sector is responsible for all financing, construction, and management of the business once the construction is complete. However, it can also be used when the business is restricted to the management and maintenance of the highway. Moreover, the model would be applicable to the case of public infrastructure financing, changing some requirements linked to the aims of private companies, such as obtaining profits. It also applies both when tolls are paid directly by the user and in cases where "shadow tolls" are charged.

We start this paper with section two, which defines the objective functions to optimize in order to obtain the optimum volume of cash flows to be securitized for each period. In section three, we define, develop and justify the constraints related to the problem. Once the bi-objective optimization problem has been set, in section four we approach the resolution of the optimization problem. To resolve this programme, we have opted for multiobjective fuzzy programming. Finally we end this paper by presenting our conclusions.

# 2 The Selection of the Portfolio of Securitized Cash Flows

## 2.1 Defining the Objective Functions

One of the main problems to be solved when the originator is going to securitize future cash flows is the optimum volume of cash flows to transfer in each period. Our approach to determining this amount of securitized future cash flows and their timing is that the originator aims to maximize the resources to be obtained from the project and these should be obtained as quickly as possible. However, before formulating these goals, we must realize the value of these free cash flows.

We define cash flows of the investment after securitization ($CF$) as:

$$CF_t = EBITDA_t(1 - z_t) + z_t(AC_t + INT_t) - TIT_t \qquad (1)$$

Where:

$EBITDA_t$ = Earnings before interest, taxes, depreciation, and amortization for the period $t$.
$z_t$ = tax rate for the period $t$.
$AC_t$ = Amortization for the period $t$.
$INT_t$ = Securitization costs for the period $t$.
$TIT_t$ = Securitized cash flows for the period $t$.

So $CF_t$ is the cash flow from investment after meeting all payments arising from the operation of the company, including tax, and after transferring the securitized cash flows for the period $t$. The tax included in (1) does not consider the tax savings generated by the financial costs of "conventional" debt, as this will be reflected in the cost at which the cash flows will be discounted. The advantage of this measure of the cash flows is the ability to deduct income securitized in full, $TIT_t$, considering the tax savings generated by the securitization in each period, $z_t$ $INT_t$, without specifying in each $TIT_t$ the amount related to the repayment of the debt and the amount related to financial costs.

To calculate the present value of the $CF_t$ we use the cost of capital of the company, which depends on the cost of debt and the cost of equity. We assume that the decision on the capital structure has already been taken, and is being implemented in other investment projects, and that only the decision to be taken regarding the proportion of future cash flows to be securitized will determine the volume of financial resources obtained through securitization.

Through securitization, the originator transfers a series of future receipts in exchange for an amount of funding. There is no specification as to which part of each securitized cash flow corresponds to remuneration or repayment of the funding received. Securitization involves a number of costs inherent in the process, in addition to the interest received by the bondholders. Nevertheless, unlike other sources of external financing, securitization explicitly stipulates to the originator which part of the financial costs corresponds to formalizing the process costs, credit enhancement, or interest received by the bondholders or other securitization fund creditors. That is why it is important to use a measure that does not require this breakdown. The proposed solution is to recognize the result of applying the effective rate of the operation to the difference between the outstanding amounts of principal and interest and the amount of outstanding expenses charged to income as an interest expense in each period. We therefore recognize the financial cost of the operation applied to the amount of the mathematical reserve of the operation as financial expenses accrued each year.

In the $CF_t$ as we defined it in (1), the tax savings will be applied to the borrowing costs incurred in each period $INT_t$, calculated by applying the effective rate of the financial cost of securitization, $k_T$, to the reserve at the beginning of each period $t$, $R_{t-1}$.

$$INT_t = k_T R_{t-1} = k_T \sum_{i=t}^{n} \frac{TIT_i}{(1+k_T)^{i-t+1}} \tag{2}$$

After defining the cash flows, we define their present value. We calculate the present value of free cash flows to shareholders $(VACFL)$ as expressed in (1). The originator will attempt to maximize this value:

$$VACFL = -FC + \sum_{t=1}^{n} \frac{CF_t}{(1+k)^t} \tag{3}$$

$FC$ is the financing from equity and traditional debt, without taking into account the resources obtained by securitization. The discount rate $k$ is the cost of capital obtained from the costs of equity and traditional debt. Thus the $VACFL$ reflects the present value of the surplus that is available to shareholders after deducting all payments from both investment and funding from the incomes.

If we reformulate the expression of $VACFL$, regrouping and isolating $TIT_t$:

$$CF_t = A_t + B_t \tag{4}$$

With:

$$A_t = EBITDA_t(1 - z_t) + z_t AC_t \tag{5}$$

$$B_t = z_t INT_t - TIT_t \tag{6}$$

$A_t$ for each period includes cash flows that would generate investment regardless of funding. In $B_t$ we group the variables related to financing through securitization, therefore including payments generated by the securitization in each period, $TIT_t$, and the tax savings generated by the financial costs of securitization, $z_t INT_t$. $B_t$ can be written by developing the part corresponding to tax savings in period $t$, for which we need to use (2):

$$B_t = z_t \text{INT}_t - TIT_t = z_t k_T \sum_{i=t}^{n} \frac{TIT_i}{(1+k_T)^{i-t+1}} - TIT_t = z_t k_T \sum_{i=t}^{n} \frac{TIT_i}{(1+k_T)^{i}} (1+k_T)^{t-1} - TIT_t$$

$$(7)$$

$k_T$ is the securitization cost including $k'_T$, the interest for the bondholders, and $g$, which includes the cost allocable to the fund management, legal costs, insurances, etc.

With the breakdown between the variables related to investment and financing, we can rewrite the *VACFL*, starting from (5) and (6) as follows:

$$VACFL = -FC + \sum_{t=1}^{n} \frac{A_t}{(1+k)^t} + \sum_{t=1}^{n} \frac{B_t}{(1+k)^t} = VACFI + VACFT \qquad (8)$$

Where *VACFI* comprises the part of *VACFL* that includes revenues and expenses generated by the investment project and its corresponding funding before securitization, i.e.:

$$VACFI = -FC + \sum_{t=1}^{n} \frac{A_t}{(1+k)^t} \qquad (9)$$

*VACFT* comprises cash flows arising from securitization. With (7) and (8) we rewrite:

$$VACFT = \sum_{t=1}^{n} \frac{B_t}{(1+k)^t} = \sum_{t=1}^{n} \frac{z_t k_T}{(1+k)^t} \sum_{i=t}^{n} \frac{TIT_i}{(1+k_T)^i} (1+k_T)^{t-1} - \sum_{t=1}^{n} \frac{TIT_t}{(1+k)^t} \quad (10)$$

And finally, we can express *VACFT* as:

$$VACFT = \sum_{t=1}^{n} \frac{TIT_t}{(1+k_T)^t} k_T \sum_{i=1}^{t} \frac{z_i}{(1+k)^i} (1+k_T)^{i-1} - \sum_{t=1}^{n} \frac{TIT_t}{(1+k)^t} \qquad (11)$$

Finally, the present value of all payments and tax savings generated by the cash flows of the securitization, *VACFT*, can be expressed as:

$$VACFT = \sum_{t=1}^{n} TIT_t \left[ \frac{k_T}{(1+k_T)^t} \sum_{i=1}^{t} \frac{(1+k_T)^{i-1} z_j}{(1+k)^i} - \frac{1}{(1+k)^t} \right] \qquad (12)$$

We call $\alpha_t$ the coefficient multiplying the $TIT_t$ variable in each period, so finally the *VACFL* can be expressed as:

$$VACFL = -FC + \sum_{t=1}^{n} \frac{A_t}{(1+k)^t} + \sum_{t=1}^{n} \alpha_t TIT_t \qquad (13)$$

Besides the requirement to maximize the free cash flow to shareholders, and since time is a key variable in the investment analysis, the question is whether to also evaluate the degree of liquidity of the project. We understand liquidity to be a measure of the distribution of cash flows indicating the time required for the investment to recover the initial outlay it requires. The recovery period will be longer the greater the capital cost, and the viability of the project will be subject to recovery is given during the life of the project. The most commonly used criterion for measuring liquidity is the payback period. A criticism traditionally aimed at this approach is that it disregards the cash flows generated by the project after the payback period. An alternative for measuring the liquidity of a project that over-comes these limitations is the calculation of duration. Numerous studies support the correlation between the duration of a project and its payback period (Blocher and Stickney 1979; Boardman et al. 1982; Hawley and Malone 1989). The duration measures the weighted financial life of the project. Since the calculation of the duration involves all the cash flows generated by the project, it eliminates the problem of the payback period as the duration takes into account the cash flows generated by the project after the payback period has ended.

The duration provides additional information. The duration measures the interest rate sensitivity of present value. In our case, this aspect of the duration may be particularly significant, since when assimilating the duration to the payback period our goal is to minimize it. We will thereby complete the initial objective of max-imizing $VACFL$ and pursue the minimization of risk that could result from changes in the cost of traditional debt and $k$. We try to provide protection against rises in interest rates that affect the cost of capital. We are to some extent immunizing the investment against future changes of interest rates. The liquidity requirement for the valuation of our project can be formulated by the goal of minimizing the duration ($D$) of the project, defined as:

$$D = \frac{\sum_{t=1}^{n} \frac{t CFL_t}{(1+k)^t}}{\sum_{t=1}^{n} \frac{CFL_t}{(1+k)^t}} \tag{14}$$

We can express the numerator of the duration (14) according to (4) and (12) as:

$$\sum_{t=1}^{n} \frac{t CF_t}{(1+k)^t} = \sum_{t=1}^{n} \frac{t A_t}{(1+k)^t} + \sum_{t=1}^{n} \frac{tz_t k_T \sum_{i=t}^{n} \frac{TIT}{(1+k_T)^i}(1+k_T)^{t-1}}{(1+k)^t} - \sum_{t=1}^{n} \frac{t TIT_t}{(1+k)^t} \tag{15}$$

We regroup the terms in which each $TIT_t$ appears. We consider the positive value of the tax savings generated up to time $t$ by each $TIT_t$ for all the cash flows generated up to time $t$. We also consider the negative value that $TIT_t$ payment represents at time $t$. The present value of the tax savings generated by a $TIT_t$ up to time $t$, which positively weights the period when it is generated, can be expressed as:

$$\sum_{i=1}^{t} \frac{iTIT_t \frac{(1+k_T)^{i-1}}{(1+k_T)^t} k_T z_i}{(1+k)^i} = \frac{k_T TIT_t}{(1+k_T)^t} \sum_{i=1}^{t} \frac{i(1+k_T)^{i-1} z_i}{(1+k)^i}.$$

In addition, each payment derived from the securitization generates the corresponding cash outflow. As a result, in overall terms, the present value of all payments and tax savings generated by all securitization cash flows, pondering the moment in time when it is generated can be expressed from (12) as:

$$\sum_{t=1}^{n} TIT_t \left[ \frac{k_T}{(1+k_T)^t} \sum_{i=1}^{t} \frac{i(1+k_T)^{i-1} z_i}{(1+k)^i} - \frac{t}{(1+k)^t} \right]$$

We denominate $\beta_t$ the coefficient multiplying $TIT_t$ in each period, which is therefore, $\beta_t = \frac{k_T}{(1+k_T)^t} \sum_{i=1}^{t} \frac{i(1+k_T)^{i-1} z_i}{(1+k)^i} - \frac{t}{(1+k)^t}$.

The denominator of the expression of the duration is simply the present value of the resources generated by the investment after meeting payments of securitization, regardless of the initial contribution to the project of equity and traditional debt. From the $VACFL$ definition as given in (13), we have:

$$\sum_{t=1}^{n} \frac{CF_t}{(1+k)^t} = \sum_{t=1}^{n} \frac{A_t}{(1+k)^t} + \sum_{t=1}^{n} \alpha_t TIT_t \tag{16}$$

And we can define the duration as:

$$D = \frac{\sum_{t=1}^{n} \frac{tA_t}{(1+k)^t} + \sum_{t=1}^{n} \beta_t TIT_t}{\sum_{t=1}^{n} \frac{A_t}{(1+k)^t} + \sum_{t=1}^{n} \alpha_t TIT_t} \tag{17}$$

We can see that the duration depends on the cash flows generated by the investment project, but also the amounts of revenues securitized in each period. The $\beta_t$ coefficient multiplied by $TIT_t$ increases the weighting of the periods when $TIT_t$ generates tax savings, and reduces the weight of the periods in which the securitized cash flows are paid.

## 2.2 Adding the Constraint Functions

The solution proposed to the bi-objective optimization problem must satisfy a number of constraints that we will define, develop and justify below. First, we must consider that the initial contribution to the project of equity and traditional debt (*FC*) plus the resources obtained by securitization, should enable the originator company to meet the series of initial payments associated with investment (*DI*). The first constraint is therefore:

$$FC + \sum_{t=1}^{n} \frac{TIT_t}{(1+k_T)^t} = DI \tag{18}$$

On the other hand, we assume that the initial contribution to the project by equity and traditional debt will be limited by a minimum and maximum resources amount.

$$FC \geq FC_{Min} \tag{19}$$

$$FC \leq FC_{Max} \tag{20}$$

$FC_{Min}$ is given by the fact that resources obtainable through securitization are limited, meaning that it will be necessary to have a minimum amount of conventional financing. Meanwhile, $FC_{Max}$ indicates the maximum amount that conventional financing can reach for the project, assuming that it will be limited.

It may also be desirable to provide a minimum for the securitized cash flows in each period. When the special purpose vehicle is constituted, on its debit side, we find the issue or issues of asset-backed securities and may also establish alternative sources of external financing. It is likely that the suppliers of these resources, which are the liabilities of the fund, require some form of periodic payment and do not allow excessive grace periods. A reasonable option would be that the value of the minimum amount of securitized cash flows in each period is allocated to financial costs generated by the securitization in each period. We will build on this case, since we think it may be one of the most common and in turn adapted to the reality of financial practice, without ruling out the possibility of adapting this restriction to any other minimum payments structure required by the liabilities of the fund. Thus, the fourth set of constraints is given by Eq. (21) and is:

$$TIT_t \geq \lambda_t, \ t = 1, 2, \ldots, n \tag{21}$$

Where we can establish that $\lambda_t = INT_t = k_T \sum_{i=t}^{n} \frac{TIT_i}{(1+k_T)^i} (1+k_T)^{t-1}$.

A key financial indicator for the success of the securitization process is the debt coverage ratio. This is the ratio of cash available for debt servicing, and gives a good estimation of the solvency of the project and therefore the financial quality of the bonds issued. Although this ratio will be included as a parameter in our optimization problem, we want to emphasize that is not obtained absolutely independent of interest rates $k_T$ y $k'_T$. A higher volatility or risk of incomes demands a greater debt coverage ratio to achieve a certain rating. The debt coverage ratio therefore shows the level of financial charges that the infrastructure can support given its revenues. As a result, for the securitization process to be feasible, the project revenues should provide a sufficient level of coverage regarding the value of securitized cash flows in each period. We denominate $COB_t$ the debt coverage ratio at time $t$, and $COB_t = \frac{EBITDA_t}{TIT_t}$. We denominate $COB_t^*$ the required debt coverage ratio at time $t$. The fifth set of constraints is:

$$RGE_t - COB_t^* TIT_t \geq 0, \ t = 1, 2, \ldots, n \tag{22}$$

Finally, the optimization problem proposed is expressed from (13), (17)–(22) as:

$$\text{Max:} \quad VACFL = -FC + \sum_{t=1}^{n} \frac{A_t}{(1+k)^t} + \sum_{t=1}^{n} \alpha_t TIT_t$$

$$\text{Min:} \quad D = \frac{\sum_{t=1}^{n} \frac{tA_t}{(1+k)^t} + \sum_{t=1}^{n} \beta_t TIT_t}{\sum_{t=1}^{n} \frac{A_t}{(1+k)^t} + \sum_{t=1}^{n} \alpha_t TIT_t}$$

Subject to:

$$FC + \sum_{t=1}^{n} \frac{TIT_t}{(1+k_T)^t} = DI$$

$$FC \geq FC_{Min}$$

$$FC \leq FC_{Max}$$

$$TIT_t \geq \lambda_t, \ t = 1, 2, \ldots, n$$

$$RGE_t - COB_t^* TIT_t \geq 0, \ t = 1, 2, \ldots, n$$

Where the decision variables are $TIT_t$, for $t = 1, 2, \ldots, n$ and $FC$.

# 3 Solving the Optimization Problem by Fuzzy Programming

We aim to solve a bi-objective optimization problem, in which it is likely that there is a conflict between the objectives to optimize, the present value of the project and its liquidity. The first step in determining the best solution for the decision-maker usually requires a prior analysis of the sensitivity of the programme, and determining the positive ideal solutions (PIS) and negative ideal solutions (NIS) of the objective functions ($VACFL$ and $D$). Given the constraints proposed, these ideal solutions delimit the area of efficient solutions or Pareto nondominated solutions.

The PIS of each objective function is its optimal value in cases where other objectives are obviated in the final decision-making. For the present value of free cash flows to shareholders, we will denominate this solution $VACFL_{PIS}$ and for duration $D_{PIS}$. $VACFL_{PIS}$ is obtained by maximizing $VACFL$ without considering any other goal. $D_{PIS}$ is the minimum value of the duration function, without considering the present value of free cash flows to shareholders generated by the infrastructure. NIS is the worst value of the objective functions for each value of the vector of decision variables. $VACFL_{NIS}$ is the present value of free cash flow to shareholders generated by the investment project, in cases where the combination of securitized cash flows allow the duration of the project to be minimized. Likewise, $D_{NIS}$ is the duration of the project where only the objective of maximizing the present value of free cash flow to shareholders generated by the project is taken into account in the design of the structure of cash flows to be securitized. As $VACFL$ should be maximized, $VACFL_{PIS} \geq VACFL_{NIS}$. As $D$ should be minimized, $D_{NIS} \geq D_{PIS}$.

Having determined these positive and negative ideal solutions, we must proceed to the resolution of the optimization problem. To resolve this programme, we opted for multiobjective fuzzy programming from among the numerous alternatives. We considered solving the programme with fuzzy programming because it allows multiobjective optimization problems to be solved when the coefficients are in principle certain (Zimmermann 1986). Moreover, the use of fuzzy programming has advantages over other methods in solving our optimization problem. Fuzzy programming provides the decision-maker with an easy interpretation of the methodology to be followed in its resolution (Lai et al. 1996), because it enables easy interaction and establishes goals to be fulfilled for each objective in a flexible way, which is closer to the way decisions are made in practice. As we will also see, transforming objective functions into goals that can be achieved with a degree of intensity that ranges from 0 (absolute failure) to 1 (full compliance), using membership functions, avoids the problem of the heterogeneous magnitudes of objective functions. It is also interesting to note that the objectives are treated symmetrically, and are all introduced as constraints.

Decision-makers can trim their preferences by checking the goal for $VACFL$, specifying two $VACFL-$ and $VACFL +$ values, where $VACFL-\leq VACFL+$. For the duration, the decision-maker checks an interval bounded by $D-$ and $D +$, where $D^- \leq D^+$. It seems logical that if these values define the aspiration level of the decision-maker in terms of the compliance of each objective, these values should be feasible and efficient. They must be delimited by the positive and negative ideal solutions of the objective functions. As a result, $VACFL^-$ should be greater or equal to the $VACFL_{NIS}$ value. The $VACFL_{NIS}$ value is the profitability of the project when the minimum duration, i.e. the best one, is reached. On the other hand, $VACFL^+$ should be less than or equal to the maximum achievable $VACFL$ ($VACFL_{PIS}$). Similarly, $D-$ should be greater than or equal to the minimum that can be achieved

($D_{PIS}$), and $D^+$ should be less than or equal to the duration corresponding to the maximum achievable VACFL value ($D_{NIS}$).

In determining $D^-$, $D^+$, $VACFL^-$ and $VACFL^+$, the decision maker must reflect the importance to be attached to achieving each objective. The resulting intervals are close to the PIS if priority is given to these objectives, and next to the NIS if considered less important.

We apply fuzzy programming to solve the programme. Specifically, we apply the measurement of satisfaction to fulfilling the goals through membership functions. For the construction of the membership functions we must consider that for the return goal, we will obtain maximum satisfaction when $VACFL \geq VACFL^+$. However, for the liquidity goal, the greatest satisfaction will be obtained when $D \leq D^-$. Furthermore, all $VACFL < VACFL^-$ and $D > D^+$ would not be acceptable values and generate a minimum level of satisfaction. We will assign the value 1 to the membership function solutions that maximize satisfaction, while 0 is to the minimum level.

To assign a gradation to intermediate values of the membership function, we use the following equations:

$$\frac{VACFL - VACFL^-}{VACFL^+ - VACFL^-}, \text{ to } VACFL \text{ and } \frac{D^+ - D}{D^+ - D^-}, \text{ to duration.}$$

We can interpret the different levels achieved as a function of ordinal utility. If we use $\mu_V$ to denominate the membership function of VACFL, and $\mu_D$ to denominate the membership function of the duration, these functions can be expressed as follows:

$$\mu_V = \begin{cases} 1 & VACFL > VACFL^+ \\ \frac{VACFL - VACFL^-}{VACFL^+ - VACFL^-} & VACFL^+ \geq VACFL \geq VACFL^- \\ 0 & VACFL < VACFL^- \end{cases} \qquad (23)$$

$$\mu_D = \begin{cases} 1 & D < D^- \\ \frac{D^+ - D}{D^+ - D^-} & D^+ \geq D \geq D^- \\ 0 & D > D^+ \end{cases} \qquad (24)$$

Eventually, VACFL will be contained in the interval $[VACFL^-, VACFL^+]$ and the duration will be contained in the interval $[D^-, D^+]$. Our aim is to solve an optimization problem to determine the values of $TIT_t$ and FC to simultaneously reach the defined objectives. Both the VACFL and D depend on the values of $TIT_t$ and FC. By assigning a membership function to each objective function, which relates the values these can be taken with satisfaction obtained, the investor can overcome the handicap of the heterogeneity of magnitudes of the objective functions. In turn, the membership functions allow greater refinement of the satisfaction that investors associate with each possible outcome.

We call the coefficient we want to maximize $\gamma$, and from the following expression we determine:

$$\gamma = Min\{\mu_V, \mu_D\} \qquad (25)$$

In defining the coefficient $\gamma$ as the minimum value of the membership functions corresponding to VACFL and D, $\gamma$ reflects the overall satisfaction that the investor obtains from the joint fulfilment of the two objectives on a scale of 0-1. This satisfaction level exactly matches the value of the satisfaction achieved with the objective that achieves the least satisfaction. We therefore opt for the more conservative result.

From (20)–(25), our programme will be written as:

$$\text{Max } f = \gamma$$

Subject to:

$$\frac{VACFL - VACFL^-}{VACFL^+ - VACFL^-} \geq \gamma$$

$$\frac{D^+ - D}{D^+ - D^-} \geq \gamma$$

$$FC + \sum_{t=1}^{n} \frac{TIT_t}{(1 + k_T)^t} = DI$$

$$FC \geq FC_{Min}$$

$$FC \leq FC_{Max}$$

$$TIT_t \geq \lambda_t, \ t = 1, 2, \ldots, n$$

$$RGE_t - COB_t^* TIT_t \geq 0, \ t = 1, 2, \ldots, n$$

$$0 \leq \gamma \leq 1$$

Where the decision variables are $\gamma$, FC and $TIT_t$, for $t = 1, 2, \ldots,$ n.

## 4  Conclusions

In this paper, we propose a methodology that can facilitate the design of the quantitative aspects of a process of securitization to finance toll roads. Specifically, we address the problem of determining the optimal amount of securitization that meets the expectations of investors, focused on maximizing cash flows generated by the investment and minimizing the time required for recovering the initial outlay. We assume an infrastructure project in which private initiative is responsible for all financing and its construction as well as managing the business. The proposed methodology would apply whether tolls are paid directly by the user or in cases where a "shadow toll" system is applied.

We aim to determine the optimal level of funding to be obtained through securitization and the optimal timing of securitized cash flows. To do this, we propose a bi-objective programme in which the company aims to maximize the surplus generated by the project (*VACFL*), and simultaneously maximize liquidity (minimizing the duration). We also consider a number of constraints related to both infrastructure projects and securitization.

Finally, have proposed the programme, we consider its resolution using fuzzy programming, and specifically the measurement of satisfaction in terms of fulfilling the goals through membership functions. The ultimate goal is expressed as the maximization of a $\gamma$ coefficient, reflecting the overall satisfaction on a scale of 0–1, which gives the investor the joint fulfilment of the two initial objectives. This level of satisfaction will correspond to the value of the satisfaction that provides the least satisfaction.

Although it has been applied to financing toll roads, the methodology described could be useful in many securitization processes, and especially those related to infrastructure funding. It can also be applicable to publicly financed infrastructures, simply by changing some requirements linked to the aims of private companies, such as obtaining profits. Securitization is still a financing instrument meriting consideration, especially in infrastructure projects, which have considerable initial outlay costs and long-term financing requirements.

# References

Boardman, C.M., Reinhart, W.J., Celec, S.E.: The role of the payback period in the theory and applications of duration to capital budgeting. J. Bus. Finance Acc. **9**(4), 165–167 (1982)

Blocher, E., Stickney, C.: Duration and risk assessments in capital budgeting. Acc. Rev. LIV (1), 180–188 (1979)

Gorton, G.: The subprime panic. Eur. Fin. Manage. **15**(1), 10–46 (2009)

Hawley, D.D., Malone, R.P.: The relative performance of duration in the capital budgeting selection process. Eng. Econ. **41**(4), 67–74 (1989)

Lai, Y.J., Hwang, C.L.: Possibilistic linear programming for managing interest rate risk. Fuzzy Sets Syst. **54**(2), 135–146 (1993)

Lai, Y.J., Hwang, C.L.: Fuzzy Multiple Objective Decision Making. Springer, Berlin (1996)

Nadauld, T.D., Sherlund, S.M.: The impact of securitization on the expansion of subprime credit. J. Financ. Econ. **107**(2), 454–476 (2013)

Sardá, S.: La titulización de derechos de credito futuros: propuesta de una metodologia aplicada a los peajes de autopista. Tesis doctoral. Universitat Rovira i Virgili (2005)

Zimmermann, H.J.: Multicriteria decision making in crisp and fuzzy environments. In: Jones, A., Kaufmann, A., Zimmermann, H.J. (eds.) Fuzzy Sets Theory and Applications, D, pp. 233–256. Reidel Publishing Company, Dordrecht (1986)

# SC: A Fuzzy Approximation for Nonlinear Regression Optimization

Sergio-Gerardo de-los-Cobos-Silva,
Miguel-Ángel Gutiérrez-Andrade, Eric-Alfredo Rincón-García,
Pedro Lara-Velázquez, Roman-Anselmo Mora-Gutiérrez
and Antonin Ponsich

**Abstract** Nonlinear regression is a statistical technique widely used in research which creates models that conceptualize the relation among many variables that are related in complex forms. These models are widely used in different areas such as economics, biology, finance, engineering, etc. These models are subsequently used for different processes, such as prediction, control or optimization. Many standard regression methods have proved to produce misleading results in certain data sets; this is especially true in ordinary least squares. In this paper a novel system of convergence (SC) is presented as well as its fundamentals and computing experience for some benchmark nonlinear regression optimization problems. An implementation using a novel PSO algorithm with three phases (PSO-3P): stabilization, generation with broad-ranging exploration, and generation with in-depth exploration, is presented and tested on 27 databases of the NIST collection with different degrees of difficulty. Numerical results show that the PSO algorithm provides better results when the SC criterion is used, compared to evaluate the usual objective function.

S.-G. de-los-Cobos-Silva (✉) · M.-Á. Gutiérrez-Andrade · P. Lara-Velázquez
Departamento de Ingeniería Eléctrica, Universidad Autónoma Metropolitana – Iztapalapa,
Av. San Rafael Atlixco 186. Col. Vicentina. Del Iztapalapa,
Mexico D.F CP. 09340, Mexico
e-mail: cobos@xanum.uam.mx

M.-Á. Gutiérrez-Andrade
e-mail: gamma@xanum.uam.mx

P. Lara-Velázquez
e-mail: plara@xanum.uam.mx

E.-A. Rincón-García · R.-A. Mora-Gutiérrez · A. Ponsich
Departamento de Sistemas, Universidad Autónoma Metropolitana – Azcapotzalco,
Av. San Pablo 180. Col. Reynosa Tamaulipas, Mexico, Mexico
e mail: rigaeral@correo.azc.uam.mx

R.-A. Mora-Gutiérrez
e-mail: ing.romanmora@gmail.com

A. Ponsich
e-mail: aspo@correo.azc.uam.mx

© Springer International Publishing Switzerland 2015
J. Gil-Aluja et al. (eds.), *Scientific Methods for the Treatment of Uncertainty
in Social Sciences*, Advances in Intelligent Systems and Computing 377,
DOI 10.1007/978-3-319-19704-3_33

407

**Keywords** Particle swarm optimization · Nonlinear regression optimization · Fuzzy approximation

# 1 Introduction

Nonlinear least squares criterion, unlike linear regression under no collinearity, does not guarantee a procedure that can find exact solutions. Even for some linear regression problems where the associated matrices have near zero elements, it is difficult to find solutions using statistical packages, as can be seen in (de-los Cobos-Silva et al. 2014).

The National Institute for Standards and Technology (NIST 2015) states that: "nonlinear least squares regression problems are intrinsically hard, and it is generally possible to find a dataset that will defeat even the most robust codes", and continues: "The datasets provided here are particularly well suited for such testing of robustness and reliability".

There are several challenges in nonlinear regression to ensure that certain methods such as Levenberg-Marquart, Gauss-Newton, etc. can not find the optimal parameter values (Kapanoglu et al. 2007), these challenges consider: (1) increased functional complexity due to non-convexity of the search space, (2) unavailable derivative information, and (3) unfavorable initial solution.

This general theoretical argumentation is supported by the experimental results (Krivý et al. 2000) where 14 regression functions have been considered. In (Kriý et al. 2000) it was shown that the standard algorithms from the statistical packages NCSS, SYSTAT, S-PLUS, SPSS for the large percentage of random starting points failed to find the global minimizer.

Different heuristic methods have been used to deal with nonlinear regression problems, such as: artificial bee colony (ABC) (de-los Cobos-Silva et al. 2013a), differencial evolution (DE), (Tvrdík and Krivý 2004; Tvrdík 2009), genetic algorithm (GA), (Kapanoglu et al. 2007), particle swarm optimization (PSO) (de-los Cobos-Silva et al. 2014, 2013b; Schwaab et al. 2008; Cheng et al. 2013).

Particle Swarm Optimization is population based algorithm that belong to a set of techniques called swarm intelligence heuristics. This kind of algorithms has been successfully implemented to solve different optimization problems. The main characteristic of these heuristics is that manage a population of individuals, where each individual represents a solution to the target problem, and some operations that modify the position of each individual exploring the solution space. Thus, each individual in a population can be moved to better solutions.

In recent decades several heuristics in optimization methods have been developed. These methods are able to find solutions close to the optimum. This paper presents a novel approach that guarantees the solution of the above points using only one algorithm. In practice, it can be observed that the proposed SC along with the three-phase PSO algorithm, allows to escape from suboptimal entrapments in many difficult instances. Moreover, SC is an alternative to the classical criteria, that allows the use of a weight aggregation function to find different solutions, even for non-convex problems.

The objective of this work is to compare numerically SC-PSO-3P on nonlinear regression functions that are well known (NIST 2015). This algorithm was tested on 27 databases, from which 8 are considered highly difficult, 11 with a medium difficulty level and 8 with a low difficulty level.

The work is divided as follows: the nonlinear regression problem considered as optimization problem is presented in the second section; background of fuzzy numbers are presented in the third section; in the fourth section the proposed Convergence System SC, as well as their definitions and fundamentals are presented. PSO is presented in the fifth section. The general guidelines of novel PSO with 3 phases is presented in the sixth section. Numerical examples are presented in the seventh section. Finally, conclusions and future research are presented in section eight.

## 2   Nonlinear Regression

Given two variables $x$ and $y$, observed on $n$ objects, where $x$ is an explanatory variable and $y$ is a variable to explain that depends on $x$, we want to describe the dependence of $y$ with respect to $x$ by a function $f$, i.e. we want to establish the functional relation $y = f(x) + \varepsilon$ where $\varepsilon$ is an error term that is a random variable with mean zero (in this work is not assumed that $\varepsilon$ follows some particular distribution). The function $f$ generally depends on certain parameters, which we denote by vector $\theta$ so the regression function can be written as:

$$y = f(x, \theta) + \varepsilon \tag{1}$$

In the regression problem, we want to find the numeric values of the parameter $\theta = (\theta_1, \theta_2, \ldots, \theta_p)$ such that it optimize some criteria, in particular we want to minimize the least squares criterion, which measures the quality of the functional approximation proposed by minimizing the sum of squared differences:

$$s(\theta) = |y - f(x, \theta)|^2 = \sum_i [y_i - f(x_i, \theta)]^2 \tag{2}$$

Where $x_1 = (x_{1i}, x_{2i}, \ldots, x_{mi})^t$,   $i = 1, 2, \ldots n$, and $y = (y_1, y_2, \ldots, y_n)^t$ are the vectors of observations of the variables, and $|\cdot|$ is the Euclidean norm. From the above we can define a global optimization problem as follows.

Given an objective function: $S: D \rightarrow \Re$, we wish to find (Tvrdík 2009):

$$\theta^* = argmin_{\theta \in D} S(\theta) \tag{3}$$

The point $\theta^*$ is called the global minimum, the search space $D$ is defined as:

$$D = \prod_i [\theta_{min_i}, \theta_{max_i}] \tag{4}$$

where $\theta_{min_i} < \theta_{max_i}$, $i = 1, 2, \ldots, p$. This specification of $S$ is known as box restriction.

Except when $f(x, \theta)$ is a linear function with respect to the parameter vector $\theta$, there is no general solution to this problem. For the nonlinear case, there are deterministic iterative algorithms, Levenberg-Maquardt or Gauss-Newton for instance, that can be used in statistical packages, but they generally fail to find the optimal solutions to the problem.

Therefore, when we want to find a nonlinear regression model of best fit using the criterion of least squares we got a general continuous programming problem, which is a problem difficult to solve (c.f. Tvrdík 2009).

## 3 Fuzzy Numbers

In this section, we introduce the basic concepts of fuzzy numbers based on (Dubois and Prade 1978). A fuzzy number: $A = (a, b, c, d; w)$ is defined as a fuzzy subset of the real line $\Re$ with membership function $h_A$ such that:

1. $h_A$ is continuous mapping from $\Re$ to the closed interval $[0, w]$
2. $h_A = 0, \forall x \in [-\infty, a]$
3. $h_A$ is a strictly increasing function on $[a, b]$.
4. $h_A = w, \forall x \in [b, c]$ where $w$ is a constant and $0 \le w \le 1$.
5. $h_A$ is strictly decreasing on $[c, d]$.
6. $h_A = 0, \forall x \in [d, +\infty]$

Where $0 \le w \le 1$ and $a$, $b$, $c$, $d$ are crisp numbers, and $a \le b \le c \le d$. If $w = 1$, the generalized fuzzy number $A$ is called a normal trapezoidal fuzzy number denoted as $A = (a, b, c, d)$. If $a = b$ and $c = d$ then $A$ is a crisp interval. If $b = c$ then $A$ is a generalized triangular fuzzy number. If $a = b = c = d$, then $A$ is a real number. The membership function $h_A$ of $A$ can be expressed as:

$$h_A = \begin{cases} h_A^L(x), & a \le x \le b \\ w, & b \le x \le c \\ h_A^R(x), & c \le x \le d \\ 0 & \text{otherwise} \end{cases}$$

Where: $h_A^L(x): [a, b] \to [0, w]$ and $h_A^R(x): [c, d] \to [0, w]$ are continuous, $h_A^L(x)$ is strictly increasing and $h_A^R(x)$ strictly decreasing. The inverse functions of $h_A^L(x)$ and $h_A^R(x)$ are denoted by $g_A^L(x)$ and $g_A^R(x)$, respectively. Let be two trapezoidal fuzzy numbers $\tilde{B}_1 = (a_1, b_1, c_1, d_1; w)$, $\tilde{B}_2 = (a_2, b_2, c_2, d_2; w)$ and $c \in \mathfrak{R}$ then: 1. $c\tilde{B}_1 = (ca_1, cb_1, cc_1, cd_1; w)$ 2. $\tilde{B}_1 + \tilde{B}_2 = (a_1 + a_2, b_1 + b_2, c_1 + c_2, d_1 + d_2; w)$.

# 4 SC: System of Convergence

This section is based on (de-los-Cobos-Silva 2015). Let:
(i) $0 \leq f_1, f_2, f_3 \leq 1$, (ii) $f_1 + f_2 + f_3 = 1$, (iii) $0 < w \leq 1$, (iv) $a_i \in \mathfrak{R}, \alpha_i \in \mathfrak{R}$
Consider the following class of fuzzy numbers:

$$C(B) = \left\{ \tilde{B}_1 = (a_i, f_1, f_2, f_3, \alpha_i; w) \mid \text{satisfying the preceding four conditions} \right\}$$

Where:
(i) $b_i = a_i + f_1 \alpha_i$ (ii) $c_i = a_i + (f_1 + f_2)\alpha_i$ (iii) $d_i = a_i + (f_1 + f_2 + f_3)\alpha_i = a_i + \alpha_i$

**Property 1** For any two fuzzy numbers in $\tilde{B}_1, \tilde{B}_2 \in C(B)$, such that $\tilde{B}_1 = (a_1, f_1, f_2, f_3, \alpha_1; w)$ and $\tilde{B}_2 = (a_2, f_1, f_2, f_3, \alpha_2; w)$ and $\forall c \in \mathfrak{R}$, the following equations are satisfied:
(i) $\tilde{B}_1 c = (ca_1, f_1, f_2, f_3, c\alpha_1; w)$ (ii) $\tilde{B}_1 + \tilde{B}_2 = (a_1 + a_2, f_1, f_2, f_3, \alpha_1 + \alpha_2; w)$

**Definition 1** Given a function $G: C(B) \to \mathfrak{R}$, and $\tilde{B}_1, \tilde{B}_2 \in C(B)$ it is said that they are SC-equivalent if and only if $G(\tilde{B}_1) = G(\tilde{B}_2)$

**Definition 2** Given a function $G: C(B) \to \mathfrak{R}$, and a $g_G \in \mathfrak{R}$, in the codomain of $G$, the following SC-equivalence class is defined $B_{g_G} \subset C(B)$ as $B_{g_G} = \left\{ \tilde{B} \in C(B) \mid G(\tilde{B}) = g_G \right\}$

**Comments**:
(i) $B_{g_G}$ is an equivalence class. (ii) $B_{g_1} \cap B_{g_2} = \varnothing, \forall g_1 \neq g_2$ (iii) $\bigcup_{g \in \mathfrak{R}} B_g = C(B)$

**Definition 3** Let $f_1, f_2, f_3, w$ such that they satisfy:
(i) $0 \leq f_1, f_2, f_3 \leq 1$. (ii) $f_1 + f_2 + f_3 = 1$ (iii) $0 < w \leq 1$
We define:

1. $F_1 = (1 + f_2)$
2. $F_2 = \left( 2f_1^2 + 6f_1 f_2 + 3f_2^2 + 3f_3 - 2f_3^2 \right)$
3. $F_3 = \left( 3f_1^3 + 4f_2^3 + 3f_3^3 + 12f_1 f_2^2 + 12f_2 f_1^2 + 6f_3 - 8f_3^2 \right)$
   it can be proven that $F_3 > 0$.
4. $A = \frac{12}{w^3(1 + 3f_2)}$

**Theorem 1** *Let:* $G: C(B) \to \Re$ given by: $G(\tilde{B}) = A[\alpha^2 F_3 - 4\alpha F_2 + 6aF_1]$
Then $G$ is bijective on each SC-equivalence class.
*Proof.* See (de-los-Cobos-Silva 2015).

**Proposition 1** Let $\tilde{B} \in C(B), s, s^*, \varepsilon \in \Re$ such that $s = s^* + \varepsilon$, then:

$$\lim_{\varepsilon \to 0} G(\tilde{B}s) = G(\tilde{B}s^*)$$

*Proof.* See (de-los-Cobos-Silva 2015).

**Proposition 2** Let $\tilde{B}_i \in C(B)$, and $s_i, \varepsilon_i \in \Re$ such that $s_i = s_i^* + \varepsilon_i, i = 1, 2, \ldots, n,$
then:

$$\lim_{\{\varepsilon_i\}_{i=1}^n \to 0} G\left(\sum_{i=1}^n \tilde{B}_i s_i\right) = G\left(\sum_{i=1}^n \tilde{B}_i s_i^*\right)$$

*Proof.* See (de-los-Cobos-Silva 2015).

## 5 Particle Swarm Optimization

The particle swarm optimization is a subset of what is known as swarm intelligence and has its roots in artificial life, social psychology, engineering and computer science. PSO differs from evolutionary computation (c.f. Kennedy et al. 2001) because the population members or agents, also called particles, are "flying" through the problem hyperspace.

PSO is based on the use of a set of particles or agents that correspond to states of an optimization problem, where each particle moves across the solution space in search of an optimal position or at least a good solution. In PSO the agents communicate with each other, and the agent with a better position (measured according to an objective function) influences the others by attracting them to it. The population is started by assigning an initial random position and speed for each element. For this article a swarm of $r$-particle solutions is considered at time $t$ given in the form: $\theta_{1t}, \theta_{2t}, \ldots, \theta_{rt}$ with $\theta_{jt} \in D, j = 1, 2, \ldots, r$, then we define a movement of the swarm of the form: The movement of the swarm is given by:

$$\theta_{jt+1} = \theta_{jt} + V_{jt+1}, \tag{5}$$

where the velocity $V_{jt+1}$ is given by:

$$V_{j,t+1} = \alpha V_{j,t} + rand(0, \phi_1) \left[ \theta'_{j,t} - \theta_{j,t} \right] + rand(0, \phi_2) \left[ \theta'_{g,t} - \theta_{j,t} \right] \qquad (6)$$

where:

- $D$: Space of feasible solutions,
- $V_{j,t}$: Speed at time $t$ of the $j$-th particle,
- $V_{j,t+1}$: Speed at time $t + 1$ in the $j$-th particle,
- $\theta_{j,t}$: $j$-th particle at time $t$,
- $\theta'_{g,t}$: The particle with the best value for all time $t$,
- $\theta'_{j,t}$: $j$-th particle with the best value to the time $t$,
- rand $(0, \phi)$: Random value uniformly distributed on the interval $[0, \phi]$
- $\alpha$: Parameters of scale.

The PSO algorithm is described in Table 1.

**Table 1** PSO algorithm

| 1. Create a population of particles distributed in the feasible space |
| --- |
| 2. Evaluate each position of the particles according to the objective function |
| 3. If the current position of a particle is better than the previous update it |
| 4. Determine the best particle (according to the best previous positions) |
| 5. Update the particle velocities $j = 1, 2, ..., r$ according to Eq. (6) |
| 6. Move the particles to new positions according to Eq. (5) |
| 7. Go to Step 2 until the termination criterion is satisfied |

# 6   SC-PSO-3P

The three-phase PSO-3P algorithm along with the SC criterion is named SC-PSO-3P. The SC-PSO-3P algorithm is described in Table 2. As can be seen in the algorithm in Table 2, the key modification is to consider the SC criterion, and the 3 phases, the rest of the algorithm remains similar. It should be mentioned, that can be used any other search algorithm instead of PSO, although the respective tests must be performed.

**Table 2** SC-PSO-3P algorithm

| |
|---|
| **1.** Set variables c, itF1, itF2, itF3, MaxIt and prop<br>**2.** Create a population of nPop particles distributed in the feasible space<br>**3.** Set cont=0 and it=1. Evaluate the position of the particles according to the function SC<br>**4.** If the current position of a particle is better (respect to SC) than the previous update it<br>**5.** Determine the best particle (according to the best previous positions against the criterion SC) If a better particle cannot be founded, let cont=cont+1<br>**6.** Update the particle velocities $j = 1, 2, ..., nPop$ according to equation (6)<br>**7.** Move the particles to new positions according to equation (5)<br>**8. (Phase 1: Stabilization)**<br>   - if it≤ itF1: go to Step 11<br>**9. (Phase 2: Generation with broad-range exploration)**<br>     - if itF1< it ≤ itF2<br>       * If cont=c<br>         • Set n=1. While n≤ nPop\*prop<br>            • Create a random particle and, with probability bigger than 0.5 sustitute randomly a particle in the swarm<br>         • Set n=n+1<br>       * Set cont=0<br>     - go to Step 11<br>**10. (Phase 3: Generation with in-depth exploration).**<br>     - if itF2≤ it ≤ itF3<br>       * If cont=c<br>         • Set n=1. While n≤ nPop\*prop<br>            • Create a random particle in a variable neighborhood of $\theta'_{g,t}$ and substitute randomly a particle in the swarm.<br>         • Set n=n+1<br>       * Set cont=0<br>     - go to Step 11<br>**11.** Select the best nPop particles according to SC criterion<br>**12.** Set it=it+1. Go to Step 3 until the termination criterion is satisfied |

# 7   Computational Results

The algorithm was implemented in Matlab R2008a and was run in an Intel Core i5-3210 M processor computer at 2.5 GHz, running on Windows 8. 50 runs were performed considering a population of 120 particles and 500 iterations per run.

In order to evaluate the performance of the proposed SC-PSO-3P, we used 27 instances of NIST, ranging from low to high difficulty, from 2 to 8 parameters, and from 6 to 250 observed data.

In (Kapanoglu et al. 2007) states that "Except in cases where the certified value is essentially zero (for example, as occurs for the three Lanczos problems), a good nonlinear least squares procedure should be able to duplicate the certified results to at least 4 or 5 digits".

The goodness of the results of the proposed algorithms was estimated by the number of duplicate digits when compared to the certified results provided in (Kapanoglu et al. 2007), which were found using iterative deterministic algorithms (Levenberg-Maquardt and Gauss-Newton). The number of duplicate digits, denoted by $\lambda$, can be calculated via the *log* relative error (Tvrdík and Krivý 2004), using the following expression.

$$\lambda = -log_{10}\left(\frac{|w-c|}{|c|}\right) \tag{7}$$

where $c \neq 0$ denotes the certified value, and $w$ denotes the value estimated by the proposed algorithm.

In this paper, the functions for nonlinear regression optimization were taken from (Kapanoglu et al. 2007), the search ranges, are reported in (de-los Cobos-Silva et al. 2014), and the efficiency was defined as:

$$Efficiency = \begin{cases} 1 - abs((h(x) - h(x^*))/h(x^*)) & si\, h(x^*) \neq 0, \\ 1 - abs(h(x)) & si\, h(x^*) = 0 \end{cases}$$

Where $h(x^*)$ is the optimum value of $h$, and $h(x)$ is the best value found by the proposed algorithm.

When the efficiency was greater than 0.999999, it was rounded to 1. The distance of the parameters found respect to certificated values was defined as:

$$Dist(\theta) = \sum_i \left[\theta_i - \theta_1^*)\right]^2 \tag{8}$$

Where $\theta_1^*, \theta_2^*, \ldots, \theta_p^*$ are the certified numeric values of the parameters, and $\theta_1, \theta_2, \ldots, \theta_p$ are the numeric values of the parameters found by the algorithm.

In Table 3 we resume the performance of the proposed SC-PSO-3P algorithm, with 120 particles and 500 iterations per run. The first column provides the level of difficulty (L = lower, M = average, H = Higher). In column 2 the model of classification is presented (Mi. = Miscellaneous, Ex. = Exponential, Ra. = Rational). Column 3 includes the name of the database. In column 4 the dimension of the problem is included (D). The fifth column provides the certified values for the residual sum of squares. The sixth column includes the best value achieved by SC-PSO-3P of the residual sum of squares. In column 7, the efficiency is presented. In column 8, we include the distance (Dist.) of the parameters found respect to certificated values. Finally column 9 provides the corresponding $\lambda$.

**Table 3** SC- PSO- 3P results using 120 particles and 500 iterations

| Dif. | T | Data Base | D | Res. sum of sqrt. opt. | Res. sum of sqrt. found | Efficiency | Dist. | λ |
|---|---|---|---|---|---|---|---|---|
| H | Mi. | Bennett5 | 4 | 5.24047440073E-4 | 5.2419235419E-4 | 1.000E+0 | 3E+03 | 3.5 |
| L | Ex. | Misala | 2 | 1.24551388940E-1 | 1.24551388944E-1 | 1.000E+0 | 2.E-11 | 10 |
| M | Mi. | ENSO | 9 | 7.88539786668E+2 | 7.8856911105E+2 | 1.000E+0 | 2.E-03 | 4.4 |
| H | Ex. | BoxBod | 2 | 1.16800887662E-3 | 1.1680088765E+3 | 1.0000E+0 | 7.90-15 | 10.4 |
| H | Ra. | Thurber | 7 | 5.642708239TE+3 | 5.6523777107E+3 | 1.0017E+0 | 4.E+01 | 2.77 |
| H | Ex. | Rat42 | 3 | 8.0565229338E+0 | 8.0565229338E+0 | 1.0000E+0 | 4.9E-15 | 11.8 |
| H | Ra. | MGH09 | 4 | 3.07505603850E-4 | 3.07505603849E-4 | 1.0000E+0 | 4.2E-14 | 11.6 |
| H | Ex. | rat43 | 4 | 8.78640490080E+3 | 8.7864049083E+3 | 1.0000E+0 | 7.5E-08 | 10.3 |
| H | Ex. | Eckerle4 | 3 | 1.46358874872E-3 | 1.46358874872E-3 | 1.0000E+0 | 1.0E-15 | 10.7 |
| H | Ex. | MGH10 | 3 | 8.7945855171E+1 | 8.7945875816E+1 | 1.0000E+0 | 1.4E-04 | 6.63 |
| M | Ex. | Gauss3 | 8 | 1.24448463600E+3 | 1.24454231709E+3 | 1.000E+0 | 2.0E-2 | 4.3 |
| L | Ex. | Lanczo3 | 6 | 1.61171935940E-8 | 2.90116362017E-6 | 1.800E+2 | 4.6E-1 | -2.2 |
| L | Ex. | Chwirut2 | 3 | 5.13048029410E+2 | 5.13048029409E+2 | 1.000E+0 | 2.E-17 | 11 |
| L | Ex. | Chwirut1 | 3 | 2.38447713930E+3 | 2.38447713930E+3 | 1.000E+0 | 6.E-17 | 11 |
| L | Ex. | Gauss1 | 8 | 1.315822243200E+3 | 1.31583573270E+3 | 1.000E+0 | 2.9E-4 | 5.0 |
| L | Ex. | Gauss2 | 8 | 1.247528209200E+3 | 1.24753546735E+3 | 1.000E+0 | 3.0E-4 | 5.2 |
| L | Mi. | DanWod | 2 | 4.317308408300E-3 | 4.31730840829E-3 | 1.000E+0 | 8.E-17 | 11 |
| L | Mi. | Misra1b | 2 | 7.546468153300E-2 | 7.546468153337E-2 | 1.000E+0 | 1.E-14 | 11 |
| M | Ra. | Kirby2 | 5 | 3.905073962400E+0 | 3.90517660952E+0 | 1.000E+0 | 2.4E-5 | 4.5 |
| M | Ra. | Hahn1 | 7 | 1.53243828540E+0 | 2.70366469911E+1 | 1.764E+1 | 1.6E-2 | -1.2 |
| M | Ex. | Nelson | 3 | 3.79768331760E+0 | 3.79768332335E+0 | 1.000E+0 | 2.E-11 | 8.8 |
| M | Ex. | MGH17 | 5 | 5.46489469750E-5 | 5.46574549274E-5 | 1.000E+0 | 4.2E-4 | 3.8 |
| M | Ex. | Lanczos1 | 6 | 1.43078677216E-25 | 1.00086316183E-19 | 6.995E+5 | 3.E-18 | -5.8 |
| M | Ex. | Lanczos2 | 6 | 2.22994281250E-11 | 1.66789625130E-9 | 7.479E+1 | 5.7E-4 | -1.8 |
| M | Mi. | Misra1c | 2 | 4.09668369710E-2 | 4.09668369706E-2 | 1.000E+0 | 1.E-10 | 11 |
| M | Mi. | Misra1d | 2 | 5.64192952830E-2 | 5.64192952826E-2 | 1.000E+0 | 1.E-13 | 11 |
| M | Mi. | Rozman1 | 4 | 4.94848473310E-4 | 4.94848473141E-4 | 1.000E+0 | 1.1E-7 | 11 |

In Table 4 we present the performance of the proposed SC-PSO-3P algorithm, with 12 particles and 500 iterations per run. In this case, we only report the data base where a $\lambda \geq 4$ was reached. The first column provides the level of difficulty (L = lower, M = average, H = Higher). In column 2 the name of the database is included. Column 3 provides the certified values for the residual sum of squares. In column 4 the best value achieved by SC-PSO-3P of the residual sum of squares is provided. In column 5, the efficiency is presented. Column 6 includes the distance (Dist.) of the parameters found respect to certificated values. Finally column 7 provides the corresponding $\lambda$.

**Table 4** SC- PSO- 3P results using 12 particles and 500 iterations

| Dif. | Data base | Res. sum of sqrt. opt. | Res. sum of sqrt. found | Efficiency | Dist. | $\lambda$ |
|------|-----------|------------------------|-------------------------|------------|-------|-----------|
| L | Misala | 1.245513889400E-01 | 1.245513889444E-01 | 1.000000 | 1.77E-12 | 10.5 |
| H | BoxBOD | 1.168008876600E+03 | 1.168008876556E+03 | 1.000000 | 4.09E-14 | 10.4 |
| H | Rat42 | 8.056522933800E+00 | 8.056522973223E+00 | 1.000000 | 8.36E-08 | 8.3 |
| H | MGH09 | 3.075056038500E-04 | 3.075056038493E-04 | 1.000000 | 6.45E-15 | 11.6 |
| L | Chwirut2 | 5.130480294100E+02 | 5.130480294069E+02 | 1.000000 | 1.06E-16 | 11.2 |
| L | Chwirut1 | 2.384477139300E+03 | 2.384477139309E+03 | 1.000000 | 3.10E-15 | 11.4 |
| L | DanWood | 4.317308408300E-03 | 4.317308409207E-03 | 1.000000 | 2.47E-12 | 9.7 |
| M | Nelson | 3.797683317600E+00 | 3.797759410441E+00 | 1.000020 | 4.81E-07 | 4.7 |
| M | Misra1d | 5.641929528300E-02 | 5.641929529547E-02 | 1.000000 | 3.64E-08 | 9.7 |

The MGH09 is a special instance, because of its high degree of difficulty. For this data base, a specific comparative study was realized. A total of 6000 evaluations of the objective function were performed using 3, 6 and 12 particles with 2000, 1000 and 500 iterations, respectively, per run for PSO, PSO and SC-PSO-3P. The results are presented in Table 5. The first column provides the number of particles (part.) and iterations (iter.). In column 2 the name of the algorithm used is included. Column 3 provides the certified values for the residual sum of squares. In column 4 the best value achieved by each algorithm of the residual sum of squares is presented. In column 5, the efficiency is provided. In column 6, the distance (Dist.) of the parameters found respect to certificated values is included. Finally column 7 provides the corresponding $\lambda$.

**Table 5** SC- PSO- 3P results for MGH09 using different parameters

| | | Res. sum of sqrt. opt. | Res. sum of sqrt. find | Efficiency | Dist. | $\lambda$ |
|----------------------|---------|------------------------|------------------------|------------|-------|-----------|
| 12 pt. × 500 iter. | SCPSO3P | 0.000307505604 | 0.000307505604 | 1.00000 | 6E-1 | 11 |
| | PSO-3P | 0.000307505604 | 0.000309635729 | 1.00692 | 1E-3 | 2 |
| | PSO | 0.000307505604 | 0.000599860548 | 1.95073 | 1E-1 | 0 |
| 6 pt. × 1000 iter. | SCPSO3P | 0.000307505604 | 0.000307536742 | 1.00010 | 8E-6 | 4 |
| | PSO-3P | 0.000307505604 | 0.000309838989 | 1.00758 | 5E-4 | 2 |
| | PSO | 0.000307505604 | 0.000599860548 | 1.95073 | 1E-1 | 0 |
| 3 pt. × 2000 iter. | SCPSO3P | 0.000307505604 | 0.000307544407 | 1.00012 | 2E-5 | 4 |
| | PSO-3P | 0.000307505604 | 0.000412511764 | 1.34147 | 4E-2 | 0 |
| | PSO | 0.000307505604 | 0.004861744550 | 15.81023 | 1E-1 | -1 |

# 8  Conclusions and Further Research

In this work, a novel criterion called System of Convergence (SC) is proposed, which was implemented using a novel PSO based algorithm with three phases: stabilization, generation with broad-ranging exploration and generation with in-depth exploration, SC-PSO-3P, which finds estimated parameters of well known nonlinear regression problems. The results show that this algorithm found the global minimum in most tasks where gradient algorithm fails.

This algorithm was tested in a set of benchmark instances. The empirical evidence shows that SC-PSO-3P is very efficient in low and highlevel of difficulty instances.

According to (Tvrdík 2009) for the database given in (Kapanoglu et al. 2007), except for the case where the certified value is essentially zero, a good procedure for non-linear least squares allows duplicate at least 4 digits of the certified values. It was observed that the method SC-PSO-3P provides very good results.

As seen from Table 5, the solutions obtained by SC-PSO-3P are significantly greater than those obtained by SC-PSO and PSO, so the proposed hybrid is very robust.

We suggest SC-PSO-3P algorithm approach as a tool for solving problems involving complex nonlinear relations, i.e. the use of SC-PSO-3P is an alternative to find the values of the parameters that provides good results in nonlinear regression problems.

As an important remark, a deeper study concerning the boundary conditions of the parameters $a$ and $\alpha$ of SC, as well as their implementation with an algorithm different from PSO must be performed.

It is also necessary to conduct a further study about the topology generated by SC. It was observed that in all the cases studied, SC can reach the global optimum or being very close to it with shorter iteration times and less iterations. Worth mentioning that in a later work a more extensive study will be conducted for each type of optimization models.

**Acknowledgments**  S.G., would like to thank to D.Sto., and to P.V. Gpe. for their inspiration, and to Ma, Ser, Mon, Chema, and to his Flaquita for all their support.

# References

Cheng, S., Zhao, C., Wu, J., Shi, Y.: Particle swarm optimization in regression analysis: a case study. Lect. Notes Comput. Sci. **7928**, 55–63 (2013)

de-los-Cobos-Silva, S.G., Gutiérrez-Andrade, M.A., Rincón-García, E.A., Lara-Velázquez, P., Aguilar-Cornejo M.: Estimación de parámetros de regresión no lineal mediante colonia de abejas artificiales. Revista de Matemática: Teoría y Aplicaciones **20**(1), 49–60 (2013a)

de-los-Cobos-Silva, S.G., Terceño-Gómez, A., Gutiérrez-Andrade, M.A., Rincón-García, E.A., Lara-Velázquez, P., Aguilar-Cornejo M.: Particle swarm optimization an alternative for parameter estimation in regression. Fuzzy Econ. Rev. **18**(2), 19–32 (2013b) de-los Cobos-Silva, S.G., Gutiérrez-Andrade, M.A., Rincón-García, E.A., Lara-Velázquez, P., Aguilar-Cornejo, M.: Colonia de Abejas Artificiales y Optimización por Enjambre de Partículas para la Estimación de Parámetros de Regresión No Lineal. Revista de Matemática: Teoría y Aplicaciones 21(1), 107–126 (2014)

de-los-Cobos-Silva, S.G.: SC: System of Convergence, Theory and Fundaments. Revista de Matemática: Teoría y Aplicaciones, to appear (2015)

Dubois, D., Prade, H.: Operations on fuzzy numbers. Int. J. Syst. Sci. 9, 613–626 (1978)

Kapanoglu, M., Koc, O., Erdogmus, S.: Genetic algorithm in parameter estimation for nonlinear regression models: an experimental approach. J. Stat. Comput. Simul. 77(10), 851–867 (2007)

Kennedy, J., Eberhart, R.C., Shi, Y.: Swarm Intelligence. Morgan Kaufmann, San Francisco (2001)

Krivý, I., Tvrdík, J., Krepec, R.: Stochastic algorithms in nonlinear regression. Comput. Stat. Data Anal. 33, 277–290 (2000)

NIST: National Institute of Standards and Technology (2005). http://www.itl.nist.gov/div898/strd/index.html

Schwaab, M., Biscaia Jr, E.C., Monteiro, J.L., Pinto, J.C.: Nonlinear parameter estimation through particle swarm optimization. Chem. Eng. Sci. 63, 1542–1552 (2008)

Tvrdík, J., Krivý, I.: Comparison of Algorithms for nonlinear regression estimates. Antoch, J. (ed.) COMSTAT, Physica-Verlag, Heidelberg, New York, 1917–1924, (2004)

Tvrdík, J.: Adaptation in differential evolution: a numerical comparison. Appl. Soft Comput. 9(3), 1149-11

# Winding Indexes at Specific Traveling Salesman Problems

Raúl O. Dichiara and Blanca I. Niel

**Abstract** The resolutions of the different Shortest and Longest Euclidean Hamiltonian Path Problems on the vertices of N-Gons, by means of a geometric and arithmetic algorithm allow us to define winding indexes for Hamiltonian and Quasi-Hamiltonian cycles. New statements characterize orientation of non necessarily regular cycles on N-Gons and deal with the existence or absence of reflective bistarred Hamiltonian tours on vertices of coupled N-Gons.

**Keywords** Reflective hamiltonian cycles · Winding indexes · N-Gons · Bistarred hamiltonian cycles · Extremal geometric problems · Networks · Regular figures

## 1 Introduction

We consider the network $N\left(K_{n2p+1}\left(e^{i\pi}\sqrt[n]{1}\right), (d_{i,j})_{nxn}\right)$, where $K_{n2p+1}\left(e^{i\pi}\sqrt[n]{1}\right)$ is the complete graph with vertices on the $n$th roots of the unity $\sqrt[n]{1} = e^{i\cdot\frac{2k\pi}{n}} = \frac{\cos(2k\pi)}{n} + i\frac{sen\,2k\pi}{n}$, for $k \in \{0, 1, 2, \ldots n-1\}$, and $D = (d_{i,j})$ is the

Mathematics Subject Classification, MSC 2010: 05C12 Distance in graphs, 05C38 Paths and Cycles, 05C22 Signed and weighted graphs, 51M20 Polyhedra and Polytopes; Regular figures, division of spaces, 51F15 Reflection groups, Reflection geometries, 20F55 Reflection and Coxeter groups, 05C35 Extremal problems, 90B10 Network models, deterministic.

R.O. Dichiara (✉)
Department IIESS Instituto de Investigaciones Económicas y Sociales del Sur (CONICET-UNS), 12 de Octubre 1098, 8vo Piso, 8000 Bahía Blanca, Argentina
e-mail: dichiara@criba.edu.ar

B.I. Niel
Departamento de Matemática, Universidad Nacional del Sur, Bahía Blanca, Argentina
e-mail: biniel@criba.edu.ar

© Springer International Publishing Switzerland 2015
J. Gil-Aluja et al. (eds.), *Scientific Methods for the Treatment of Uncertainty in Social Sciences*, Advances in Intelligent Systems and Computing 377,
DOI 10.1007/978-3-319-19704-3_34

$n \times n$ matrix of the Euclidean distances between nodes (Buckly and Harary 1990). In these structures we dealt with the Longest Cyclic and Non-Cyclic Euclidean Hamiltonian Path Problems in vertices of $2p + 2$-Gons (Applegate et al. 2006). We struggled with the analogous extremal geometric problems on the vertices of $2p + 1$-Gons in Niel et al. (2010). In this disclosure, we posed at Sect. 3 procedures that associate winding indexes which interpretate orientations in Euclidean Hamiltonian and Quasi-Hamiltonian Cycles (Barvinok et al. 2003). Particularly, the methodology comes forth an extension of Lemma 2.6 and Theorem 2.7 in Niel (2012). Furthermore, Sect. 4 settles the conditions for the existence of Reflective Bistarred Hamiltonian Cycles in the architectures of $N(K_{N=2n}(e^{i\pi}\sqrt[n]{1}, re^{i\pi}\sqrt[n]{1}), d_{NxN}^E)$ networks, wherein the structural parameter $r$ is a fixed positive real number in (0, 1). This approach emerges from a generalized scope of the variational geometric rays at the Quasi-Spheric Mirror construction in Niel (2006) given birth from Hamilton thoughts in Hamilton (1833), whose central geometric optics concept has been quoted at page 15 in Niel (2012).

Precisely, Sect. 2 is devoted to the recapitulation of intrinsic concepts inherent of the vertex locus of a regular N-Gon. Task divided in the contents of *"Natural evaluations of Euclidean path lengths"* and *"Resuming terms and definitions"*. Theorem 3.1 and Theorem 3.2 are the main results of Sect. 3. Herein, illustrative examples highlight and interpret the winding indexes appended to any Euclidean Hamiltonian cycles in $N\left(K_{n2p+1}(e^{i\pi}\sqrt[n]{1}), (d_{i,j})_{nxn}\right)$ networks. Section 4 confirms in Theorem 4.1.1 the necessary conditions for the existence of Reflective Bistarred Euclidean Hamiltonian cycles passing through the vertices of regular coupled Nodd-Gons. Result that leaves an open door to the regular shapes of *"the reflective maximum density bistarred Euclidean Hamiltonian cycles"* as the solution of Max. TSP at $N(K_{N=2nodd}(e^{i\pi}\sqrt[n]{1}, re^{i\pi}\sqrt[n]{1}), d_{NxN}^E)$ net arquitectures (Watts 1999). Contrariwise, Theorem 4.2.1 explicits the unfeasibility of any reflective bistarred Hamiltonian cycle resolve the Max. TSP in the vertices of regular coupled Neven-Gons. Appendix A display elemental sentences to obtain each one of the Euclidean Hamiltonian cycles in an exhaustive exploration in the nets $N\left(K_n(e^{i\pi}\sqrt[n]{1}), (d_{i,j})_{nxn}\right)$, with $n$ less than or equal to ten, because of the running required time. Conclusions and relevant references end this paper.

# 2   Recapitulation of Concepts

In this work we regain concepts, terms and definitions from previous papers, in particular from Niel (2012), although the example illustrations are pictured at the vertices of $2p + 1$-Gons. Worthy of mention is that the discussed aspects have, in general, evenness independence.

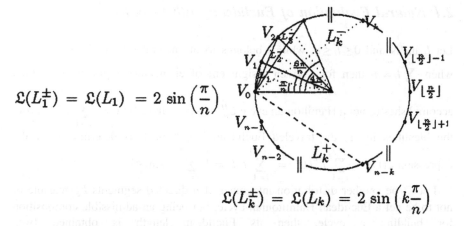

$$\mathcal{L}(L_1^{\pm}) = \mathcal{L}(L_1) = 2 \sin\left(\frac{\pi}{n}\right)$$

$$\mathcal{L}(L_k^{\pm}) = \mathcal{L}(L_k) = 2 \sin\left(k\frac{\pi}{n}\right)$$

**Fig. 1** $L_k$ lengths in vertices of a $N_{even}$-Gon

From geometric argument of the vertex locus follows (1), the recursive formulæ (2), (3) and (4) which determine the lengths of the $L_k$ segments and their differences, see Fig. 1.

Since,

$$\zeta(L_k^{\pm}) = \zeta(L_k^-) = \zeta(L_k^+) = \zeta(L_k) \cong L_k, \quad 1 \leq k \leq \lfloor n/2 \rfloor \tag{1}$$

$$\zeta(L_1^{\pm}) = \zeta(L_1) = 2 \sin\left(\frac{\pi}{n}\right)$$

$$\zeta(L_k^{\pm}) = \zeta(L_k) = 2 \sin\left(\frac{k\pi}{n}\right) \tag{2}$$

$$\zeta(L_k) = \zeta(L_{k-1}) + 4 \sin\left(\frac{\pi}{2n}\right) \cos\left[\frac{(2k-1)\pi}{2n}\right], \quad 1 \leq k \leq \lfloor\frac{n}{2}\rfloor$$

$$\zeta(L_k) = \zeta(L_{k-2}) + 4 \sin\left(\frac{\pi}{2n}\right)\left[\cos\left[\frac{(2k-3)\pi}{2n}\right] + \cos\left[\frac{(2k-1)\pi}{2n}\right]\right], \quad 1 \leq k \leq \lfloor\frac{n}{2}\rfloor \tag{3}$$

$$\zeta(L_k) = \zeta(L_1) + 4 \sin\left(\frac{\pi}{2n}\right)\left[\cos\left[\frac{3\pi}{2n}\right] + \cos\left[\frac{5\pi}{2n}\right] + \cdots + \cos\left[\frac{(2k-1)\pi}{2n}\right]\right],$$
$$1 \leq k \leq \lfloor\frac{n}{2}\rfloor \tag{4}$$

The bracket term inferior and superior bounds in (4) are $(k-1)\cos\left[\frac{(2k-1)\pi}{2n}\right]$ and $(k-1)\cos\left(\frac{3\pi}{2n}\right)$.

## 2.1 Natural Evaluation of Euclidean Path Lengths

Let $l_i$ in IN and $0 \le l_i \le n$, while $i$ belongs to the natural set $\{1, 2, \cdots, \lfloor n/2 \rfloor\}$, when $\sum_{i=1}^{n} l_i = n$, then for a given assignment of elementary segments $L_i^{\pm}$ which accomplishes or not a Hamiltonian cycle $C_n^H$, is symbolized by $C = \sum_{i=1}^{\lfloor n/2 \rfloor} l_i L_i^{\pm}$. From the identities in (1) the traveled Euclidean length of $C$ is determined for the expression $\zeta(C) = \sum_{i=1}^{\lfloor n/2 \rfloor} l_i \zeta(L_i^{\pm}) = \sum_{i=1}^{\lfloor n/2 \rfloor} l_i\, L_i = 2 \sum_{i=1}^{\lfloor n/2 \rfloor} l_i\, \sin\left(\frac{i\pi}{n}\right)$.

Let $C'$ be another distribution integrated of $n$ directed segments $L_k^{\pm}$ feasible or not to be get a Euclidean Hamiltonian cycle, but being an admissible composition for building a cycle, then its Euclidean length is obtained from $\zeta(C') = \sum_{i=1}^{\lfloor n/2 \rfloor} l'_i \zeta(L_i^{\pm}) = \sum_{i=1}^{\lfloor n/2 \rfloor} l'_i\, L_i = 2 \sum_{i=1}^{\lfloor n/2 \rfloor} l_i\, \sin\left(\frac{i\pi}{n}\right)$. Consequently, the traveled length comparison of $C$ and $C'$ results $\zeta(C') - \zeta(C') = 2 \sum_{i=1}^{\lfloor n/2 \rfloor} (l_i - l'_i)\, \sin\left(\frac{i\pi}{n}\right)$.

**Example 2.1** Let $\zeta(C') = \zeta\left(3L_1^{\pm} + 3L_2^{\pm} + L_3^{\pm} + 2L_4^{\pm}\right)$ and $\zeta(C) = \zeta\left(3L_1^{\pm} + 2L_2^{\pm} + L_3^{\pm} + 3L_4^{\pm}\right)$. Both Euclidean traveled lengths are attained by the Euclidean Hamiltonian cycles $C_H^{'9} = 3L_1^{-} + L_4^{-} + L_2^{+} + L_3^{-} + 2L_2^{+} + L_4^{+}$ and $C_H^{'9} = L_1^{-} + L_3^{-} + L_1^{-} + L_2^{+} + 3L_4^{-} + L_2^{-} L_1^{-}$, respectively.

$$\zeta(C_H^{9}) - \zeta(C'9_H) = 2\left[\sin\left(\frac{4\pi}{9}\right) - \sin\left(\frac{2\pi}{9}\right)\right] > 0$$

First one $\zeta(C')$ ranked at 53rd and the second one $\zeta(C)$ at 71st position in the spectra of traveled Euclidean Hamiltonian cycles at vertices of the regular 9-Gon. Information renders by a run of an exhaustive exploration code. Particularly, each one of these lengths were nine hundred times achieved by some Euclidean Hamiltonian tour.

## 2.2 Resuming Terms and Definitions

In this section, we regain concepts and terms from the paper (Niel 2012), hence we decide not to overwhelm with therein disclosed details.

Let be $V_o, \ldots, V_{n-1}$ the points of $e^{i\pi} \sqrt[n]{1}$ set and let them be clockwise enumerated for the integers modulo $n$, $Z_n$, from the vertex $V_o = (-1, 0)$. For each $0 \le k \le \lfloor n/2 \rfloor$ and each $j \, \varepsilon \, Z_n$, let $L_{\bar{k},j}^{-}$ denote the segment that joins $V_j$ with $V_{j+k}$, while $L_{k,j}^{+}$ denotes the one that joins $V_j$ to $V_{j+(n-k)} = V_{j-k}$. From now onwards, $L_k^{-}$

and $L_k^+$ denote $L_{k,0}^-$ and $L_{k,0}^+$, respectively. Let $l_{max}$ be the diameter, it joins the vertex $V_j$ with its opposite $V_{j+\frac{n}{2}}$, only if $n$ is even. In addition, $P_n$ symbolizes a regular $n$-Gon inscribed in the unitary circle and with vertices in $V_o, \ldots, V_{n-1}$.

**Fig. 2** $L_k^{\mp}$ in vertices of a $N_{odd}$-Gon

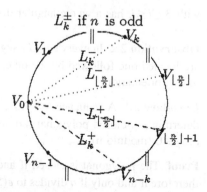

**Definition 2.1** For each integer n, $L$ is a "$L_k$ segment" if for any $k$ such that $0 \le k \le \lfloor n/2 \rfloor$, and for any $j \, \varepsilon \, Z_n$, is $L = L_{k,j}^-$ or $L = L_{k,j}^+$, see Fig. 2.

**Definition 2.2** If $L$ is an $L_k$ segment, "the associated integer to $L$", which is denoted by $e(L)$ is defined as:

$$e(L) = \begin{cases} k & if \quad L = L_{k,j}^- \\ n - k \equiv -k (mod \, ulo \, n) & if \quad L = L_{k,j}^+ \end{cases} \tag{5}$$

**Definition 2.3** If $S = \{L_1, \ldots, L_j\}$ is a sequence of $L_k$ segments, "the integer associated to the pathway determined by $S$" denoted $e(S)$ is given by $e(S) = \sum_{i=1}^{j} e(L_i) \pmod{n}$.

**Observation 2.1** The central angle that determines a $L_k$ segment in the unitary circumference is $e(L_k)\frac{2\pi}{n}$, i.e.

$$\angle(L_k) = \begin{cases} k\frac{2\pi}{n} & if \quad L_k = L_k^- \\ (n-k)\frac{2\pi}{n} & if \quad L_k = L_k^+ \end{cases} \qquad 1 \le k \le \lfloor \frac{n}{2} \rfloor.$$

**Example 2.2** $L_k$ be directed segments with vertices in the $n^{th}$ roots of the unity $\{V_0 = (-1,0), \ldots, V_{n-1}\}$, see Fig. 2. Then, their respective angular advances are:

1. $\angle(L_1^-) = \frac{2\pi}{n}$;
2. $\angle(L_1^+) = (n-1)\frac{2\pi}{n}$;
3. $\angle\left(L_{\lfloor\frac{n}{2}\rfloor}^+\right) = \left(\lfloor\frac{n}{2}\rfloor + 1\right)\frac{2\pi}{n_{odd}}$

**Proposition 2.1** If $S = \{L_{k1}, L_{k2}, \ldots, L_{kj}\}$ is a sequence of $L_k$ segments. The polygonal that is built by placing the first $L_{k1}$ segment from $V_o$, the second $L_{k2}$ segment from the vertex $V_{o + e_{(Lk1)}}$, and so on with the rest of the $L_{k1}$ segments in $S$ with $3 \leq i \leq j$, has a whole angular displacement given by $\sum_{i=1}^{j} e(L_i) \frac{2\pi}{n} = e(S) \frac{2\pi}{n}$.

**Observation 2.2** In every case $L_k$ segments of a sequence $S = \{L_{k1}, L_{k2}, \ldots, L_{kj}\}$, are located one followed by the other by considering the correlative angles always in cw. orientation.

**Corollary 2.1** A sequence $S = \{L_{k1}, L_{k2}, \ldots, L_{kj}\}$, with $2 \leq j \leq n$, of $L_k$ segments determines a closed polygonal (no necessarily a Hamiltonian cycle) if and only if $e(S) \equiv 0$ (modulo n).

**Proof** The polygonal is closed if and only if $e(S) \frac{2\pi}{n}$ is an integer multiple of $2\pi$, therefore if and only if $n$ divides to $e(S)$, i.e. if and only if $e(S) \equiv 0$ (modulo n).   □

# 3   Statements for the Winding Indexes

Let us consider the network $N(K_n(e^{i\pi} \sqrt[n]{1}), D)$, where $K_n(\sqrt[n]{1})$ is the complete graph with vertices on the $n$th roots of the unity and $D = (d_{i,j})$ is the $n \times n$ matrix of the Euclidean distances between nodes. In other words, these network architectures are in congruence with the Euclidean complete connectivity over the vertices of a regular $N$-Gon (Kirillov 1999). Here, we deal with Quasi-Hamiltonian or Hamiltonian Cycles in $N(K_{n=2p+1}(e^{i\pi} \sqrt[n]{1}), d_{(i,j)nxn})$ networks. However, in general, the results have validity for any closed polygonal built on vertices of any arbitrary regular $N$-Gon (Treatman et al. 2000).

**Theorem 3.1** Let $S = \{L_{k1}, L_{k2}, \ldots, L_{kn}\}$ be a sequence of $n$ segments $L_k, 1 \leq k \leq \lfloor n/2 \rfloor$, which determines a cycle C in $N(K_{n=2p+1}(e^{i\pi} \sqrt[n]{1}) d_{(i,j)nxn})$ networks, then exists a unique integer $z(C)$, $1 \leq z(C) \leq (n-1)$, such that

$$e(S) = \sum_{i=1}^{n} e(L_i) = z(C) \cdot n \equiv 0 \text{ (modulo n)}.$$

**Proof** From $1 \leq e(L_{ki}) \leq (n-1)$, and for $1 \leq i \leq n$, then $n \leq \sum_{i=1}^{n} e(L_{ki}) \leq n(n-1)$

$\Leftrightarrow 1 \leq n^{-1} \sum_{i=1}^{n} e(L_{ki}) \leq (n-1)$. Since C is a cycle $\sum_{i=1}^{n} e(L_{ki}) \equiv 0$ (modulo n), i.e. this sum is a multiple of n. Therefore $z(C) = n^{-1} \sum_{i=1}^{n} e(L_{ki})$ is the integer.

Next definition follows from Proposition 2.1 and Corollary 2.1. □

**Definition 3.1** Let $C$ be a Euclidean Hamiltonian cycle in $N(K_{n=2p+1}$ $(e^{i\pi} \sqrt[n]{1}), d_{(i,j)nxn})$ networks, the integer $z(C)$ is named the index of the cycle $C$, and it represents the number of turns of the cycle $C$ around the origin (or around of the centre of the unitary circumference).

**Remark 3.1** In particular, Theorem 3.1 asserts that the maximum index for a Euclidean Hamiltonian cycle with vertices in a regular N-Gon is $(n-1)$.

Direct from the definition of the associated integer to a directed segment $e(L_k^\pm)$, it is less than or equal to $(n-1)$ and only $e(L_1^+) = (n-1)$.

**Example 3.1** Winding indexes in the minimum density Euclidean Hamiltonian cyclic polygonals on the vertices of a regular N-Gon.

In reference to the first and second cases exhibited at Example 2.2 pg. 4, results that:

1. $S_1 = \left\{ \underbrace{L_1^-, \ldots, L_1^-}_{n} \right\}$. In this case $C_i$ is the Euclidean Hamiltonian cyclic

polygonal built upon the vertices of a regular N-Gon, which is in correspondence with the nth roots of the unity. Hence $nL_1^-$ travels cw. once around the centre since $\sum_{i=1}^{n} e(L_1^-) = n$. Therefore, the index of the cycle is $z(C_1) = 1$, e.g. see Fig. 4, left.

2. $S_1 = \left\{ \underbrace{L_1^+, \ldots, L_1^+}_{n} \right\}$. Here, $z(C_2) = (n^{-1}) \sum_{i=1}^{n} e(L_1^+) = n-1 \equiv -1$ (modulo n).

Since the angular displacement of each $L_1^+$ segment is $(n-1) \cdot \frac{2\pi}{n}$, the consecutive collocation of the $n$ segments $L_1^+$ implies the accomplishment of $n-1$ complete turns around the origin. Therefore, the sides of $P_n$ are located in a contrary way with respect to those in the previous item. See Fig. 4, right. In other words, the congruence $n-1 \equiv -1$ (modulo $n$) means that $C_2$ construction describes $n-1$ turns cw. It is equivalent to the performing of a single ccw. turn for the Euclidean Hamiltonian cyclic pathway $nL_1^+$, e.g. see Fig. 4, right.

The minimum density Euclidean Hamiltonian cycles $\left\{ \underbrace{L_1^-, \ldots, L_1^-}_{n} \right\}$ and

$\left\{ \underbrace{L_1^+, \ldots, L_1^+}_{n} \right\}$ determine cyclic trajectories which move in the same geometric

place, however they have opposite orientation, the first one is cw. while the second one travels in ccw. Deserving of notice is that the index of the first one is $z(C_1) = 1$ and the index corresponding to $C_2$ is $z(C_2) = n-1$. Hence, $z(C_2) \equiv -z(C_1)$ (modulo $n$). Furthermore, $1 \leq z(C_2) \leq \lfloor n/2 \rfloor$ and $\lfloor n/2 \rfloor \leq z(C_2) \leq n-1$.

**Example 3.2** Winding indexes in the maximum density starred Euclidean Hamiltonian cyclic polygonals on the vertices of a regular 2p + 1-Gon.

$$S_3 \left\{ \overbrace{L^-_{\lfloor n/2 \rfloor}, \ldots, L^-_{\lfloor n/2 \rfloor}}^{n=2p+1} \right\}. \text{ In } N\left(K_{n2p+1}\left(e^{i\pi}\sqrt[n]{1}\right), (d_{i,j})_{nxn}\right) \text{ networks, } S_3 \text{ defines}$$

a cycle denoted by $C_3$, which is a Euclidean Hamiltonian cyclic pathway $C_H^{2p+1}$, [6, 12]. Its index is $z(C_3) = \lfloor n/2 \rfloor$. Moreover, $n \cdot L^-_{\lfloor \frac{n}{2} \rfloor}$ accomplishes the Max. TSP. These trajectories travel $\lfloor n/2 \rfloor$ times around the origin in cw. angular advance.

$$S_4 \left\{ \overbrace{L^+_{\lfloor n/2 \rfloor}, \ldots, L^+_{\lfloor n/2 \rfloor}}^{n=2p+1} \right\}. \text{ In this case } z(C_4) = \lfloor n/2 \rfloor + 1. \text{ This Euclidean Hamil-}$$

tonian cycle $nL^+_{\lfloor n/2 \rfloor}$ also accomplishes the Max. TSP on the vertices of a regular $2p+1$-Gon, but herein in ccw. Orientation.[1] For an illustrative example, see Fig. 4, 3rd frame from left to right.

The starred polygonals of maximum density living in $N\left(K_{n2p+1}\left(e^{i\pi}\sqrt[n]{1}\right), (d_{i,j})_{nxn}\right)$ networks, are the Hamiltonian cycles $C_3$ and $C_4$ defined by the respective sequences of directed segments $S_3$ and $S_4$ and moving on the same geometric regular and perfect shape (Coxeter 1963). It is easy to verify that the linear segments in the first pathway determine a cw. orientation, in contrast with the second trajectory which travels in ccw. movement around the origin of the circumference. Moreover, the indexes $z(C_3)$ and $z(C_4)$ verify $1 \leq z(C_3) \leq \lfloor n/2 \rfloor$ and $\lfloor n/2 \rfloor < z(C_4) \leq n-1$, respectively.

Hence, $z(C_4) = \lfloor n/2 \rfloor + 1 \equiv -\lfloor n/2 \rfloor$ (modulo $n$), i.e. $z(C_4) = -z(C_3)$ (modulo $n$).

**Theorem 3.2** If $S = \{L^{\mp}_{k1}, L^{\mp}_{k2}, \ldots, L^{\mp}_{km-1}, L^{\mp}_{km}\}$ in $N(K_{n2p+1}\left(e^{i\pi}\sqrt[n]{1}\right), (d_{i,j})_{nxn})$ networks determines a cycle $C$ such that $1 \leq z(C) \leq \lfloor n/2 \rfloor$, then its opposite $\bar{C}$ cycle defined by $\bar{S} = \{L^{\mp}_{kn}, L^{\mp}_{km-1}, \ldots, L^{\mp}_{k2}, L^{\mp}_{k1}\}$ verifies that $\lfloor n/2 \rfloor \leq z(\bar{C}) \leq n-1$, and reciprocally. Moreover $z(\bar{C}) \equiv -z(C)$ (modulo n).

**Proof** From Definition 2.2 of $e\left(L^{\mp}_k\right)$ arises that $e\left(L^{\mp}_k\right) = n - e\left(L^{\mp}_k\right)$ whichever be the $L_k$ segment, $1 \leq k \leq \lfloor n/2 \rfloor$. Consequently $z(C) = n^{-1} \sum_{i=1}^{n} e\left(L^{\mp}_{k_i}\right)$ and

$$z(\bar{C}) = n^{-1} \sum_{i=1}^{n} e\left(L^{\mp}_{k_{n+1-i}}\right) = n^{-1} \sum_{i=1}^{n} \left[n - e\left(L^{\mp}_{k_i}\right)\right] = n^{-1}(n - \sum_{i=1}^{n} 1 - \sum_{i=1}^{n} e\left(L^{\mp}_{k_i}\right)) = \frac{n^2 - z(C)n}{n} = n - z(C) \equiv -z(C)$$

(modulo $n$). Therefore, $z(\bar{C}) \equiv n - z(C) \equiv -z(C)$ (modulo $n$).

---

[1]Ref. to Theorem 2.1.1 at p. 69, specially item viii) if $\beta = -\pi$ and if $n$ is odd, at p. 70 in Niel (2006).

In   addition   $1 \le z(C) \le \lfloor n/2 \rfloor \Rightarrow n - \lfloor n/2 \rfloor \le n - z(C) \le n - 1$,   and   since
$n = \lfloor n/2 \rfloor + \lfloor n/2 \rfloor + 1$, then $\lfloor n/2 \rfloor + 1 \le n - z(C) \equiv -z(C)$ (modulo $n$) $\le n - 1$, i.e.
$\lfloor n/2 \rfloor < z(\bar{C}) \le n - 1$. Reciprocally, if $\lfloor n/2 \rfloor < z(\bar{C}) \le n - 1$ results $1 \le n - z(\bar{C}) <$
$\lfloor n/2 \rfloor + 1$, therefore $1 \le n - z(\bar{C}) \le \lfloor n/2 \rfloor$, hence $1 \le z(C) \le \lfloor n/2 \rfloor$.                  □

**Definition 3.2** Given   a   Euclidean   Hamiltonian   cycle   C   in   $N\left(K_{n2p+1}\right.$
$\left(e^{i\pi} \sqrt[n]{1}\right), (d_{i,j})_{nxn})$   networks   if   the   index   $z(C)$   is   a   integer   such   that
$1 \le z(C) \le \lfloor n/2 \rfloor$, then $C$ has cw. orientation. On the contrary, if the index $z(C)$
varies following $\lfloor n/2 \rfloor \le z(C) \le n - 1$, the cycle travels in ccw. orientation around
the origin.

**Remark 3.2** Let $C$ be an Euclidean Hamiltonian cycle over the vertices of a regular
$2p + 2$-Gon, for any $C_H^n$ in $N\left(K_{n2p+1}\left(e^{i\pi} \sqrt[n]{1}\right), (d_{i,j})_{nxn}\right)$ networks with $z(C) =$
$\lfloor n/2 \rfloor = \frac{n}{2}$, its opposite cycle $z(\bar{C})$ verifies that $z(\bar{C}) = n - z(C) = n - \frac{n}{2} = \frac{n}{2} = z(C)$.
Therefore $z(C) + z(\bar{C}) = n \equiv 0$ in $Z_n$, that is $z(\bar{C}) = -z(\bar{C})$ (modulo $n$). Hence,
Definition 3.2 does not allow the distinctions of cw. and ccw. Orientation for the
cycle $C$ and its opposite $\bar{C}$.

**Note 3.1** In a regular $2p + 2$-Gon, that is if $n$ is even $\angle\left(L_{n/2}^-\right) = \angle\left(L_{n/2}^+\right)$, then
$e\left(L_{n/2}^-\right) = e\left(L_{n/2}^+\right) \equiv \frac{n}{2}$.

**Example 3.3** $S = \left\{L_2^-, L_1^-, \ldots, L_2^+, L_2^+\right\}$ defines a Euclidean Hamiltonian cycle
$C$ in the regular 4-Gon, i.e. $C$ lives in $C_H^4$ at $N\left(K_{n2p+1}\left(e^{i\pi} \sqrt[n]{1}\right), (d_{i,j})_{4x4}\right)$ network
such that $z(C) = 2$. While $-S = \left\{L_1^-, L_2^-, \ldots, L_1^+, L_2^+\right\}$ has the same index,
$z(\bar{C}) = 2$. Hence, Definition 3.2 does not allow the distinctions of cw. and ccw.
movements for the cycles built by $S$ and $\overleftarrow{S}$.

It is high time to point out that the acronym TSP stands for Traveling Salesperson Problem, in reference to optimum cyclic Hamiltonian pathway. Furthermore, TSPP represents Traveling Salesperson Path Problem, in correspondence
with optimum non-cyclic Hamiltonian pathway.

**Example 3.4** $\tau_k$ cycles   defined   by   (6)   in   the   networks   $N(K_{n=2p+2}$
$\left(e^{i\pi} \sqrt[n]{1}\right), (d_{i,j})_{nxn})$, accomplish Remark 3.2, since $z\left(C_H^n\right) = z\left(C_H^{\bar{n}}\right) = n/2$.

$$\bar{\Gamma}_k: \underbrace{l_{q_{max}}^-}_{i}; l_{max}; \underbrace{l_{q_{max}}^+}_{i+1}; \underbrace{l_{q_{max}}^-, l_{q_{max}}^+}_{n/2-k-1}; \underbrace{l_{q_{max}}^+}_{k-1-i}; l_{max} \underbrace{l_{q_{max}}^-}_{k-1-i1} \tag{6}$$

For each $k$, if $1 \le k \le n/2 - 1$, the trajectory $\Gamma_k$ built by the previous sequences
are Euclidean Hamiltonian cycles in $N\left(K_{n=2p+2}\left(e^{i\pi} \sqrt[n]{1}\right), (d_{i,j})_{nxn}\right)$ networks with
initial and final ending points at $V_0(-1, 0)$. Furthermore, these pathways solve the $\frac{n}{2}$
different Max. TSPPs, as we have proved in Niel (2014).

**Example 3.5** In the networks $N\left(K_{n=2l+1}\left(e^{i\pi}\sqrt[n]{1}\right),\left(d_{i,j}\right)_{nxn}\right)$ live $\widetilde{T}_k^{n\,odd}$ cycles defined by (7), (8), (9), (10) and (11) which have rotation indexes $z(C_H^n)\neq z(\bar{C}_H^n)$, since $z(C_H^n)=\lfloor n/2\rfloor$ while $z(\bar{C}_H^n)=\lfloor n/2\rfloor+1$.

The trajectories $\widetilde{T}_k^{n\,odd}$, for each $k$, $1\leq k\leq\lfloor n/2\rfloor-1$ and $p=\lfloor n/2\rfloor-k$, built by the sequences below are Euclidean Hamiltonian cycles in $N\left(K_{n=2l+1}\left(\sqrt[n]{1}\right),D\right)$ networks with initial and final ending points at $V_0(-1,0)$. The proofs to determine that each $\widetilde{T}_k^{n\,odd}$ pathway, defined by the expressions (7)–(11), are Euclidean Hamiltonian cycles result from direct applications of the Arithmetic Algorithm proposed by the authors. Furthermore, these pathways solve the $\lfloor n/2\rfloor$ different Max. TSPPs over the nodes congruent to the vertices of a regular $2p+1$-Gon, as we have proved in previous works.

$$\widetilde{T}_k^{n\,odd}\ :\ L_{k=\lfloor n/2\rfloor-p_{odd}}^-,\ \underbrace{L_{\lfloor n/2\rfloor}^++L_{\lfloor n/2\rfloor-1}^+}_{\lfloor p_{odd}/2\rfloor}+\underbrace{L_{\lfloor n/2\rfloor}^+}_{2}+\underbrace{L_{\lfloor n/2\rfloor-1}^-+L_{\lfloor n/2\rfloor}^-}_{\lfloor p_{odd}/2\rfloor+1}+(n_{odd}-2p_{odd}-3)L_{\lfloor n/2\rfloor}^-$$

$$(7)$$

$$\widetilde{T}_k^{n\,odd}\ :\ L_{k=\lfloor n/2\rfloor-p_{odd}}^-,\ \underbrace{L_{\lfloor n/2\rfloor}^++L_{\lfloor n/2\rfloor-1}^+}_{\lfloor p_{odd}/2\rfloor-i}+\underbrace{L_{\lfloor n/2\rfloor}^+}_{2}+\underbrace{L_{\lfloor n/2\rfloor-1}^-+L_{\lfloor n/2\rfloor}^-}_{\lfloor p_{odd}/2\rfloor-i}+(n_{odd}-2p_{odd}-3)L_{\lfloor n/2\rfloor}^-$$

$$+L_{\lfloor n/2\rfloor-1}^-,\ \underbrace{L_{\lfloor n/2\rfloor}^-+L_{\lfloor n/2\rfloor-1}^-}_{i-1}+\underbrace{L_{\lfloor n/2\rfloor}^-}_{2}+\underbrace{L_{\lfloor n/2\rfloor-1}^++L_{\lfloor n/2\rfloor}^+}_{i}\qquad 1\leq i\leq\lfloor p_{odd}/2\rfloor$$

$$(8)$$

$$\widetilde{T}_k^{n\,odd}\ :\ L_{k=\lfloor n/2\rfloor-p_{even}}^-,\ \underbrace{L_{\lfloor n/2\rfloor}^++L_{\lfloor n/2\rfloor-1}^+}_{\lfloor p_{even}/2\rfloor}+\underbrace{L_{\lfloor n/2\rfloor}^-}_{2}+\underbrace{L_{\lfloor n/2\rfloor-1}^-+L_{\lfloor n/2\rfloor}^-}_{\lfloor p_{even}/2\rfloor}+(n_{odd}-2p_{odd}-3)L_{\lfloor n/2\rfloor}^-$$

$$(9)$$

$$\widetilde{T}_k^{n\,odd}\ :\ L_{k=\lfloor n/2\rfloor-p_{even}}^-,\ \underbrace{L_{\lfloor n/2\rfloor}^++L_{\lfloor n/2\rfloor-1}^+}_{\lfloor p_{even}/2\rfloor-i-1}+\underbrace{L_{\lfloor n/2\rfloor}^-}_{2}+\underbrace{L_{\lfloor n/2\rfloor-1}^-+L_{\lfloor n/2\rfloor}^-}_{\lfloor p_{even}/2\rfloor-2-i}+L_{\lfloor n/2\rfloor-1}^-$$

$$+(n_{odd}-2p_{odd}-3)L_{\lfloor n/2\rfloor}^-+0\leq i\leq p_{evend}/2^{-2}+L_{\lfloor n/2\rfloor-1}^-+\underbrace{L_{\lfloor n/2\rfloor}^-+L_{\lfloor n/2\rfloor-1}^-}_{i}+\underbrace{L_{\lfloor n/2\rfloor}^-}_{2}+\underbrace{L_{\lfloor n/2\rfloor-1}^-+L_{\lfloor n/2\rfloor}^+}_{i+1}$$

$$(10)$$

$$\widetilde{T}_k^{n\,odd}\ :\ L_{k=\lfloor n/2\rfloor-p_{even}}^-,\ \underbrace{L_{\lfloor n/2\rfloor}^++L_{\lfloor n/2\rfloor-1}^+}_{\lfloor p_{evend}/2\rfloor-i-1}+\underbrace{L_{\lfloor n/2\rfloor}^+}_{2}+\underbrace{L_{\lfloor n/2\rfloor-1}^-+L_{\lfloor n/2\rfloor}^-}_{\lfloor p_{even}/2\rfloor-2-i}+L_{\lfloor n/2\rfloor-1}^-$$

$$+(n_{odd}-2p_{odd}-3)L_{\lfloor n/2\rfloor}^-+0\leq i\leq p_{evend}/2^{-1}+L_{\lfloor n/2\rfloor-1}^-+\underbrace{L_{\lfloor n/2\rfloor}^-+L_{\lfloor n/2\rfloor-1}^-}_{i}+\underbrace{L_{\lfloor n/2\rfloor}^-}_{2}+\underbrace{L_{\lfloor n/2\rfloor-1}^++L_{\lfloor n/2\rfloor}^+}_{1}$$

$$(11)$$

## 3.1 Winding Indexes at the Euclidean Quasi-Hamiltonian Cycles in N-Gons

Now onwards, we consider as a Euclidean Quasi-Hamiltonian cycle of order $m$, $m < N$, and denote by $C_{Q-H}^m$ any cyclic polygonal passing only once time through $m$ of the $N$ vertices of a regular N-Gon. Herein, the idea of the rotation indexes is rationally extended to Euclidean Quasi-Hamiltonian cycles built on certain $m$ vertices of a regular N-Gon.

**Theorem 3.1.1** Let $m$ be an integer, odd or even, such as $2 \leq m \leq n$ and let $S'$ be a subset of $m$ vertices $S' = \{V_0, V_{i_1}, \ldots, V_{i_{m-1}}\}$ of the set $S = \{V_0, V_{i_1}, \ldots, V_{i_{n-1}}\}$ in congruence with the $n$ roots of the unity $e^{i\pi}\sqrt[n]{1}$, ordered in cw. orientation from vertex $V_0(-1,0)$. If $C'$ is a Quasi-Hamiltonian cycle of order $m$, i.e. $C_{Q-H}^m$, with vertices on the points of $S'$ and determined by the segments $\{V_{k_1}, V_{k_2}, \ldots,$ $V_{k_{m-1}}, V_{k_m}\}$ then its rotation index $z(C') = n^{-1} \sum_{i=1}^{m} e(L_{k_i})$ verifies that $1 \leq z(C') \leq$ $m-1$. Meanwhile, the winding index of the Quasi-Hamiltonian opposite cycle $\bar{C}$ satisfies $z(C') = m - z(C') \equiv -z(C')$ (modulo m).

**Proof** Since $C'$ is $C_{Q-H}^m$ a cycle with vertices in $S'$ its $m$ directed segments are, in particular, $m$ segments $L_k$ in $S$, which are designated with the same symbols $\{L_{k_1}, L_{k_2}, \ldots, V_{k_m}\}$ and therefore $\sum_{i=1}^{m} e(L_{k_i}) \equiv 0$ (modulo n). In this circumstance, it is known that exists $z(C')$ such that $\sum_{i=1}^{m} e(L_{k_i}) = z(C')$ with $1 \leq z(C') \leq n-1$. This integer counts the number of turns that the cycle $C'$ accomplishes around of the center. If $\{V_0, V_1^*, \ldots, V_{m-1}^*\}$ denotes the vertices on the $m$-Gon, their are congruent to the $m$ roots of the unity $e^{i\pi}\sqrt[m]{1}$, ordered cw. from $V_0$. Therein, since the points in $S'$ are ordered in the same way from $V_0$, a natural vis-`a-vis correspondence is established, as follows $V_0 \to V_0$, $V_{i_*} \to V_1^*$; $V_i \to V_1^*$ and so on until $V_{i_{m-1}} \to V_{m-1}^*$. Hence, the segments $\{L_{k_1}, L_{k_2}, \ldots, V_{k_m}\}$ of the Euclidean Quasi Hamiltonian cycle $C'$ are converted vis-`a-vis in the segments $\{L_{k_1}^*, L_{k_2}^*, \ldots, V_{k_m}^*\}$ which correspond to the Hamiltonian cycle $C^*$. The nature of the transformation of $C'$ in $C^*$, warrants that both cyclic polygonals $C'$ and $C^*$ describe the same number of turns around the origin. However $C^*$ is a $C_H^m$ then by Theorem 3.2 results that $z(C') = n^{-1} \sum_{i=1}^{m} e(L_{k_i}) = z(C^*) = m^{-1} \sum_{i=1}^{m} e(L_{k_i}^*)$. In addition, if $\overleftarrow{C^*}$ is the opposite cycle of $C^*$ its index verifies $z(\overleftarrow{C^*}) = m - z(C^*) \equiv -z(C^*)$ (modulo m). Since $\overleftarrow{C^*}$ is in correspondence with $C'$, therefore $z(\overleftarrow{C'}) = m - z(\overleftarrow{C^*}) = m - z(\overleftarrow{C^*}) = m - z(C') \equiv z(C')$ (modulo m). $\qquad \square$

**Example 3.1.1** Let $S' = \{V_0, V_{i_1}, V_{i_2}, V_{i_3}, V_{i_4}\} \subseteq \{V_0, V_1, \ldots, V_8\}$, herein $V_{i_1} = V_1$, $V_{i_2} = V_2$, $V_{i_3} = V_4$ and $V_{i_5} = V_5$. The directed segments $L_{k_1} = L_4^-$, $L_{k_2} = L_2^+$, $L_{k_3} = L_3^-$, $L_{k_4} = L_4^+$ and $L_{k_5} = L_1^+$ determine a quasi-cycle $C'$, i.e. the $C_{Q-H}^5$ with vertices $S'$. In this case, the transformed $C^*$, is the $C_H^5$ of the quasi-cycle $C'$ defined by: $\left\{ L_{k_1}^* = L_2^+, \ L_{k_2}^* = L_1^+, \ L_{k_3}^* = L_2^-, \ L_{k_4}^* = L_2^-, \ L_{k_5}^* = L_1^+ \right\}$, see Fig. 3.

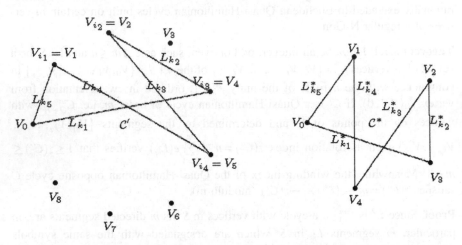

**Fig. 3** Example of Theorem 3.1.1 transformation from a $C_{Q-H}^5$ in the 9-Gon to a $C_H^5$ in the 5-Gon

Then $\overleftarrow{C'} : \{L_1^-, L_4^-, L_3^+, L_2^-, L_4^+\}$ and $\overleftarrow{C^*} : \{L_1^-, L_2^-, L_2^+, L_2^-, L_2^-\}$. The rotation indexes of the quasi-cycle and cycle result: $z(C') = \frac{4+7+3+5+8}{9} = 3$

$z(C^*) = \frac{3+4+2+2+4}{5} = 3$ $\quad z(C') = \frac{1+4+6+2+5}{9} = 2$ $\quad z\left(\overleftarrow{C^*}\right) = \frac{1+3+3+1+2}{5} = 2 \equiv -3$

(modulo 5), that is $1 \le z\left(\overleftarrow{C^*}\right) = 2 \le \lfloor 5/2 \rfloor$ and $\lfloor 5/2 \rfloor < 3 = z\left(\overleftarrow{C'}\right) \le 4 = 5 - 1$.

Therefore $z(C') = z(C^*) = 3$ ccw. and $\left(\overleftarrow{C'}\right) = 2 \equiv -3$ (*modulo* 5) cw. Hence,

$C'$ compasses three circles ccw. while its opposite $\overleftarrow{C^*}$ encircles three times the center although in cw. orientation.

Based on the previous analysis and the algorithmic procedure in §2.2. Fundamental statements: Lemma 2.6 and Theorem 2.7 in Niel (2012) ensue Corollary 3.1.1.

**Definition 3.1.1** Let $S = \left\{ L_{k1}^\pm, \ldots, L_{kn}^\pm \right\}$, $n \ge 3$, be a sequence corresponding to an Euclidean Hamiltonian cyclic trajectory upon the vertices of a regular polygon of odd number of sides. Meanwhile $l_R$ is the integer determined as

$1 \le l_R \le n^{-1} \sum_{i=1}^{n} e(L_{k_i}) \le n - 1$. If $1 \le l_R \le \lfloor n/2 \rfloor$, then $S$ in $C_H^n$ cw. with orientated rotation index $\gamma = l_R$. If $\lfloor n/2 \rfloor \le l_R \le n - 1$, then $S$ in $C_H^n$ ccw. with orientated rotation index $\tilde{\gamma} = -(n - l_R)$, $\tilde{\gamma} \equiv -l_R$ (modulo n).

**Observation 3.1.1** If $n$ is even the previous definition kept if $1 \le l_R < \lfloor n/2 \rfloor$ or if $\lfloor n/2 \rfloor < l_R \le n - 1$.

**Example 3.1.2** Given $S\{5L_1^-\}$ and $\overleftarrow{S}\{5L_1^+\}$ in $C_H^5$, see Fig. 4. With associated integers $e(S) = 5$ and $e\left(\overleftarrow{S}\right) = 20$, therefore $l_R(S) = 1$ and $l_R\left(\overleftarrow{S}\right) = 4$, i.e. $\gamma(S) = 1$ and $\tilde{\gamma}\left(\overleftarrow{S}\right) = -1$ (modulo 5).

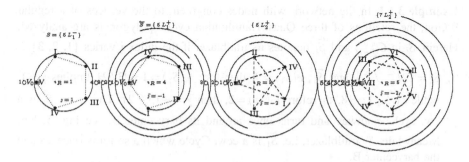

**Fig. 4** Rotation indexes $\gamma$ and $\widetilde{\lambda}$, e.g. $5L_1^-$, $5L_1^+$, $5L_2^+$ and $7L_2^+$

Let $S$ and $\overleftarrow{S}$ be in $C_H^5$ given as $S = \{5L_2^-\}$ and $\overleftarrow{S} = \{5L_2^+\}$, then $e(S) = 10$ and $e\left(\overleftarrow{S}\right) = 15$ therefore $l_R(S) = 2$ while $l_R\left(\overleftarrow{S}\right) = 3$, i.e. $\gamma(S) = 2$ and $\gamma\left(\overleftarrow{S}\right) = -2$, see 3rd frame at Fig. 4.

Let $S$ and $\overleftarrow{S}$ live in $C_H^5$ composed by $S = \{7L_2^-\}$ and $\overleftarrow{S} = \{7L_2^+\}$, then $e(S) = 14$ and $e\left(\overleftarrow{S}\right) = 35$, therefore $l_R(S) = 2$ and $l_R\left(\overleftarrow{S}\right) = 2$, i.e. $\gamma(S) = 2$ and $\gamma\left(\overleftarrow{S}\right) = -2$, see Fig. 4, right.

**Corollary 3.1.1** Let $S = \{L_{k1}^{\pm}, \ldots, L_{kn}^{\pm}\}$ and $\overleftarrow{S} = \{L_m^{\pm}, \ldots, L_1^{\pm}\}$, with $3 \le m \le n$, be Euclidean Quasi-Hamiltonian cycles upon $m$ of the $n$ vertices of a regular n-Gon of odd number of sides and such that both trajectories start at $V_0(-1, 0)$. That is, $S$ and $\overleftarrow{S}$ in $C_{Q-H}^m$, but they differ in the orientation.

When $1 \le l_R(S) = n^{-1} \sum_{i=1}^{m} e\left(L_{k_i}^{\mp}\right) \ne l_R\left(\overleftarrow{S}\right) = n^{-1} \sum_{i=1}^{m} e\left(L_{k_i}^{\mp}\right) 1$, exists two possibilities: $1 \le l_R \le \lfloor m/2 \rfloor$ in which case $S$ in $C_{Q-H}^m$ is cw. with rotation index $\gamma = l_R$ and

$\overleftarrow{S}$ in $C_{Q-H}^m$ is ccw. with rotation index $\widetilde{\gamma} = -\gamma$. Contrariwise, if $\lfloor m/2 \rfloor \le l_R \le m-1$ the cyclic polygonal $S$ in $C_{Q-H}^m$ is ccw. oriented, meanwhile $\overleftarrow{S}$ in $C_{Q-H}^m$ is cw. traveled.

**Proof** Directly from Theorem 3.1.1

If $S$ in $C_{Q-H}^m$, i.e. $S$ is a Quasi-Hamiltonian cycle of order $m$, $m < n_{odd}$ and let $\overleftarrow{S}$ be its trailblazer cycle, then their respective barycenter rotation indexes $\gamma$ and $\widetilde{\lambda}$ verify the next identity $\gamma + \widetilde{\gamma} = m < n_{odd}$.

If a Quasi-Hamiltonian cycle $C_{Q-H}^m$ does not close the centre C of the $P_n$ in which it is inscribed, the rotation index around of its barycenter B is the same that the index around the unitary circumference.                                                        □

**Example 3.1.1** In the network with nodes congruent to the vertices of a regular 9-Gon, the rotations of three Quasi-Hamiltonian cyclic polygonals are analyzed. Herein, $S_i$ in $C_{Q-H}^m$ and $\overleftarrow{S}_i$ denotes its trailblazer if the index i varies $\{1, 2, 3\}$.

1. $S_1 = \{2L_2^-, L_4^+\}$ and $\overleftarrow{S}_1$, $e(S_1) = 9$, therefore $l_R(S_1) = 1 \le \lfloor 3/2 \rfloor$ implies $\gamma = 1$, while $e\left(\overleftarrow{S}_1\right) = 18$, therefore $l_R(S_1) = 2$, then $\widetilde{\gamma} = -(3-2)$. Therefore, $S_1$ is a cw. cyclic pathway and turns once around its barycenter B, see Fig. 5, left. Meanwhile, $S_1$ trailblazer, i.e. $\overleftarrow{S}_1$ is a ccw. Cycle which also turns once around the barycenter B.

2. $S_2 = \{L_1^+, L_4^+, L_4^-, L_2^-\}$ and $\overleftarrow{S}_2$, $e(S_2) = 18$, $e\left(\overleftarrow{S}_2\right) = 18$ therefore $l_R(S_2) = l_R\left(\overleftarrow{S}_2\right)$. In this case, see centre of Fig. 5, Theorem 3.1.1 does not infer about the orientation of the cycles.

3. $S_3 = \{3L_4^+, 2L_3^+\}$, then $e(S_3) = 27$ and $l_R(S_3) = 3$. Herein $m = 5$ and $\lfloor 5/2 \rfloor < l_R(S_3) = 3 \le 4$, therefore from Definition 3.1.1 follows that $S_3$ turns two times ccw. around the barycenter. While its trailblazer $\overleftarrow{S}_3 = \{2L_3^-, 3L_4^-\}$, since $e\left(\overleftarrow{S}_3\right) = 18$ and $l_R(S_3) = 2$ encircles cw. twice. See non directed cycle $\overset{\leftrightarrow}{S}_3$ at Fig. 5, right.

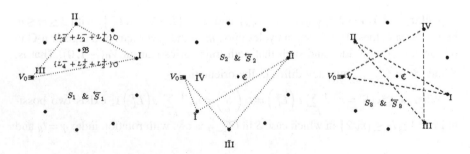

**Fig. 5** Rotations in Quasi-Hamiltonian cycles: e.g. $n = 9$, $m = 3$, $m = 4$ and $m = 5$

## 4 Bistarred Hamiltonian Cycles in Coupled N-Gons

Let $N\left(K_{N=2n}\left(e^{i\pi}\sqrt[n]{1}, re^{i\pi}\sqrt[n]{1}\right), d_{NxN}^{E}\right)$ symbolize the structures of the networks built in the vertices of two regular N-Gons, in the manner that one of them has its vertices on the nth roots of the unity and the other at the radial projections of the previous points over the inner concentric circumference of radio $r$, i.e. $0 < r < 1$. $K_{N=2n}$ symbolizes complete connectivity amongst the $N = 2n$ nodes $\left(e^{i\pi}, re^{i\pi}\sqrt[n]{1}\right)$, where $\sqrt[n]{1} = e^{i\cdot\frac{2k\pi}{n}}$, with $k$ in $\{0, 1, \ldots, n-1\}$. The linkage are weighted by the Euclidean distance between nodes, which are denoted in the $2n \times 2n$ matrix array $d_{NxN}^{E}$ (Witte et al. 2012). Specifically, the geometric concepts and illustrative examples, herein consider the nodes cw. located from $V_0(-1, 0)$, $V_0'(-r, 0)$, $V_0(-1, 0)$, e.g. see Fig. 7. A Hamiltonian tour at the generalized Petersen graphs, normally its corresponding Euclidean weighted non regular and non reflective cyclic shape, will not capture our attention in this section, e.g. the Hamiltonian cycle in the Petersen graph $P(16, 4)$ (Raucher 2014) lives in the $N\left(K_{2.16}\left(e^{i\pi}\sqrt[16]{1}, re^{i\pi}\sqrt[16]{1}\right), d_{NxN}^{E}\right)$ net. Moreover, the architectures of $N\left(K_{N=2n}\left(e^{i\pi}\sqrt[n]{1}, re^{i\pi}\sqrt[n]{1}\right), d_{NxN}^{E}\right)$, when $r = 0$ become the nets $N\left(K_n\left(e^{i\pi}\sqrt[n]{1}\right), d_{NxN}^{E}\right)$, which have been studied in a concatenated series of works (Dichiara et al. 2009).

**Conjecture 4.1** ¿ Are Max. TSP in $N\left(K_{N=2nodd}\left(e^{i\pi}\sqrt[nodd]{1}, re^{i\pi}\sqrt[nodd]{1}\right), d_{NxN}^{E}\right)$ networks solved by the perfect and regular shapes of the reflective maximum density bistarred cyclic polygonals as the reflective maximum density starred cyclic polygonals do in $N\left(K_{nodd}\left(e^{i\pi}\sqrt[nodd]{1}\right), d_{NxN}^{E}\right)$ structures ?

**Example 4.1** Figure 6 shows the reflective maximum density bistarred Hamiltonian cycles accomplished in $N\left(K_{N=2nodd}\left(e^{i\pi}\sqrt[nodd]{1}, re^{i\pi}\sqrt[nodd]{1}\right), d_{NxN}^{E}\right)$ networks for coupled nodd-Gons. Specifically, coupled 3-Gon, 5-Gon and 9-Gon architectures when the radius of the inner circumference is $r = 0.5$.

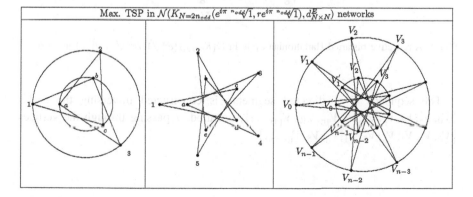

**Fig. 6** Bistarred reflective Hamiltonian Cycles in vertices of Coupled Nodd-Gons

## 5  Existence of Bistarred Hamiltonian Polygonals in Coupled Nodd-Gons

We deal with the existence or absence of the bistarred reflective Hamiltonian cyclic polygonals, in the architecture of complete connectivity in $N\big(K_{N=2n}\big(e^{i\pi}\sqrt[n]{1}$ $,re^{i\pi}\sqrt[n]{1}\big),d^{E}_{N\times N}\big)$ networks, built in the following way: For each $s$, such that $1\le s\le\lfloor n/2\rfloor$ the directed segments are of the form $\overrightarrow{V_0,V_s'}$, $\overrightarrow{V_s',V_{2s}}$, $\overrightarrow{V_{2s},V_{3s}'}$, wherein the subindices $ks$ belong to $Z_n$ (Lord 2008).

**Notation 4.1.1** *Let* $_pL_q$ *and* $_rL_t'$ *be directed segments defined as:* $_pL_q=V_q,V_q'$ *and* $_rL_t'=V_r',V_t$.

**Example 4.1.1** In $N\big(K_{N=2\times7}\big(e^{i\pi}\sqrt[7]{1},re^{i\pi}\sqrt[7]{1}\big),d^{E}_{N\times N}\big)$ network, it is considered the reflective bistarred Hamiltonian cyclic polygonal for $s=2$, see the explanation in Fig. 7.

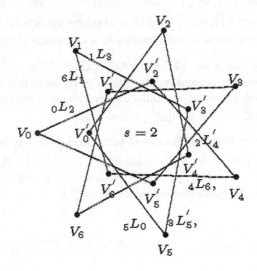

**Fig. 7** A reflective bistarred Hamiltonian cycle in $N\big(K_{N=2\times7}\big(e^{i\pi}\sqrt[7]{1},re^{i\pi}\sqrt[7]{1}\big),d^{E}_{N\times N}\big)$ network

The sequence $S$ of directed segments is constructed traversing the nodes $\{V_0,V_2',V_4,V_6',V_8,V_8,V_{10}',V_{12},V_{14}',\ldots\}$, i.e. mod. 7 passing through the vertices $\{V_0,V_2',V_4,V_6',V_8,V_1,V_3',V_5,V_0',\ldots\}$

$$S: \left\{ \overrightarrow{V_0, V_2'}, \overrightarrow{V_2', V_4}, \overrightarrow{V_4, V_6'}, \overrightarrow{V_6', V_1}, \overrightarrow{V_1, V_3'}, \overrightarrow{V_3', V_5}, \overrightarrow{V_5, V_0'}, \overrightarrow{V_0', V_2}, \overrightarrow{V_2, V_4'}, \overrightarrow{V_4', V_6}, \overrightarrow{V_6, V_1'}, \overrightarrow{V_1', V_3}, \overrightarrow{V_3, V_5'}, \overrightarrow{V_5', V_0} \right\} =$$

$$S: \left\{ 0L_2, 2L_4', 4L_6, 6L_1', 1L_3, 3L_5', 5L_0, 0L_2', 2L_4, 4L_6', 6L_1, 1L_3', 3L_5, 5L_0' \right\}$$

**Example 4.1.2** In the structure of the $N\left(K_{N=2 \times 15}\left(e^{i\pi} \sqrt[15]{1}, re^{i\pi} \sqrt[15]{1}\right), d_{N \times N}^E\right)$ networks, with $r$ in $(0, 1)$, Fig. 8 shows reflective bistarred Quasi-Hamiltonian cyclic polygonals in coupled 15-Gons, for $n = 15$ and $s$ no relatively primes, specifically when s is chosen in the integer set $\{3, 5, 6\}$, see Fig. 8.

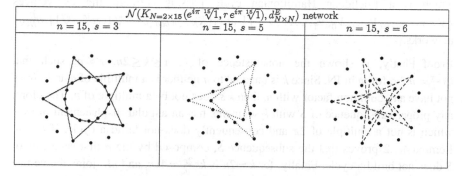

| $\mathcal{N}\left(K_{N=2 \times 15}\left(e^{i\pi} \sqrt[15]{1}, re^{i\pi} \sqrt[15]{1}\right), d_{N \times N}^E\right)$ network | | |
|---|---|---|
| $n = 15, \ s = 3$ | $n = 15, \ s = 5$ | $n = 15, \ s = 6$ |

**Fig. 8** Reflective Bistarred Quasi-Hamiltonian cyclic polygonals in coupled 15-Gons

In general, the directed segments $_pL_q$ and $_rL_t'$ are of the form $_{(i-1)s}L_{is} = \overrightarrow{V_{(i-1)s}, V_{is}'}$ and $_{is}L_{(i+1)s}' = \overrightarrow{V_{is}', V_{(i+1)s}}$ with $1 \leq i \leq 2n - 1$.

**Lemma 4.1.1** For $n$ and $s$ positive integers such that $1 \leq s \leq \lfloor n/2 \rfloor$ the corresponding directed segments $_{(i-1)s}L_{is}$ and $_{is}L_{(i+1)s}'$ have the same length and determine an angular displacement equal to $\angle\left(_{(i-1)s}L_{is}\right) = \angle\left(_{is}L_{(i+1)s}'\right) = s\frac{2\pi}{n}$.

**Proof** In both cases from the direct verification of basic geometric arguments. $\square$

**Lemma 4.1.2** If $n = 2k + 1$ in IN and if $s$ is such that $1 \leq s \leq \lfloor n/2 \rfloor$, the sequence of directed segments

$$S = \left\{ _{(i-1)s}L_{is}, \ _{is}L_{(i+1)s}' \Big|_{1 \leq i \leq 2n - 1} \right\} \text{ has the proper subsequence } \overline{S} \text{ of } n \text{ directed}$$

segments $\overline{S} = \left\{ _0L_s, \ _sL_{2s}', \ _{2s}L_{3s}, \ _{3s}L_{4s}', \ \cdots, \ _{(n-2)s}L_{(n-1)s}', \ _{(n-1)s}L_{ns} \right\}$ built by the first $n$ segments in $S$, which starts at $V_0(-1, 0)$ and ends at the vertex $V_0(-r, 0)$. In addition, the angular displacement determined for $\overline{S}$ is $s 2\pi$. That is, the polygonal constructed by $\overline{S}$ performs s turns around the origin.

**Proof** It comes by the observation that the segments in $S$ are consecutively collocated and the first one $_0L_s$ joins the vertice $V_0$ with $V_s'$ while the last segment of $\overline{S}$ links $V_{(n-1)s}$ to $V_{ns}'$. That is, $V_{n-s}$ with $V_0'$, since $ns \equiv 0$ and $(n-1)s \equiv -s \equiv n-s$ (modulo n). Moreover, the angular displacement of the polygonal determined by $\overline{S}$ is $ns\frac{2\pi}{n} = s2\pi$. $\qquad\square$

**Theorem 4.1.1** If $n$ and $s$ in IN, let $n$ be odd, $s$ does not divide $n$, i.e. $s$ and $n$ are relatively primes, with $1 \leq s \leq \lfloor n/2 \rfloor$, then the sequence of directed segments $S = \left\{ {}_{(i-1)s}L_{is}, \; {}_{is}L_{(i+1)s}' \Big|_{1 \leq i \leq 2n-1} \right\}$ determines a reflective bistarred polygonal which is a Euclidean Hamiltonian cycle passing through the $2n$ vertices $\left\{ V_0, V_1, \ldots, V_{n-1}, V_0', V_1', \ldots, V_{n-1}' \right\}$ of $\mathrm{N}\left( K_{N=2n \; odd}\left( e^{i\pi \; {}^{nodd}\!\sqrt{1}}, re^{i\pi \; {}^{nodd}\!\sqrt{1}} \right), d_{NxN}^E \right)$ networks.

**Proof** Firstly, it shown the nonexistence of $k$, $1 \leq k \leq 2n$, $k \neq n$, such that $ks\frac{2\pi}{n} = m2\pi$, for $m$ in IN. Since $k$ is unequal to $n$ neither is a multiple of $n$ and $s$ does not have any common factor with $n$, then $k\,s$ does not be a multiple of $n$. Therefore, any proper subsequence of $S$ with $k$ segments has an angular displacement of $ks\frac{2\pi}{n}$ which is not a multiple of $2\pi$ and consequently does not build a cycle. If $k = n$, Lemma 4.1.2 proves that the subsequence $\overline{S}$, composed by the $n$ first segments of $S$ does not build a cycle. Finally, for $k = 2n$ is $ks\frac{2\pi}{n} = 4s\pi$, and $_0L_s$ joins the vertice $V_0$ to $V_s'$ and $_{(2n-1)s}L_{2ns}' = {}_{n-s}L_0'$ unites $V_{n-s}'$ with $V_0$, then $S$ is a Hamiltonian cycle of order $2n$ upon the set of vertices $\left\{ V_0, V_1, \ldots, V_{n-1}, V_0', V_1', \ldots, V_{n-1}' \right\}$ (Fig. 9). $\qquad\square$

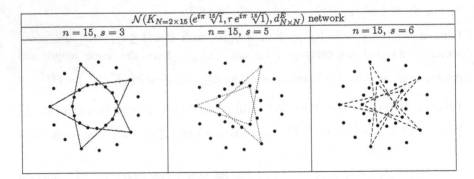

| $\mathcal{N}\left( K_{N=2\times15}\left( e^{i\pi\,\sqrt[15]{1}}, r\,e^{i\pi\,\sqrt[15]{1}} \right), d_{N\times N}^E \right)$ network | | |
|---|---|---|
| $n = 15, \; s = 3$ | $n = 15, \; s = 5$ | $n = 15, \; s = 6$ |

**Fig. 9** Bistarred reflective Hamiltonian cyclic polygonals in coupled 7-Gons

**Remark 4.1.1** In coupled regular Nodd-Gons exists the maximum density reflective bistarred Hamiltonian cycle. Then, their regular shapes could solve the Max. TSP in $\mathrm{N}\left( K_{N=2n \; odd}\left( e^{i\pi \; {}^{nodd}\!\sqrt{1}}, re^{i\pi \; {}^{nodd}\!\sqrt{1}} \right), d_{NxN}^E \right)$ networks.

## 5.1 Existence of Bistarred Quasi-Hamiltonian Cycles in Coupled Neven-Gons

Next result establishes the absence of reflective bistarred Euclidean Hamiltonian cycles of order $2n$ passing through each vertice of coupled Neven-Gons.

**Theorem 4.2.1** If $n$ in IN, is even, and $s$ in IN, such that $1 \le s \le n/2$, then it does not exist any bistarred reflective Hamiltonian cyclic polygonal $S = \{_{(i-1)s}L_{is},$ $_{is}L'_{(i+1)s}|_{1 \le i \le 2n-1}\}$ upon the $2n$ vertices $\{V_0, V_1, \ldots, V_{n-1}, V'_0, V'_1, \ldots, V'_{n-1}\}$ of $N\left(K_{N=2n\ evem}\left(e^{i\pi\ \ ^{evem}\sqrt{1}}, re^{i\pi\ \ ^{evem}\sqrt{1}}\right), d^E_{NxN}\right)$ networks.

**Proof** Let $S$ be the sequence $S = \{S_1; S_2\}$ with: $S_1 = \{_0L_s, {}_sL'_{2s}, {}_{2s}L'_{3s}, \cdots, {}_{(n-2)s}L_{(n-1)s}, {}_{(n-1)s}L_{ns}\}$ and $S_2 = \{_{ns}L_{(n+1)s}, {}_{(n+1)s}L'_{(n+2)s}, \cdots, {}_{(2n-2)s}L_{(n-1)s}, {}_{(2n-1)s}L'_{2ns}\}$. $S$ has the property that its subsequence $S_1$, built by the first $n$ segments of $S$, accomplishes an angular displacement of $ns\frac{2\pi}{n} = s2\pi$. Moreover, the first segment $_0L_s$ of $S_1$ connects $V_0$ to $V'_s$ and the last one $_{(n-1)s}L'_{ns} = {}_{n-s}L'_0$ links $V'_{n-s}$ to $V_0$. Hence, the polygonal associated to the subsequence $S_1$ is a cycle of $n$ segments, which starts and ends at $V_0$ if $s$ and $n$ are coprime integers. Therefore, $S$ sequence does not determine a Hamiltonian cycle on the $2n$ vertices $\{V_0, V_1, \ldots, V_{n-1}, V'_0, V'_1, \ldots, V'_{n-1}\}$. Furthermore, the directed segments of the second half of $S$ repeat exactly and in the same order the $n$ segments in the first half $S_1$ and therefore the cyclic pathway once more describes the cycle defined by $S_1$, that is, $_{ns}L'_{(n+1)s} = {}_0L_s$, $_{(n+1)s}L'_{(n+2)s} = {}_sL'_{2s}$, $_{(n+2)s}L'_{(n+3)s} = {}_{2s}L'_{3s}$, and so on $_{(2n+2)s}L'_{(2n+1)s} = {}_{(n-2)s}L'_{(n-1)s}$, until $_{(2n-1)s}L'_{2ns} = {}_{(n-1)s}L'_{ns}$. Figure 10 shows the cases $n = 8$ for $s = 1$ and $s = 3$. However, if $s$ divides $n$ the circumstances are slightly different because of cyclic polygonals are conformed by $\frac{n}{s}$ directed segments and repeated $s$ in the sequence $S_1$ and over by the $n$ congruent links of $S_2$, e.g. see central illustration of Fig. 10, it represents the case $n = 8$ for $S = 2$, in which four directed connectors composed the cyclic pathways twice repeated by ordering

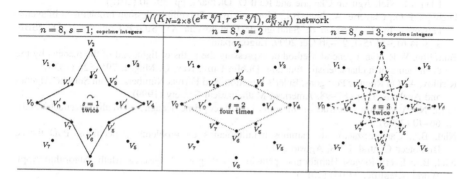

**Fig. 10** Reflective Bistarred Quasi-Hamiltonian cyclic polygonals in coupled 8-Gons

at $S_1$ and vis-`a-vis replicated according with $S_2$. Illustrative case is $s = n/2$. It consists in the replication of $n/2$ digons $V_0 \rightarrow V_n/2 \rightarrow V_0$ following the directed links in $S_1$. Meanwhile, its twin sequence $S_2$ mimics the $n$ forthcoming rebounds.    □

# 6  Conclusion

This paper is a study concatenated to previous researches that we have done in the treatment of Specific Traveling Salesman Problem on complete connected networks built over vertices of the regular $N$-Gons. Herein, the results are advocated to determine the rotation or winding indexes of relevant cyclic pathways. Moreover, the bistarred reflective Euclidean Hamiltonian and Quasi-Hamiltonian clockwise cycles are singled out at the coupled $N$-Gons in the network architectures of $2N$ nodes. Furthermore, since the Maximum Density Bistarred Reflective Hamiltonian Cycle exists in paired $N$odd- Gons, next paper will be dedicated to prove if these regular polygonals solve the Max. TSP at $\mathrm{N}\left(K_{N=2n\,odd}\left(e^{i\pi\ ^{nodd}\!\sqrt{1}},\right.\right.$ $\left.\left.re^{i\pi\ ^{nodd}\!\sqrt{1}}\right), d^E_{NxN}\right)$ networks. It is time to emphasize that, Euler's totient function determines the number of Hamiltonian and Quasi-Hamiltonian cycles living at these coupled $N$-Gons net structures, i.e. $\mathrm{N}\left(K_{N=2n}\left(e^{i\pi\ ^n\!\sqrt{1}}, re^{i\pi\ ^n\!\sqrt{1}}\right), d^E_{NxN}\right)$.

# References

Applegate, D., Bixby, R.E., Chavatal, V., Cook, W.J.: Traveling Salesman Problem: A Computational Study. Princeton University Press, Princeton (2006)

Barvinok, A.I., Gimadi, E.Kh., Serdyukov, A.I.: The Maximum Traveling Salesman Problem (eds. Gutin, G., Punnen, A.) (2002)

Buckly, F., Harary, F.: Distance in Graphs. Addison-Wesley Publishing Co., New York (1990)

Coxeter, H.S.M.: Introduction to Geometry. Wiley Inc., New York (1963)

Witte, H., Jedlinski, M., Dichiara, R.O. (eds.): Sustainable Logistics. edi-UNS 2012. Section One: Theoretical Foundations, Chapter IV: Traveling Salesman Problems in Bilayered Networks Blanca I. Niel, Agustín Claverie and Raúl O. Dichiara, pp. 39–40 (2012)

Dichiara, R.O., Claverie, A., Niel, B.I.: Min. and max. Traveling salesman problems in bilayered networks. In: Universidad de Santiago de Compostela XV SIGEF Congress. C.D.: ISBN 978-84-613-5575-4, 29–30 Oct 2009, Lugo, Spain

Hamilton, W.R.: On a general Method of expressing the paths of light, and of the planets, by the coefficients of a characteristic function. Dublin Univ. Rev. Q. Mag. I, 795–826 (1833)

Kirillov, A.: On Regular Polygons, Euler's Function, and Fermat Numbers. $K$vant Selecta: Algebra and Analysis, I, pp. 87–98. American Mathematical Society (1999)

Lord, N.: A uniform construction of some infinite coprime sequences. Math. Gazette **92**(523), 66–70 (2008)

Niel, B.I.: Caracterización de caminos hamiltonianos en problemas específicos. PhD thesis. Biblioteca Central. UNS, Agosto 2014

Niel, B.I.: Every longest Hamiltonian path in even N-gons. J. Discrete Math. Algorithm Appl. World Scientific. **04**(04) (2012)

Niel, B.I., Reartes, W.A., Brignole, N.B.: Every longest Hamiltonian path in odd N-gons. In: SIAM Conference on Discrete Mathematics, Austin, Texas, 14–17 June 2010

Niel, B.I.: Geometry of the Euclidean Hamiltonian suboptimal and optimal paths in the $N\left(K_n\left(e^{i\pi}\sqrt{[n]}\,1\right), \left(d_{i,j}\right)_{nxn}\right)$ Networks. In: Proceedings of the VIII "Dr. Antonio A. R. Monteiro", Congress of Mathematics, pp. 67–84, 2006, Mathematics Department, Universidad Nacional del Sur, 8000 Bahía Blanca, Argentina, http://inmabb.criba.edu.ar/cm/actas/pdf/09−niel.pdf

Rauscher, R.: Ein neuer Ansatz zur Ermittlung von Hamiltonkreisen in verallgemeinerten Petersenschen

Graphen, P. (n, k).: Lingener Studien zu Management und Technik. LIT Verlag, Berlin (2014)

Treatman, S., Wickham, C., Bradley, D.M.: Constructible approximations of regular polygons. Am. Math. Monthly **107**(10), 911–922 (2000) (Serge Tabachnikov, Editor, AMS)

Watts, D.J.: Small Worlds: The Dynamics of Networks Between Order and Randomness. Princeton Studies in Complexity (1999)

# Author Index

© Springer International Publishing Switzerland 2015

443

J. Gil-Aluja et al. (eds.), *Scientific Methods for the Treatment of Uncertainty in Social Sciences*, Advances in Intelligent Systems and Computing 377, DOI 10.1007/978-3-319-19704-3

Printed in the United States
By Bookmasters